Reaction Engineering, Catalyst Preparation, and Kinetics

Reaction Engineering, Catalyst Preparation, and Kinetics

Jorge M. Marchetti

CRC Press is an imprint of the
Taylor & Francis Group, an **informa** business

First edition published 2021
by CRC Press
6000 Broken Sound Parkway NW, Suite 300, Boca Raton, FL 33487-2742

and by CRC Press
2 Park Square, Milton Park, Abingdon, Oxon, OX14 4RN

© 2021 Taylor & Francis Group, LLC

CRC Press is an imprint of Taylor & Francis Group, LLC

The right of Jorge M. Marchetti to be identified as author of this work has been asserted by him in accordance with sections 77 and 78 of the Copyright, Designs and Patents Act 1988.

Reasonable efforts have been made to publish reliable data and information, but the author and publisher cannot assume responsibility for the validity of all materials or the consequences of their use. The authors and publishers have attempted to trace the copyright holders of all material reproduced in this publication and apologize to copyright holders if permission to publish in this form has not been obtained. If any copyright material has not been acknowledged please write and let us know so we may rectify in any future reprint.

Except as permitted under U.S. Copyright Law, no part of this book may be reprinted, reproduced, transmitted, or utilized in any form by any electronic, mechanical, or other means, now known or hereafter invented, including photocopying, microfilming, and recording, or in any information storage or retrieval system, without written permission from the publishers.

For permission to photocopy or use material electronically from this work, access www.copyright.com or contact the Copyright Clearance Center, Inc. (CCC), 222 Rosewood Drive, Danvers, MA 01923, 978-750-8400. For works that are not available on CCC please contact mpkbookspermissions@tandf.co.uk

Trademark notice: Product or corporate names may be trademarks or registered trademarks and are used only for identification and explanation without intent to infringe.

Library of Congress Cataloging-in-Publication Data
Names: Marchetti, Jorge Mario, author.
Title: Reaction engineering, catalyst preparation, and kinetics / by Jorge Marchetti.
Description: First edition. | Boca Raton, FL : CRC Press, 2021. | Includes bibliographical references and index. | Summary: "This textbook covers reaction engineering, catalyst preparation, and kinetics. It features a section of fully solved examples as well as end of chapter problems. It includes coverage of catalyst characterization and its impact on kinetics and reactor modeling. It presents simpler cases as well as fully developed complicated scenarios"-- Provided by publisher.
Identifiers: LCCN 2020055172 (print) | LCCN 2020055173 (ebook) | ISBN 9781138605985 (hardback) | ISBN 9780367764234 (ebook)
Subjects: LCSH: Chemical processes. | Chemical reactors. | Catalysts. | Chemical kinetics.
Classification: LCC TP155.7 .M364 2021 (print) | LCC TP155.7 (ebook) | DDC 660/.2995--dc23
LC record available at https://lccn.loc.gov/2020055172
LC ebook record available at https://lccn.loc.gov/2020055173

ISBN: 978-1-138-60598-5 (hbk)
ISBN: 978-0-367-76423-4 (pbk)
ISBN: 978-0-429-46684-7 (ebk)

DOI: 10.1201/9780429466847

Typeset in Times
by SPi Technologies India Pvt Ltd (Straive)

Visit the Support Material: https://www.routledge.com/
Reaction-Engineering-Catalyst-Preparation-and-Kinetics/Marchetti/p/book/9781138605985

To my wife, Anna.
To my kids, Björn and Lukas
In the memory of my mother, Zulema.

Contents

Preface xi
Acknowledgements xiii
Author xv

Chapter 1 Catalysis Preparation and Characterization 1

    1.1 Introduction: Basic Concept and Origin of Catalysis 1
    1.2 Cataloging Catalytic Material 2
        1.2.1 Homogeneous Catalysis 2
        1.2.2 Heterogeneous Catalysis 2
            1.2.2.1 Nanoscale Catalytic Materials 4
            1.2.2.2 Porous Materials 4
            1.2.2.3 Supported Catalysts 5
        1.2.3 Biocatalysis 6
        1.2.4 Photocatalysis 8
    1.3 Catalyst Preparation Techniques 9
        1.3.1 Precipitation 10
        1.3.2 Coprecipitation 10
            1.3.2.1 Theory of Nucleation 12
        1.3.3 Impregnation 12
        1.3.4 Sol Gel Method 13
        1.3.5 Chemical Vapor Deposition 14
        1.3.6 Ion Exchange 15
        1.3.7 Immobilization of Catalysts 16
            1.3.7.1 Covalent Bonding 17
            1.3.7.2 Entrapment 17
            1.3.7.3 Adsorption 18
            1.3.7.4 Ionic Bonding 18
        1.3.8 Other Alternatives 18
    1.4 Catalyst Characterization Techniques 19
        1.4.1 X-ray Diffraction (XRD) 19
        1.4.2 Porosity Measurements 21
        1.4.3 Scanning Electron Microscopy (SEM) 23
        1.4.4 Transmission Electron Microscopy (TEM) 24
        1.4.5 Infrared Spectroscopy (IR) 25
        1.4.6 Thermal Analysis 26
        1.4.7 Raman Spectroscopy 29
        1.4.8 X-ray Photoelectron Spectroscopy (XPS) 30
    1.5 Industrial Catalytic Processes 33
        1.5.1 Synthesis Gas 34

		1.5.2	Ammonia Synthesis 35
		1.5.3	Selective Oxidation 36
		1.5.4	Fischer–Tropsch Process 36
		1.5.5	Biodiesel Production 37
	References 38

Chapter 2 Reactor Design: Mole and Energy Balance 43
    2.1 Introduction 43
    2.2 Types of Reactors 43
        2.2.1 Batch Reactor 45
        2.2.2 Semi-Batch Reactor 49
            2.2.2.1 Balance for Reactant A 51
            2.2.2.2 Balance for Reactant B 52
        2.2.3 Continuous Reactors 57
            2.2.3.1 Continuous Stirred Tank Reactor (CSTR) 58
            2.2.3.2 Plug Flow Reactor (PFR) 62
            2.2.3.3 Pack Bed Reactor (PBR) 70
            2.2.3.4 Membrane Reactor (MR) 71
    2.3 Conversion 73
    2.4 Reactors in Series 79
    2.5 Reactors in Parallel 81
    2.6 Moles Balance Calculations 82
    2.7 Reaction in Gas Phase 86
    2.8 Pressure Drop 94
    2.9 Multiple Reactions 100
        2.9.1 Series Reactions 101
        2.9.2 Parallel Reactions 104
            2.9.2.1 Scenario 1 105
            2.9.2.2 Scenario 2 106
        2.9.3 Combined Reactions 110
    2.10 Equilibrium Reactions 113
    2.11 Non-Isothermal Reactors 117
        2.11.1 Adiabatic 122
            2.11.1.1 Adiabatic Process for a Reaction
                       at Equilibrium 130
        2.11.2 Plug Flow with Heat Added/Removed at a
            Constant Outside Temperature 134
        2.11.3 Co-Current Heating/Cooling Fluid System 141
        2.11.4 Counter Current Heating/Cooling Fluid System 142
        2.11.5 CSTR with Heat Transfer 147
    2.12 Non-Isothermal, Non-Steady State Reactors 152
        2.12.1 Batch Reactor 153
        2.12.2 Semi-Batch Reactor 154

Contents ix

          2.13   Multiple Reaction System in a Plug Flow Reactor 159
          2.14   Multiple Reactions in a Batch or Semi-Batch Reactor 166
          Notes 171
          References 172
          Problems 172

Chapter 3   Reaction Kinetics 185

          3.1   Introduction 185
          3.2   Elementary Reactions 185
          3.3   Non-Elementary Reactions 186
          3.4   Multiple Reactions 188
          3.5   Evaluation of Experimental Data 188
                 3.5.1   Integration Method 189
                 3.5.2   Differential Method 190
                 3.5.3   Generic Method 194
          3.6   Kinetics Modeling for Simple Reactions 194
                 3.6.1   1-Phase Irreversible Reaction 195
                 3.6.2   1-Phase Reversible Reaction 196
                         3.6.2.1   Equilibrium Method 196
                         3.6.2.2   PSSH Method 199
                 3.6.3   2-Phase Irreversible Reaction 200
                         3.6.3.1   PSSH Method 200
                 3.6.4   2-Phase Reversible Reaction 202
                         3.6.4.1   Equilibrium Method 202
                         3.6.4.2   PSSH Method 207
          3.7   Kinetic Modeling of Complex Systems 209
                 3.7.1   Homogeneous Systems 210
                 3.7.2   Heterogeneous Systems 219
          3.8   Catalyst Deactivation 225
          3.9   Mass Transfer Limitations 228
                 3.9.1   Internal Mass Transfer Limitations 231
                         3.9.1.1   Cylindrical Pore 233
                         3.9.1.2   Spherical Pore 235
                 3.9.2   Overall Mass Transfer Limitations 238
          Note 249
          References 249
          Problems 250

Chapter 4   Completely Solved Example 259

          4.1   Introduction 259
          4.2   Description of the Problem 260
          4.3   Laboratory Equipment Employed 260
          4.4   Experimental Procedure 261

4.5	Sample Analysis and Errors		262
4.6	Data Evaluation		263
4.7	Mathematical Model 1		269
4.8	Comparison of Data and Model 1		274
4.9	Mathematical Model 2		281
4.10	Comparison of Data and Model 2		291
4.11	Final Expression for Kinetics		303
4.12	Simulation of an Isothermal Plug Flow Reactor Using Kinetics from 4.11		304
4.13	Simulation of an Adiabatic Plug Flow Reactor with the Kinetics from 4.11		307
4.14	Simulation of a Constant Heat Transfer Plug Flow Reactor with Kinetics from 4.11		310
4.15	Simulation of a Co-Current Heat Transfer Flow in a Plug Flow Reactor with Kinetics from 4.11		314
4.16	Simulation of a Counter-Current Heat Transfer Flow in a Plug Flow Reactor with Kinetics from 4.11		318
4.17	Comparison For A Gas Phase System with Pressure Drop		323
	4.17.1	Isothermal	323
	4.17.2	Adiabatic	329
	4.17.3	Constant External Temperature	336
	4.17.4	Co-Current External Flow	343
	4.17.5	Counter-Current External Flow	349
	4.17.6	Comparison of Previous Cases	356
References			360
Index			361

Preface

This book is an introduction to the universe of Reaction Engineering and Catalysis. My intention with this book is to give some basic concepts and tools that students can use to solve problems related to this topic. The book organization is based on what I have encountered as challenges while being a student, working on the topic as well as from my students and their struggles.

This book is designed for students of both undergraduate and graduate levels within the field of chemical reaction engineering. Each chapter of the book is built in such a way that we go from the simple ideas and concepts toward the more complex and realistic cases and situations. The tools given within this textbook should provide the students and readers with sufficient knowledge to tackle real problems in industry.

This book is divided into four main chapters, each dealing with one specific topic. This division has been made based on my experience when working within this field and when teaching reaction engineering for over 15 years.

Chapter 1 is an introduction to the world of catalysis; it is a brief summary of the different types of catalysts that can be found, their properties, characteristics, and uses. The chapter includes a subsection on the preparation methodologies, with a short description of each of them, as well as advantages and disadvantages, a subsection on the characterization methods, not all of them but those that are considered the most relevant and more used and what property can be measured in each case and what information can be provided by each method. Finally, this chapter has five examples of industrial cases where different types of catalysts are being used.

Chapter 2 is related to reactor design; it contains the fundamentals for the development and use of the mole and energy balances for batch, semibatch, and continuous reaction system (including membrane systems). It deals with the variation of temperature and pressure and their effect on the conversion and yields. Along the chapter, the complexity of the problem increases in order to combine all the different effects from variations in the reactor's temperature, pressure, volume, etc. However, for this chapter, the reaction rates used have been given; this simplification has led to the development of Chapter 3.

Chapter 3 has the main goal of introducing the reader to the world of reaction kinetics and modeling. Within this chapter, we give the reader an introduction to the different possible reactions as well as different methodologies to treat the experimental data to obtain a kinetics model that is accurate. The complexity of the development of reaction kinetics increases as we move forward within the chapter, going from irreversible to reversible systems, from one phase to two or more phases, from one single reaction to multiple reaction system. When dealing with heterogeneous reactions with a solid catalyst, this chapter also introduces the effect of deactivation of the catalyst and their effect on the kinetics model as well as some introduction to mass transfer limitation within the reactive systems.

Finally, Chapter 4 presents a fully proposed and solved problem. This chapter deals with a typical problem that is faced when performing laboratory work. It introduces a description of the equipment used, as well as the analytical tools required for sampling and measuring. The core of this chapter is the development of mathematical models and kinetics based on the experimental data. This has been done as complete as possible, with cases that did not work, to show what the reader needs to do when actually using this approach, what the different steps are, what modifications should be considered, and what solutions can be found when the final kinetics expression is found; different reactive systems (isothermal, adiabatic, co-current flow, counter current flow, constant external fluid temperature) are presented, solved, and compared. I hope that this fully developed problem could assist students for a better understanding on the different steps required, from the lab data to a fully simulated reactor using their own collected data.

Jorge M. Marchetti
Faculty of Science and Technology
Norwegian University of Life Science
Ås, Viken, Norway

Acknowledgements

Anna, Björn, and Lukas, THANK YOU.

Anna, when I came up with the idea of writing this book, 3 years before it was done, you were amazingly positive about it, very optimistic, like you always are. The writing process has not been easy, especially at the end, but you were there, giving me the space and time to finish, to reach the final line. Thank you for allowing me this time for my professional growth, for my personal development. I love you so deeply.

To my kids, Björn and Lukas, for not understanding what I was doing, for not thinking that I needed time to work, but instead for wanting me to play with you, to be there for you, to do things together. Your perspective of life, your desire to include me in your life, your games, your free time, has allowed me to relax, to balance the work and the free time, has forced me to think differently, hopefully, more creatively, and more pedagogically. Thank you for making me remember how amazing it is to lie on the floor and let the imagination go wild while playing with anything, from a marvel to a racing car.

To my colleague and friend, Dr. Mangesh Avhad. Thank you for the discussions on the topic, for the inputs, comments, suggestions, and help in gathering information for the catalysis preparation and characterization section.

To my former students that have challenged me with questions during class, that have pushed my curiosity and have given me, I hope, some inside knowledge on their learning process.

All the personnel involved at Taylor & Francis for their help and support preparing, correcting, and making this manuscript into a book, thank you. Special thanks to Allison Shatkin and Gabrielle Vernachio for helping to start this project but more important, for being there assisting me along the writing and publishing process.

To the Faculty of Life Science and Technology at the Norwegian University of Life Science for giving me the space, time, and resources to write this book.

Author

Professor **Dr. Jorge M. Marchetti** was granted his 5-year Bachelor in 2003 from Universidad Nacional del Sur, Bahía Blanca, Argentina in Chemical Engineering. In 2008, he was awarded his Ph.D. in Chemical Engineering in Biodiesel Production and in 2009, he was given his Ph.D. in Physics in the area of Hydrogen Storage; both Ph.D.s were given by Universidad Nacional del Sur in Argentina. From 2008 to 2010, Dr. Marchetti had a post-doctoral fellowship at Norwegian University of Science and Technology, Trondheim, Norway (NTNU) working on Natural Gas Refining at the Chemical Engineering Department. After that, he was appointed as an Assistant Professor at Chalmers University of Technology in Gothenburg, Sweden, at the Forrest Products and Chemical Engineering Division from the Chemical Engineering Department. From 2011, Dr. Marchetti has been an Associate Professor at the Norwegian University of Life Sciences, NMBU until July 2017 when he was promoted to Full Professor in Chemical Engineering. Currently, Prof. Marchetti is the leader of the Reaction Engineering and Catalysis Group, working on waste valorization, catalytic development and performance, reactor engineering, reaction modeling, process system engineering, and techno-economic assessments. Dr. Marchetti has over 70 per review publications, two books within the field of renewable energy, eight chapter contributions, and has been involved in more than 100 international meetings. He is involved as founder, editorial board member, and reviewer member of several index journals from Elsevier as well as Springer. Dr. Marchetti was awarded The Processing Division Student Excellence Award of The American Oil Chemists' Society, given by the American Oil Chemist Society. Prof. Marchetti was ranked within the top 1% of the scientist worldwide within the subfield of energy according the latest work published by Stanford University in cooperation with SciTech Strategies and Elsevier (https://lnkd.in/dRCSsT3) "Updated science-wide author databases of standardized citation indicators" published in October 2020.

1 Catalysis Preparation and Characterization

1.1 INTRODUCTION: BASIC CONCEPT AND ORIGIN OF CATALYSIS

The word *catalysis* is an important constituent of our daily life and contributes substantially to societal welfare. In today's circumstances, approximately 80–90% of the world's total chemical production processes use catalytic systems to accelerate various chemical reactions. A distinctive characteristic of a catalytic system is to change the rate of a chemical reaction, accelerate the formation of intermediate compounds, and generate desired products, while strongly controlling the selectivity and increasing the yield.

A catalyst is a substance that increases the rate of a chemical reaction without participating in it. The rate of a chemical reaction is mainly determined by the energetic barriers that need to be overcome to transform reactants to products. This energy, called activation energy, can be considerably decreased by using catalysts. A schematic representation can be seen in Figure 1.1.

An effective catalytic system provides high efficiency to convert raw materials into a single or multiple product through a series of elementary steps (or repeated cycles). It is regenerated to its original form at the end of each cycle during its lifetime. A catalyst only modifies reaction kinetics, and not the thermodynamics of a reaction.

The phrase *catalytic process* is believed to have been coined by Berzelius (1836). However, it is also claimed that some of the characteristics of catalysts were first presented by Fulhame in 1794, who demonstrated that the presence of water (H_2O) was required for the oxidation of carbon monoxide (CO) and that H_2O was

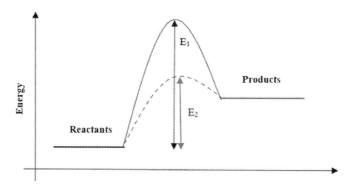

FIGURE 1.1 Energy between reactants and products without catalyst (E_1) and with a catalyst (E_2).

DOI: 10.1201/9780429466847-1

unaffected by this chemical transformation. In 1812, Kirchhoff presented a similar interpretation where starch was hydrolyzed to sugars by using dilute acids. In 1817, Sir Humphry Davy stated that combustible gases explode in the presence of air after exposure to heated platinum (Pt), at temperatures below the ignition threshold. In 1834, Faraday proposed that reactants adsorb on the surface of the catalyst. In 1877, Lemoine explained that catalysts do not affect the position of reaction equilibrium, but only alter the rate at which a reaction will progress. Thus, the proposed definition of a catalyst is that it *accelerates a chemical reaction without affecting the position of the equilibrium.*

1.2 CATALOGING CATALYTIC MATERIAL

Depending on the nature of the material, we can separate the different catalysts into homogeneous and heterogeneous types and heterogeneous can be categorized into supported and non-supported catalysts. In addition, we make a new distinction based on the size, introducing nanocatalysts. This section will discuss different types of catalysts while in Sections 1.3 and 1.4, their preparation and characterization techniques, respectively, will be provided.

1.2.1 HOMOGENEOUS CATALYSIS

A homogeneous catalyst is a catalytic material that is in the same phase as the reactants and products. For instance, for the transformation of vegetable oil with sodium methoxide to produce biodiesel, all chemicals are in liquid phase. Homogeneous catalysts can also be categorized as acidic and basic, as well as organic and inorganic. Enzymatic homogeneous catalysts have been associated here but it has been room for some debate, so they are placed into a category of their own. However, as long as a catalyst is not supported, *i.e.*, it is in liquid form, with the reaction also in liquid phase, it fits within this definition and category as well.

The advantages of homogeneous catalyst are high selectivity due to a large and defined number of active sites, and the mild reaction conditions required, in terms of temperature and time. However, by being in the same phase as the reactants, the separation of the catalyst from the rest of the reaction medium is regularly a very complicated and time consuming task, leading to a very expensive process which is less viable from an industrial perspective. Another drawback of these catalysts is that their thermal stability is lower than that of heterogeneous catalysts.

1.2.2 HETEROGENEOUS CATALYSIS

Heterogeneous catalyst implies that the reactants and the catalytic material are not in the same phase, like liquid–solid reactions or gas–solid systems. Heterogeneous catalysts are used in over 80% of the industrial chemical processes due to several benefits in comparison with homogeneous alternatives. Even though they could be less selective than homogeneous alternatives and the operational conditions are usually harsher, due to the heterogeneity of the materials, they are easily separated from the reaction medium (typically, a simple filtration is enough). Furthermore, they can be

Catalysis Preparation and Characterization

produced with different techniques and have different properties such as porosity, porous size, porous distribution, and surface area, allowing the catalysts to be tailor-made depending on the application and the reactants involved. As these materials are custom-made, their reproducibility is high, and their performance is quite stable. The major drawback of heterogeneous catalysts is that they may suffer from deactivation, *i.e.*, that after some time, the catalyst will no longer perform desirably.

A chemical reaction catalyzed by heterogeneous catalytic materials generally takes place at the interface between two phases. This demands in-depth investigation of the reaction pathway to understand the actual behavior of the material from a scientific research perspective. Figure 1.2 shows a schematic representation of a gas reaction over a catalytic surface and shows the three steps of adsorption of reactants, surface reaction, and desorption of the product.

The heart of heterogeneous catalyst lies in the surface-active sites. It is highly recommended to investigate the physicochemical properties of heterogeneous catalytic systems before its optimization for a particular catalytic process. In other words, determining an optimal number of active sites per reactor volume is very important from a reaction stoichiometry and process economics viewpoint.

In the context of development of heterogeneous catalysts, the characterization of materials is highly essential as it provides insights into the relationship between the activity of the catalyst and physicochemical properties of the material. If the structure and composition of catalysts can be correlated with its activity and selectivity, the performance of the catalyst can be understood, thus, improving reproducibility. Moreover, the field of heterogeneous catalysis is wide and highly interdisciplinary in nature, which demands the cooperation between chemists, physicists, surface scientists, material scientists, reaction engineers, and theorists and experimentalists. Additionally, based on the industry, like food, pharmaceuticals, automobiles, petrochemical, and biochemical industries, other experts are involved. With an approximation of around 80–90% of all the modern-day chemical processes using heterogeneous catalysts, consistent research efforts are under scientific attention for exploring different materials and upgrading their technical performance. The type of characterization performed on solid materials is dependent on the preparation methods and the estimated physicochemical properties. According to the International Union of Pure and Applied Chemistry (IUPAC), materials are categorized according to their average pore diameters as microporous (less than 2 nm), mesoporous (2–50 nm), and macroporous (more than 50 nm). The following sections provide a few examples of different types of solid materials that are used as catalysts in a variety of chemical reactions.

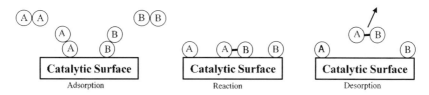

FIGURE 1.2 Scheme of a gas–solid reaction.

1.2.2.1 Nanoscale Catalytic Materials

Nanoscale catalytic materials are catalysts with particles size between 1 and 100 nm, which are seen as the future of the catalysis world because they exhibit excellent electronic, physical, mechanical, thermal, and catalytic properties. Nanomaterials can possess different structural shapes such as rods, tubes, fibers, wires, cubes, colloids, and quantum dots. Nanoscale catalysts are a subarea of the heterogeneous catalysts and are discussed in a subsection since they are used and produced differently. The main interest when developing nanoscale catalysts is to augment the number of active catalytic sites by specifically designing the structure and downsizing the catalyst particle size, based on the premise that heterogeneous materials assisted catalytic reactions are governed by the adsorption–reaction–desorption pathway. Significant success has been achieved in controlling the structural features of solid materials, even at the atomic level. The development of nanocatalysts has also been important due to the need to increase the specific surface area by minimizing its size. This, in return, has a positive impact on deciding the rate of a particular chemical reaction and obtaining the desired products with high selectivity.

Nanocatalysts are being used as a combined solution to take advantage of the best characteristics of a homogeneous catalyst, *i.e.*, high surface area and therefore, large contact area between the reactants and the catalyst, while still easy to separate as the catalyst is in solid phase, making it insoluble in the reaction mixture. In addition, nanocatalysts have high selectivity, high activity, great stability, high energy efficiency, and good atom economy, which significantly reduces waste generation [1].

In the work published by Singh and Tandon [1], they present an important figure summarizing applications of different nanocatalysts, such as the production of 1,2,3-triazole using gold nanoparticles, C–C coupling reactions using platinum nanocatalysis, and the synthesis of phenols, anilines, and thiophenols from aryl halides using nanocatalysts from copper, among others.

1.2.2.2 Porous Materials

Nowadays, many highly porous catalytic materials such as foams, and ordered structures, exist. Nevertheless, it is important to understand the difference between surface roughness and porosity, as demonstrated in Figure 1.3. A particular material is porous if the defects are deeper inside the structure, whereas roughness is usually on the external surface.

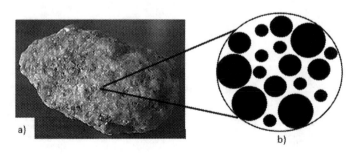

FIGURE 1.3 Surface roughness (a), porosity (b).

Furthermore, pore size distribution is beneficial for the mass transfer of the reaction species as molecular transportation is highly affected by changes in the pore dimensions. The pore size is the distance between opposite walls or the internal pore width. Pore sizes are accurately obtained only if the pore shapes of any material are precisely defined. This is very critical, especially for materials with different ordered and disordered structures, such as porous carbon-honeycomb, ink bottle like-glass, nanotubes, and new-generation silica materials with linked micropores and mesopores.

Porous materials find applications in separation processes, gas storage, smart electronic devices, ion-exchange membranes, and the catalysis sector. Industrial processes such as water treatment, indoor air quality, selective synthesis of value-added chemicals, and different types of liquid biofuels, demand materials with porosity ranging between micropore and macropore region.

As an example, the scientists at Mobil groups observed a significant breakthrough in the field of mesoporous alumino-silicate materials following the discovery of molecular sieves. These materials possess large uniform pore structures. The first pathway for the synthesis of such materials involves the organization of cationic surfactant species into lyotropic liquid crystal phases, which serves as a template for the formation of hexagonal structure. The aggregation of surfactant micelles then arranges into the hexagonal arrays of rods followed by the interaction of silicate anions with the cationic groups of the surfactant species. The condensation of silicate species results in the formation of an inorganic polymeric species. The removal of the template is performed by calcination to have hexagonally arranged inorganic hollow cylinders. The next step involves the ordering of silicate moieties in a hexagonal arrangement. The randomly distributed surfactant micelles interact with the silicate via columbic interactions leading toward the formation of multiple silica monolayers, which are then arranged into highly ordered mesoporous networks with an energetically favored hexagonal arrangement. The condensation of silicate species also occurs during this stage.

The other strategies for the synthesis of mesoporous silica are charge density matching, folded sheet mechanism, and silica liquid crystal models. The assembly of mesoporous materials can be driven by hydrogen bonds using neutral templates such as non-ionic poly-(ethylene oxide) (PEO) surfactants and inorganic precursors [2,3].

1.2.2.3 Supported Catalysts

Precious noble metals like ruthenium (Ru), rhodium (Rh), palladium (Pd), platinum (Pt), gold (Au), and silver (Ag) are well known to present excellent catalytic activities. However, these materials are too expensive to be attractive for many industrial applications, but they are used for some specific applications. For instance, Pt is commonly used for fuel cells, Pd, Pt, and Ni are used for oil hydrogenation, and Ag is commonly used for ethylene oxide and formaldehyde production [4–6]. This is possible since the amount of these expensive catalysts is reduced to a minimum by supporting small quantities of the active phase over a substrate that mainly provides mechanical structure and stability. This is achieved by controlling the particle sizes and improving the metal dispersion on the support. Figure 1.4 shows the deposition of expensive metal over a cheaper substrate. Figure 1.4a shows a pure noble metal

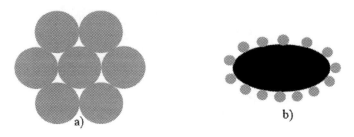

FIGURE 1.4 (a) Pure noble metal and (b) cheaper substrate with dispersed metal.

catalyst, while Figure 1.4b presents a similar sized substrate with some surface-deposited metal.

Different synthesis protocols are tested to reduce the particle size to increase the active catalytic sites on the surface, exposed to the reactant, to accelerate the chemical transformation. Furthermore, the physicochemical properties of the support play an important role in providing a synergistic effect to the catalyst. For example, Au nanoparticles present excellent activity to catalyze the CO oxidation because the reactive oxygen is supplied by a nanocrystalline ceria (CeO_2) support for Au nanoparticles. The oxygen is supplied in the form of surface η_1 superoxide species and peroxide adsorbed species, at the one-electron defect site, to the supported gold catalyst. Moreover, the material synthesis technique decides the oxygen vacancies in ceria. Raman analysis suggests that the surface properties are different for precipitated and crystalline CeO_2; nanocrystalline CeO_2 stabilizes oxygen as superoxide and peroxide species, whereas conventionally precipitated ceria tends to stabilize oxygen adsorbed species and molecular oxygen. Both cationic and metallic Au species are present in the Au catalyst on nanocrystalline CeO_2, and that the catalytic sites incorporate cationic gold [7].

1.2.3 Biocatalysis

Biocatalysts is the subarea of catalysis that studies the reactions when using a living organism to speed up the process; the most common living organisms being used are yeast, bacteria, fungi, and enzymes.

Yeast has been using since ancient Egypt to produce alcoholic beverages, and we still use this technique today. Enzymes are the typical biocatalysts due to their huge market and increasing interest. Enzymatic catalysis, which is mainly protein-based, has been used for over 100 years for different chemical purposes [8]. The main advantages of enzymes over other catalysts are that they have extremely high selectivity; the human body has over 70,000 enzymes and every single one of them has one particular task [9], and they can be supported over different substrates. Therefore, they have the main advantages of a heterogeneous catalyst (minus bleaching into the solution), and the operation conditions are very mild (low temperature). Thus, they are extremely good energy savers; some enzymes are quite stable and can be used for long periods of time. However, usually their price is too high, making some of them impossible to be used commercially. Lipases from fungi and bacteria are easy to

Catalysis Preparation and Characterization 7

produce in bulk amounts because of their extracellular nature. These microorganisms have been utilized for several different purposes and already existing enzymes have discovered new applications. Lipases are used as catalysts in food processing, detergents, pharmaceuticals, paper, cosmetics, and chemical synthesis industries [8].

Enzymes work under two main mechanisms as catalysts, *i.e.*, the lock and key and the induced-fit model. For the lock and key mechanism, the enzyme is the lock, and the substrate is the key; therefore, you must have the right key to open a lock, otherwise it will not work. This is an analogy used to show how selective the enzymes are toward a desired substrate and therefore, to produce a specific product. The difference with the induced-fit model is that the substrate plays a role in shaping the enzyme and modifying it in order to allow the substrate to interact. Then, the enzyme will return to its original shape after the interaction.

Figure 1.5 a–d shows a "lock and key model." Figure 1.5a shows the reactants approaching the enzyme, Figure 1.5b is when the reactants interacts with the enzyme and produced the desire (D) and undesired (U) products, and this can be seen in Figure 1.5c; finally, the products are desorbed from the enzyme and as it can be seen, in Figure 1.5d, there is no modification in the enzymes active sites. Figure 1.6 a–d shows the mechanism for the induced model. Figure 1.6a shows the reactants and the enzyme prior to the interactions and Figure 1.6b shows how the reactants starts to interact with the enzyme before this one is modified to induce the fitting. Figure 1.6c represents the transformation of the reactants into the desire products, while Figure 1.6d represents the desorption from the active sites of the products.

It is important to not confuse the induced-fit model with the conformational approach, which is not presented in this book. It differs from the induced-fit model in the fact that the enzyme first modifies its active site to fit the reactants and then it

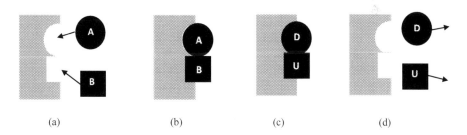

FIGURE 1.5 Representation of a lock and key model.

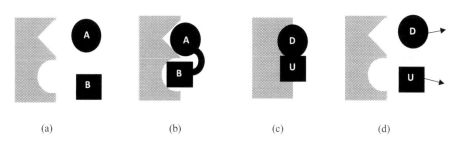

FIGURE 1.6 Representation of an induced-fit model.

starts the interaction with it, while in the induced-fit model, the modification of the active site is done due to the interaction with the reactant.

1.2.4 Photocatalysis

Photocatalysis is the sub-branch of catalysis that studies the chemical reactions that are initiated by the presence of light. Light activates the catalysis and therefore stimulates the chemical reactions. Some typical photoreactions are oxygen reduction and oxidation of organic compounds, while the most generally known photoreaction is photosynthesis, where plants use sunlight to transform CO_2 and H_2O into glucose ($C_6H_{12}O_6$) and O_2.

In photocatalysis, a semiconductor material could be used to perform a chemical reaction stimulated by photons as an energy source. An ideal photocatalyst is photostable, chemically inert, and available at an affordable price. Although a variety of semiconductors are now available, titanium oxide is still seen as the catalytic material with the most potential for practical applications. This is related to its performance, availability, robustness, cost of the material, stability under chemical conditions, and safety. TiO_2 has been used for H_2O oxidation with H_2 production, water splitting with no need of extra energy, and reduction of CO_2, among other uses [9]. The overall photocatalytic activity of TiO_2 is determined by its crystalline structure, surface area, density of surface hydroxyl groups, and adsorption/desorption characteristics. TiO_2 has three common crystalline polymorphs: anatase, rutile, and brookite. Figure 1.7 shows a schematic representation of the reduction of CO_2 over TiO_2 in the presence of water.

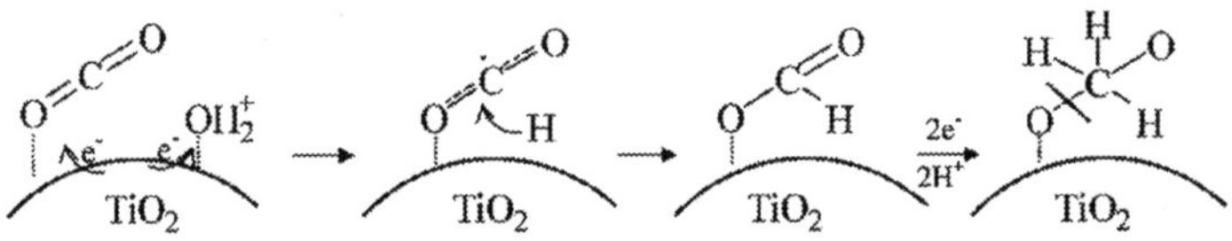

FIGURE 1.7 Proposed mechanism for the photoreduction of CO_2 in the presence of water [9].

One of the major advantages of using photocatalysis is that it involves the possibility of performing chemical reactions at ambient temperature. On the other hand, the limitation with using titanium oxide, for example, is its large band gap (3.2 eV), making it active only upon exposure to the UV light and high recombination rate of electrons and holes that makes photocatalytic reactions have poor energy efficiency. Figure 1.8 shows an example of a Z-scheme photocatalytic degradation [10].

One of the major sources for recombination of electrons and holes is due to the presence of defects in the lattice. In the case of TiO_2, the mobility of electrons is higher in comparison to the holes. The band gap can be reduced by doping TiO_2, which would increase the concentration of trap sites. The recombination rate can be slowed down by annealing the catalyst that results in the augmentation of crystallinity. The inclusion of noble metals nanoparticles (Pt, Pd, Au) are also used to improve the separation of electrons and holes wherein electrons will be transferred to the noble metals and holes remain close to the photocatalyst.

Catalysis Preparation and Characterization

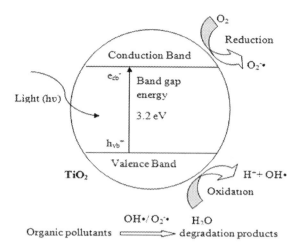

FIGURE 1.8 Schematic representation of a photocatalytic degradation over TiO_2 [10].

The photocatalyst-assisted chemical reactions can be summarized in the following steps:

- Diffusion of the reactant from bulk to the surface
- Adsorption of reactants on the surface
- Absorption of photons and generation of electron–hole pairs
- Transfer of electrons and holes to the catalytic surface and to the adsorbed species
- Reaction of adsorbed species to products
- Desorption of products
- Diffusion of products from surface to bulk

Photocatalysis is one of the technologies that holds potential and has been gaining commercial interests for air purification processes, water treatment, energy storage, and valuable chemicals production. Improving air quality, for example, is becoming increasingly important due to increasing pollution and global demand. Purifying indoor air is especially highly relevant as people spend up to 80% of their time indoors to avoid health concerns such as headaches, fatigue, and asthma symptoms.

1.3 CATALYST PREPARATION TECHNIQUES

As mentioned, catalysts have become an indispensable part of today's chemical-based society; they are involved in most products used in our daily life. How the different catalysts perform is directly correlated with the catalyst properties such as porosity, size, and surface area. These properties are linked, to some extent, with the methodology used for the preparation of the materials. Thus, different approaches will provide different final properties and therefore, different catalysts to be used in different applications. A description of some of the various techniques will be presented in this subsection in order to show these methodologies and the main effect they can have on the different materials.

1.3.1 Precipitation

The most commonly applied procedure for the preparation of heterogeneous catalysts and solid supports is the precipitation technique. The process is performed by the controlled addition of a precipitation reagent like an acid or base. The coprecipitation methodology involves more than one component. Precipitation is cheap, reliable, and easy to implement; however, as a technique itself, it requires a good control for the addition of reagents as well as a good separation procedure afterward to purify the desired catalyst and eliminate all other chemicals. Typically, a solution with the desired chemical is mixed with another reagent to reach the desired concentration, pH, and reactions conditions in general. This allows the desired new chemicals to be formed, and this new material is typically in a solid phase precipitating in the solution. On separation and purification of the solid from the solution, the desired catalysts are obtained. A typical laboratory example is the production of AgCl from $AgNO_3$ and KCl. Industrially, this technique is used to produce Fe_2O_3 (for Fischer-Tropsch reactions), TiO_2, and $AlPO_4$ (polymerizations process), among others.

When determining the final properties of the catalyst, several operational parameters need to be considered in order to achieve, for example, the right homogeneity in the final product. Ertl *et al.* [11] present a very clear description of how each different parameter affects the properties of the precipitate. For instance, the solvent is crucial for the textural properties and the crystallinity, the precipitating agent is relevant for the homogeneity, the pH for the crystal phase, and the mixing for the composition of the precipitate. Other factors include temperature, pressure and additives. The formation of the solid phase within the liquid solvent, generally takes place by two possible pathways: (i) nucleation and (ii) growth or agglomeration.

Shahinuzzaman *et al.* [12] presented a pathway on how to prepare a catalyst using the precipitation technique. Figure 1.9 shows a reproduction (with permission) of this scheme pathway.

1.3.2 Coprecipitation

The coprecipitation technique is generally used to produce two component-based catalysts (for example, metal-based catalysts). This technique has very closely related properties to the precipitation technique and it is generally introduced as one subsection. This methodology differs from precipitation mainly since it is not possible to achieve homogeneity when precipitating one solute. The final composition and structure of the catalysts or catalyst precursors will depend, as before, on several factors, such as pH, temperature, mixing rate, and the solubility among the two metal-based reagents. This solubility, together with the quantity of each of them employed, as well as the solubility of each of them with the liquid mix, will lead to different types of catalysts, with different properties. Therefore, the same metal reagents could lead to two very different catalysts with different applications based on the solution liquid. Swatiska *et al.* [13] used this approach for the preparation of a Cu-ZnO-MgO-Al_2O_3 for the direct synthesis of DME. In their work, they present a scheme for the coprecipitation of reagents; Figure 1.10 shows the pathway employed when using coprecipitation to produce the desired catalysts.

Catalysis Preparation and Characterization

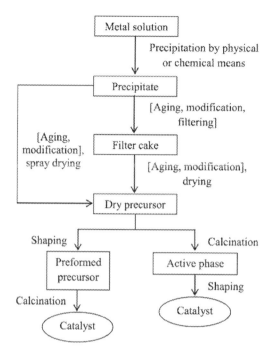

FIGURE 1.9 Preparation scheme for the precipitation methodology. (*Reproduced with permission* [12]).

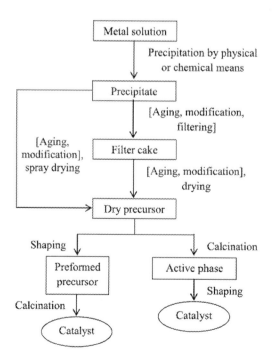

FIGURE 1.10 Copreparation scheme [13].

According to Schwarz et al. [14], there are two well-defined chemical routes when using coprecipitation: (i) sequential precipitation of separate chemicals and (ii) coprecipitation of well-known chemicals. The main difference is that for the sequential pathway, each chemical precipitate separately from each other, while in the second pathway, the chemicals interact with each other prior to or during the precipitation.

Some of the possible catalysts and their application are Cu-ZnO/Al$_2$O$_3$ for methanol synthesis, (VO$_2$)$_2$P$_2$O$_7$ for the selective oxidation of butane, and V-Mo oxides for selective oxidation of acrolein, among others.

1.3.2.1 Theory of Nucleation

Here we would like to present a brief mathematical description of the theory of nucleation, and we strongly recommend the reader to read [11] for a more detailed explanation. If we consider a spherical nucleus where the precipitation occurs only on the surface, the total free enthalpy can be calculated, taking into the interaction of the nucleus with the surface, as follows:

$$\Delta G_{total} = \frac{2}{3}\pi r\left(m_{fes} - m_{fed}\right) + 2\pi r^2 \gamma + \pi r^2 \gamma_{ss} \qquad (1.1)$$

where m_{fes} is the molecular free energy of the solid, m_{fed} is the molecular free energy of the dissolved solid, γ is the interfacial energy, and γ_{ss} is the interfacial energy with the support.

1.3.3 Impregnation

This technique relies on the impregnation of an active phase over a solid substrate. The active phase is then in contact with the substrate. By this contact, the catalytic active materials are impregnated (wet impregnation is also a name for this approach), and then, after the impregnation is finished, a drying, or calcinating process is required to eliminate all remaining volatile chemicals and the solution medium, while depositing the metal onto the target surface. Figure 1.11 shows a scheme of the steps involved in the impregnation process.

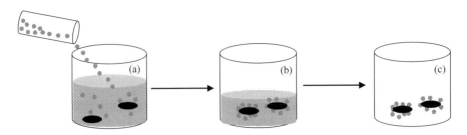

FIGURE 1.11 Impregnation process scheme showing the (a) addition of catalytic substrate, (b) removal of solution after impregnation, and (c) the final catalytic material.

Catalysis Preparation and Characterization

There are mainly two different types of impregnation: wet impregnation and capillary wet impregnation. The main difference depends in the quantity used in each case, where wet impregnation is related to an excess of solution, capillary wet impregnation is carried out with a solution amount equal to the pore volume of the solid material.

Temperature is the main factor affecting this process; it affects the solution as well as the substrate, as it can change their density, viscosity, and solubility. This approach is generally exothermic, however, there are some cases where this small increase of temperature can have negative effects; for example, if the solution has more than one precursor and they are unstable. The precipitation of one or more of them due to changing temperature, is likely. Other typical factors are related to the mass transfer limitations that appear for the flow of liquid within the pores, this is directly correlated with factors like the porous size, the diameter, and the porosity of the substrate [11].

This technique is relatively cheap; the porosity of the catalysts is not determined by the active phase but mainly by the substrate itself. It is hard to produce in large quantities and to obtain a uniform distribution of the active phase over the catalysts. Mass transfer limitations for the impregnation process can be overcome by using different substrates. Well-known catalytic supports include ceria, zirconia, zeolites, silicas, alumina, and carbon, among others. For example, Pt-Re over Y-alumina is used for the dehydrogenation of fuels to produce platform chemicals.

1.3.4 Sol Gel Method

This method is very commonly used for the preparation of carbon-based components, metal oxides, silicon-based materials, and zeolites. This method consists mainly on the transformation of monomers or solids particles that are in a liquid (sol) into a gel, which can contain both liquid and solid phases. In order to obtain the final catalytic materials, a removal of the liquid phase is required. This can lead to two main types of products, an aerogel or a xerogel, where the main difference can be seen in the shrinkage experienced by the materials and this shrinkage is related to the type of drying employed.

Figure 1.12 presents a scheme of the steps that take place in the sol gel approach.

This methodology can be used mainly to prepare three types of catalysts: (i) bulk inorganic mono or multimetallic catalysts, (ii) bulk multiphasic materials, and (iii) uni- or multiphasic coatings, films, or membranes [11].

Each step of the process presented in Figure 1.12 will be different in operation conditions, concentration, and reaction parameters, based on what type of the three above-mentioned types of catalysts is desired. We will not get into the details and recommend the *Handbook of Heterogeneous Catalysts* for further reading as well as the references within [11].

The advantages of this methodology are that it is a simple process that can be used to produce the desired catalysts, that the structure can be tailor-made based on the application where it will be used. Some applications are within the area of energy, aerospace engineering, textiles, and solar cell applications. More

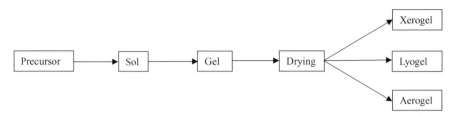

FIGURE 1.12 Process scheme for the impregnation methodology.

specifically, Nb_2O_5 over SiO_2 can be used for isomerization of butene, while Pt-Sn/Al_2O_3 can be used for propane dehydrogenation and heptane dehydrocyclization, among other applications [15].

1.3.5 Chemical Vapor Deposition

Chemical vapor deposition (CVD) is generally used to produce coatings on different surfaces, it has been a methodology generally used for the production of semiconductors, ceramics and lately for carbon nanotubes [16]. This methodology is based on a gas reacting with a hot surface producing a solid deposition on it. Figure 1.13 shows a representation of the process. This chemical reaction can produce high purity catalysts with special dopants on the external surface of the above-mentioned hot surface. This is a very versatile methodology for preparing very different types of catalysts. However, the high vapor pressure values of the precursors and use of hazardous chemicals is a significant drawback for this technique. In addition, in some cases, the by products from the precursors are also toxic and therefore, the process becomes more and more costly due to the need for neutralization of those hazardous by products.

This methodology has some advantages, *i.e.*, the surface is generally more uniformly covered, it is done under some vacuum but does not require high vacuum levels; it can be done in a continuous process which minimizes the introduction of impurities into the surface. Some of the disadvantages related to this technique are the high temperatures required, and the possibility of combustion due to some chemicals being flammable within the temperature range.

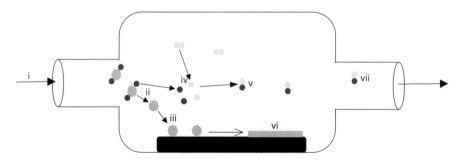

FIGURE 1.13 Process scheme for chemical vapor deposition, (•) Chemical G^1, (∘) Chemical G^2, (∗) Chemical M.

Catalysis Preparation and Characterization

Figure 1.13 shows all the steps for the coating of a surface with a metal, described as follows: (i) the chemical MG^1_2 flows into the system (the selection of the chemical is arbitrary), (ii) the precursor interacts with a different gas (G^2_2, also arbitrary selected), in order to produce (iii) a pure metal M that will be deposited on the surface and produce (iv–v) new gas by products (G^1G^2). The desposition of the metal will produce a layer over the surface, producing the desired coating (step (vi)), while the new chemical is removed from the reaction chamber in step (vii). Several different types of metals can be used such as gold, cobalt, gallium, molybdenum, platinum, and nickel, among others.

Atomic layer deposition (ALD) is a subsection of CVD, in which two different reactants are used. The vapors of the first chemical interact with the surface, producing a deposition over the surface that is used by the second chemical to get attached. This is a two-step reaction, which is usually stepwise or sequential. This technique produces a thin layer that is deposited over the desired substrate. It is a technique that can be used to deposit specific dopants over special substrates, to produce a thin layer which maximizes the catalytic properties of a material, and it can be used to have a better control of the composition of the deposited material, among other advantages. Johnson *et al.* [17] present a good review on the topic and provide a complete list of different materials that can be used for the ALD approach.

1.3.6 Ion Exchange

Ion exchange is a technique where an ion electrostatically interacts with the surface of a support and is transformed by another ion. Usually, the surface that is to be covered with a target ion, M, is first soaked into a solution with another ion, N, which covers the surface and then N is replaced by M. This process takes place until an equilibrium of the ions is achieved, as shown in Equation 1.2. The solid is then washed, filtrated, and dried and the final catalyst is ready.

$$N^+_{Solid} + M^+_{Solution} \rightleftarrows N^+_{Solution} + M^+_{Solid} \qquad (1.2)$$

Figure 1.14 shows a schematic representation of the reaction shown in Equation 1.2, where the ions are interchanged. Figure 1.14 shows step (a), where M is approaching the surface covered with N^+; in step (b), we can see how the exchange of ions takes

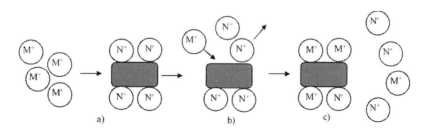

FIGURE 1.14 Ion-exchange scheme.

place, while step (c) is the final step where the equilibrium among the different ions has been achieved.

This methodology is used for the preparation of several different types of catalysts, in order to support special ions over different substrates as well as to doped different catalysts.

One of the most commonly used catalysts are ion exchange resins, which are defined as crosslinked sulfonated polystyrene structures. They are prepared by the exchange of ions (cations or anions), and when used in a catalytic reaction, these ions interchange with the reacting liquid. The most commonly used resin is a sulfonic acid resin which has several different applications in industry such as tert-butanol production, methyl tert-butyl ether, biodiesel production, water purification, metal purification, hydrolysis, and pharmaceuticals. A typical chemical structure of a resin can be seen in Figure 1.15 [18].

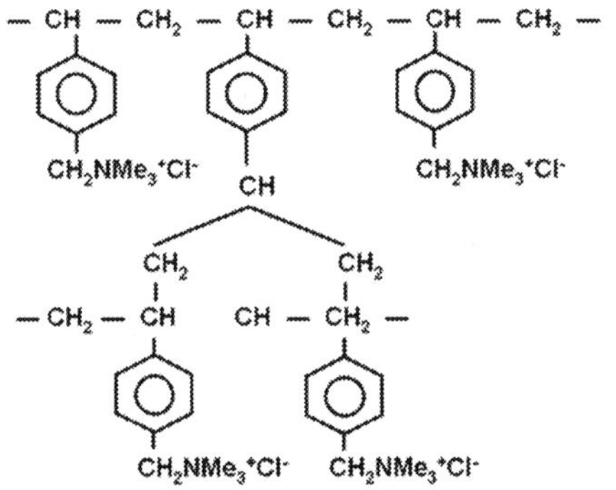

FIGURE 1.15 Polystyrene anion resin. (*Reproduced with permission* [18]).

1.3.7 Immobilization of Catalysts

Catalyst immobilization is a technique where a homogeneous catalyst is anchored over a support. This can be achieved via different techniques, like adsorption, covalent bonding, ionic bonding, crosslinking (co)polymerization, and entrapment, among others, where each of these techniques has advantages and disadvantages. They are also different based on the type of support to be used, the type of active phase to be immobilized, and the catalyst application.

Usually, this methodology is associated with enzymes; therefore, we will present some of the most common techniques used to achieve catalyst immobilization, with emphasis on enzymatic immobilization. For the case of enzymes, several advantages can be achieved when immobilized, for instance, the enzyme can be re-used, it can be used in a continuous process and separated from the product more efficiently, it is more stable and suffers less degradation which improves the process as a whole, and the enzyme performance can be improved due to the substrate. However, the number of industrial applications that can use this type of catalysts are limited, the drawbacks include the expensive process of the purification and isolation of the enzymes. The enzyme is therefore, typically immobilized over different substrates, and the most

Catalysis Preparation and Characterization

common substrates are glass, silica, ceramics, carbon, as well as synthetic and natural-based polymers.

1.3.7.1 Covalent Bonding

This technique involves the covalent attachment of catalysts to a support/carrier (organic or inorganic). Different supports have varying affinities with different types of catalysts or enzymes, and this influences the strength of the covalent bonding. This is a very simple and well-established methodology and is generally used. However, the enzyme can suffer chemical modifications that will alter its activity due to the new chemical bond with the support. Figure 1.16 presents a representation of this covalent bonding.

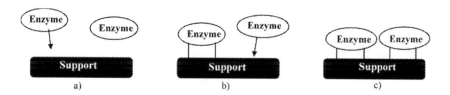

FIGURE 1.16 Schematic representation of a covalent bonding: (a) enzymes approaching the support, (b) some enzymes have covalently attached to the support, and (c) enzymes have covalently covered the support.

1.3.7.2 Entrapment

The method of entrapment is used to embed an enzyme into a porous matrix; as shown in a schematic representation in Figure 1.17. Entrapment can be carried out by embedding in a matrix or by encapsulation in a membrane and the catalyst activity is not modified. This is a highly flexible, cheap, and fast immobilization technique; however, leakage of the enzyme as well as mass transfer limitations are the main drawbacks with these types of catalysts.

Entrapment can be done in a matrix or a membrane. For matrix entrapment, the catalyst is then located in the empty spaces of a polymer-based matrix, see Figure 1.17. This can be done by submerging the support in a solution containing the biocatalysts.

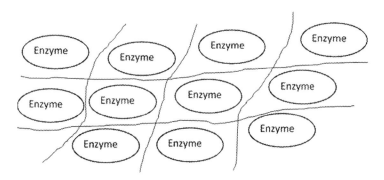

FIGURE 1.17 Schematic representation of an entrapped enzyme.

There are different methods to prepare the desired matrix, like polymerization, precipitation, and gelation. Each of these methods permits different materials to be entrapped and therefore, different catalysts to be produced [11].

When using membranes, a similar methodology applies; here, the membrane formation takes place in the presence of the biocatalysts. This approach is very good for limiting the mass transfer restrictions; however, their activity can be compromised. A second approach consisted in adding the biocatalysts to the membrane after the support has been created. It will yield larger catalyst systems with good mechanical properties.

1.3.7.3 Adsorption

Adsorption is a well-known, old technique to support different chemicals over the surface of a solid and can be used for many different applications. In this case, it is used to attach a biocatalyst, generally enzymes, to the surface of a solid support. This can take place due to different forces, like van der Waals and hydrogen interactions. These interactions determine how strongly the biocatalysts are attached to the support. These interactions are strongly dependent on some properties and characteristics of both the catalyst and the support, such as pH, pressure, temperature, and surface area. Figure 1.18 shows a representation of the adsorption process of an enzyme over a solid support.

FIGURE 1.18 Schematic representation of an adsorption process: (a) before adsorption and (b) after adsorption.

1.3.7.4 Ionic Bonding

Similar to covalent bonding, ionic bonding implies the chemical interaction of the biocatalysts with the support. These interactions are due to the difference in charge within the atoms involved in the process. This type of bonding occurs when electrons are donated from one atom to another, which generally happens when a metal and a non-metal interact.

1.3.8 OTHER ALTERNATIVES

Finally, we will like to name some other alternatives that can be found in the literature such as hydrothermal synthesis, flame hydrolysis, grafting, coimpregnation, gelation, flocculation, melt infiltration, and colloidal synthesis, among others.

1.4 CATALYST CHARACTERIZATION TECHNIQUES

For detailed physico-chemical characterization of these catalytic materials, several different properties and characteristics need to be measured and analyzed, such as the surface area, the porosity, the pore diameter and pore size distribution, the morphology, the bulk chemical composition, and the surface chemical composition. These can be achieved by different analytical techniques such as X-ray diffraction, scanning and transmission electron microscopy, infrared, Raman spectroscopy, and methods based on adsorption and desorption of probe molecules. These investigations are expected to provide information about phase structure, crystallinity, crystallite size, surface structure, textural properties, nature of active sites, morphology, particle size, nature of acidity, and other characteristic features. The structure–activity relationship can also be better understood by using these techniques, which helps to improve the activity of the catalyst for various applications.

1.4.1 X-ray Diffraction (XRD)

X-ray diffraction (XRD) is one of the fundamental techniques employed for catalyst characterization. It has been an important technique for determining the structure of materials characterized by long-range order. It is used to identify crystalline phase(s) of catalysts by means of lattice structural parameters and crystallinity. X-rays are highly intrusive electromagnetic radiations, which are electrically neutral. Their frequency lies between the ultra-violet (UV) and gamma radiations, and their wavelengths (λ) range from approximately 0.04 to 1000 Å. For diffraction applications, X-rays with only shorter wavelengths ranging from few Å to 0.1 Å (1–12 keV) are used because these are comparable to the size of the atoms. Therefore, the X-rays are ideally suited for deducing the structural arrangement of atoms and molecules for various materials. The diffraction method involves the interaction between the incident monochromatized X-rays (like Cu Kα or Mo Kα source) with the atoms of a periodic lattice. X-rays scattered by atoms in an ordered lattice interfere constructively as per Bragg's law, as presented in Equation 1.3.

$$n\lambda = 2d\sin(\theta) \quad (1.3)$$

with $n = 1, 2, 3...$, where, λ is the wavelength of the X-rays, d is the distance between two lattice planes, θ is the angle between the incoming X-rays and the normal to the reflecting lattice plane, and n is the integer called order of the reflection.

By measuring the angle 2θ, under which constructively interfering X-rays leave the crystal, the Bragg's equation gives the corresponding lattice spacing, which is characteristic for a particular compound. The width of the diffraction peaks signifies the dimension of the reflecting planes. It is known that the width of a diffraction peak increases when the crystallite size is reduced below a certain limit (<100 nm). Therefore, XRD patterns can be used to estimate the average crystallite size of very small crystallites using XRD line broadening, by applying the Scherrer formula presented in Equation 1.4:

$$t = \frac{0.9\lambda}{\beta \cos(\theta)} \quad (1.4)$$

where t is the thickness of the crystallites (in Å), λ is the wavelength of X-rays, θ is the diffraction angle, and β is the full width at half maxima of the diffraction peak.

Our group has worked on the production of biodiesel from different oils using new renewable catalysts [19–21]. We have prepared a CaO-based catalyst enhanced with glycerol. Figure 1.19 shows the variation of intensity when measuring 2θ [22].

The diffraction peaks at 2θ, i.e., 32°, 37°, 54°, 64°, and 67° are attributed to the CaO species. In the case of CaO, the absence of characteristic peaks for $CaCO_3$ at 2θ, i.e., 29.4° as well as for $Ca(OH)_2$ at 2θ, i.e., 18°, 34°, 47.2°, and 50.8° confirmed that the calcination process of M. Galloprovincialis shells at 800°C resulted in the formation of only CaO species. After calcination at 800°C, both $CaCO_3$ and calcium hydroxide ($Ca(OH)_2$) decompose. The diffraction pattern for CaDg was obtained after the reaction of CaO with glycerol, in the presence of plant oil, at 65°C. Small but detectable characteristic peaks at 2θ, i.e., 8.2°, 10.2°, 24.4°, 26.6°, 34.4°, and 36.2°, in the diffractogram confirmed the formation of CaDg species. The XRD patterns of the used CaO and CaDg catalyst were also obtained and compared with those

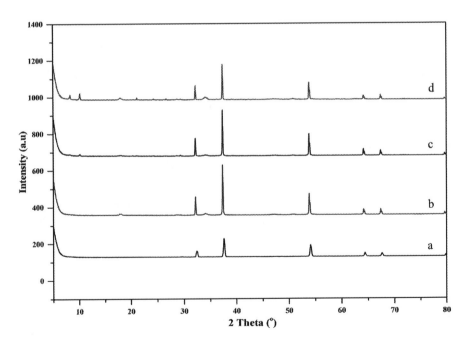

FIGURE 1.19 XRD results for CaO-based catalysts: (a) fresh CaO, (b) used CaO, (c) fresh CaDg, and (d) used CaDg. (*Reproduced with permission* [22]).

of fresh catalysts. In the case of used CaO catalyst, the appearance of diffraction peaks at 2θ, i.e., 29.4° and those at 18°, 34°, 47.2°, and 50.8°, indicated poisoning of CaO catalyst due to an adsorption of surrounding CO_2 and moisture, respectively. The hydration and carbonation reaction resulting in the formation of $CaCO_3$ and $Ca(OH)_2$ species might reduce the catalytic activity of CaO. In the case of collected CaDg catalyst, the increase in the intensity of diffraction peaks at 2θ, i.e., 8.2°, 10.2°, 24.4°, 26.6°, 34.4°, and 36.2° indicated that untransformed CaO that remained during the synthesis of CaDg reacted with the glycerol produced during the alcoholysis reaction. The consistent generation of CaO–glycerol complex during the alcoholysis reaction was anticipated to promote the activity of CaDg catalyst.

1.4.2 Porosity Measurements

Evaluation of surface area is a very generally practiced characterization as the catalytic activity is possibly linked indirectly to the total number of surface sites. Surface area is categorized into internal surface area, which is the area of pore walls, and the external surface area, which is the area of surface outside micropores and mesoporoes. The specific surface area, pore volume, pore size, and porosity of any material, are highly dependent on the synthesis protocol adapted. Therefore, the synthesis procedure is always established based on the application of the catalyst and the catalyst support.

The most accepted method of measuring surface area of catalytic materials is the one that is based on the theory developed by Brunauer, Emmett, and Teller in 1938. It considers the multilayer adsorption of a gas, generally nitrogen, under the assumptions that: (i) the adsorption energy remains constant as the coverage goes from zero to fully covered for the primary layer of the adsorbate and each successive layers above, (ii) the enthalpy of adsorption is the same for any other layer except the first one, (iii) there is no inter-molecular interaction, however they attract and retain molecules striking them from the gas phase, and (iv) a new layer can be initiated before the completion of the one under. Equation 1.5 shows the Brunauer–Emmett–Teller (BET) equation:

$$\frac{P}{\left(V(P_0 - P)\right)} = \frac{1}{(C*V_m)} + \left(\left[(C-1)\Big/(C*V_m)\right] * \left(\frac{P}{P_0}\right)\right) \tag{1.5}$$

where P_0 is saturation vapor pressure of the adsorbate at the experimental temperature, P is adsorption equilibrium pressure, V_m is the volume of adsorbate required for monolayer coverage, V is volume of gas adsorbed at pressure P and C is a constant related to the heat of adsorption and liquefaction. In order to quantify the amount of gas adsorbed (regularly is nitrogen), a linear relationship between P/P_0 and $P/(V(P_0/P))$ is required. This linear portion is only restricted to a limited portion of the entire isotherm, generally for P/P_0 of 0.05–0.30. The slope of the straight line S is equal to $(C-1)/(C*V_m)$ and the intercept I is equal to $1/(CV_m)$. Both the parameters are used to calculate the monolayer volume, V_m given by $1/(S + I)$.

The surface area of the catalyst (SBET) is related to V_m by Equation 1.6.

$$\text{SBET} = \left(\frac{V_m}{22414}\right) N_A \sigma \tag{1.6}$$

where N_A is the Avogadro number and σ is the mean cross-sectional area covered by one adsorbate molecule.

Nitrogen gas is most widely used for surface area determinations as it exhibits intermediate C values (50–250) on most solid surfaces, impeding localized adsorption or behavior as a two-dimensional gas. It has been demonstrated that the C influences the value of the cross-sectional area of an adsorbate. Therefore, the acceptable range of C for nitrogen makes it possible to calculate its cross-sectional area from its bulk liquid properties. For the hexagonal close-packed nitrogen monolayer at −196°C, the cross-sectional area σ for nitrogen is 16.2 Å.

In the case of supported metal (or metal oxide) catalysts, it is important to know what fraction of the active metal atoms are exposed. This is so because atoms located in the interior of the metal particles do not participate in surface reactions, and hence are not available for the purposes of catalytic processes. The metal dispersion, D, is defined as the fraction of metal atoms found on the surface of the active metal particles; it is expressed as a percentage of all metal atoms present in the sample. The higher the dispersion, the more exposed is the metal surface and more efficient is the catalyst. The value of D, dispersion, can be obtained using Equation 1.7.

$$D = \frac{N_m * S * M}{100L} \tag{1.7}$$

where M and L are the molecular weight and percent loading of the supported metal, S is adsorption stoichiometry, N_m is the monolayer uptake of chemisorbed gas (µmol/g).

An alternative methodology to obtain the surface area, plus the particle size distribution, is mercury porosimetry. This technique can also provide information regarding the bulk density, the fractal dimensions, and the permeability, among other properties. Mercury, as liquid, has a property of not wetting the surfaces, therefore, in order to use it to fill in different porous cavities, pressure needs to be used. This will permit to estimate the porous radius as a function of the applied pressure, surface tension, and contact angle; for mercury, they are typically 484 mN m^{-1} and 140°, respectively. This relationship can be seen in Equation 1.8.

$$r_p = -\left(\frac{2\tau}{P}\right)\cos\phi \tag{1.8}$$

where τ is the surface tension and ϕ is the contact angle.

Catalysis Preparation and Characterization

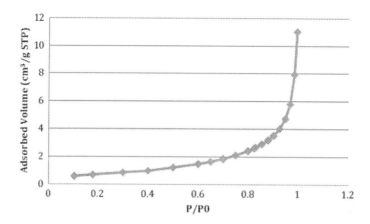

FIGURE 1.20 Nitrogen isotherm. (*Reproduced with permission* [23])

As an example, Figure 1.20 shows the results for nitrogen adsorption in a CaO catalyst [23], which shows that the mentioned catalysis is mesoporous.

1.4.3 SCANNING ELECTRON MICROSCOPY (SEM)

Scanning electron microscopy (SEM) is a versatile technique, which can provide information regarding particle size, shape, and texture of samples. SEM differs from the optical microscope regarding the use of electrons to form an image of the sample instead of light. SEM is a type of electron microscopy that images the sample surface by scanning it with a high-energy beam of electrons in a raster scan pattern. It is a straightforward technique to probe the morphological features of the materials. A beam of electrons produced by an electron gun travels vertically through the microscope held in a vacuum and is focused onto the sample. Electron and sample interactions result in the generation of X-rays, and backscattered and secondary electrons. These are collected by a detector and converted to an image of samples. SEM scans over a sample surface with a probe of electrons (5–50 eV) and detects the yield of either secondary or back-scattered electrons as a function of the position of the primary beam. Contrast is generally caused by the orientation that parts of the surface facing the detector appear brighter than parts of the surface with their surface normal pointing away from the detector. The interaction between the electron beam and the sample produces different types of signals providing detailed information about the surface structure and morphology of the sample. Secondary and backscattered electrons are used for imaging the samples. Scanning the energy of emitted X-rays (energy dispersive X-ray spectroscopy, EDX) identifies the element and also quantifies its amount in the sample. To avoid charge buildup, samples with low electronic conductivity were covered by Au-Pd layer. Charge buildup results in the static electric field over sample surface and deflects secondary electrons, which can make imaging difficult. A major advantage of SEM is that bulk samples can also be directly studied by this technique. Figure 1.21 shows SEM images of dried avocado seeds after staining with toluidine which show that the lipid bodies are intact and preserved inside the seeds.

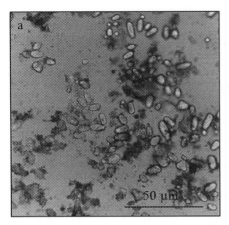 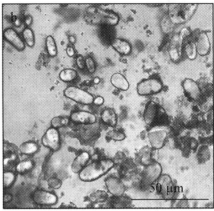

FIGURE 1.21 SEM images of avocado seeds with lipids inside [24].

1.4.4 Transmission Electron Microscopy (TEM)

Transmission electron microscopy is used for high-resolution imaging of thin films of a solid sample for compositional and micro structural analysis. The original form of electron microscopy, *i.e.*, transmission electron microscopy (TEM), involves a high voltage electron beam emitted by a cathode and formed by magnetic lenses. The electron beam that has been partially transmitted through a very thin (and thus, semi-transparent for electrons) specimen carries information about the structure of the specimen. The spatial variation in this information, *i.e.*, the image, is then magnified by a series of magnetic lenses until it is recorded by hitting a fluorescent screen, photographic plate, or light sensitive sensor such as a charge-coupled device (CCD) camera. The image detected by the CCD may be displayed in real time on a monitor or computer. The ability to determine the positions of atoms within materials has made the TEM an indispensable tool for nanotechnology research and developments

FIGURE 1.22 TEM image of CaO. (*Reproduced with permission* [23])

in the fields of heterogeneous catalysis and of semiconductor devices for electronics and photonics, among others.

Figure 1.22 shows a TEM image of the crystallographic structure of CaO which shows that the structure of the newly prepare catalysts is cubic.

1.4.5 INFRARED SPECTROSCOPY (IR)

Infrared spectroscopy is a modern spectroscopic technique that has found profound applications in the field of catalysis. This is primarily due to the fact that IR provides actual information on the structure, geometry, and orientation of practically all molecules that are present in the sample, irrespective of the physical state, temperature, or pressure. It is therefore, a useful tool to identify phases that are present in the catalyst or its precursor stages, the adsorbed species, adsorption sites, and the way in which the adsorbed species are chemisorbed on the surface of the catalyst. Infrared spectroscopy is the most common form of vibrational spectroscopy and it depends on the excitation of vibrations in molecules or in solid lattices by the absorption of photons, which occurs if a dipole moment changes during the vibration. The intensity of the infrared band is proportional to the change in dipole moment. A variety of IR techniques have been used to get information on the surface chemistry of different solids. With respect to the characterization of solid catalysts, two techniques largely predominate, namely, the transmission/absorption and the diffuse reflection techniques. In the first case, the sample consists typically of 10–100 mg of catalyst, pressed into a self-supporting disc of approximately 1 cm^2 and a few tenths of a millimeter thickness. In the diffuse reflectance mode (DRIFT), samples can be measured by simple deposition on a sample holder, avoiding the tedious preparation of wafers. This technique is especially useful for strongly scattering or absorbing samples. The infrared absorption spectrum is described by the Kubelka Munk function that is presented in Equation 1.9 [25].

$$F(R_\infty) = \frac{K}{S} = \frac{(1-R_\infty)^2}{2R_\infty} \tag{1.9}$$

where K is the absorption coefficient, which is a function of the frequency ν, S is the scattering coefficient, and R_∞ is the reflectivity of a sample of infinite thickness, measured as a function of ν.

The most common application of IR in catalysis is to identify adsorbed species and to study the way in which these species are chemisorbed on the surface of the catalyst. More specifically, IR spectroscopy has been used to study the adsorption of typical probe molecules like ammonia, pyridine and other bases, hydrocarbons, and carbon dioxide, which can monitor the basic sites on oxide catalysts. Investigation of adsorbed species in relation to their behavior in catalytic reactions is the main field of application of IR spectroscopy.

Figure 1.23 shows the IR spectra. A broad band was observed for Ba^{2+}–T–TiO^2 when it was exposed to CO_2 at T = 25°, while a clear shoulder was observed in the band for Ca^{2+}–T–TiO_2 and Sr^{2+}–T–TiO_2 [26].

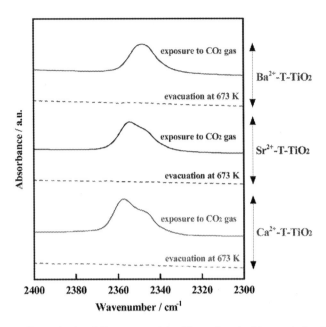

FIGURE 1.23 IR results for different materials. (*Reproduced with permission* [26])

1.4.6 Thermal Analysis

In thermal analysis, the change in certain chemical and physical property of a sample, as a function of temperature, is analyzed. Thermal analysis can be used to study phase transitions, solid-state reactions, reduction/oxidation, and decompositions of samples under different atmospheres. Thermo-analytical techniques involve the measurement of the response of any solid under study (energy or mass, released or consumed) as a function of temperature (or time) dynamically by application of a linear temperature program. Thermogravimetry is a technique which measures the change in mass of a material as a function of temperature and time, in a controlled manner. This variation in mass can be either a loss of mass (vapor emission) or a gain of mass (gas fixation). It is ideally used to assess volatile content, thermal stability, degradation characteristics, aging/lifetime breakdown, sintering behavior, and reaction kinetics. Differential thermal analysis is a technique which measures the temperature difference of the sample versus a reference, caused by thermal events in a material. It is ideally used to determine the melting point, glass transition temperature, crystallinity, degree of curing, heat capacity, impurities present, etc.

As an example, Figure 1.24 shows the variation of weight loss as a function of temperature for CaO when produced from Mussel shells [22]. Figure 1.24 shows that a weight loss of 3.1% was observed after heating the sample from 25 to 800°C. The two minor weight losses found at 100°C, and between 325 and 375°C could be due the removal of the physiosorbed moisture and decomposition/oxidation of organic components, respectively.

There are three major temperature program-based techniques that are generally used when characterizing solid catalysts: (i) temperature programmed desorption (TPD), (ii) temperature programed reduction (TPR), and (iii) temperature-programmed oxidation

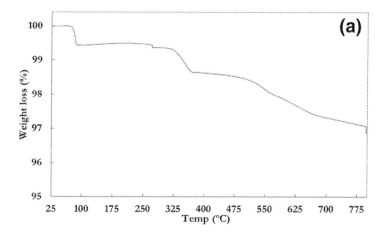

FIGURE 1.24 Weight lost as a function of temperature for CaO produce from Mussel shells. (*Reproduced with permission* [22])

(TPO). These techniques have become highly relevant since they are generally used for the solid–gas reaction studies. Typically this is a case that can be found in a pyrolysis reactor, in a solid–gas catalytic system for deposition of material (fertilizer production, medication production), in a fuel cell, and in a battery, among others.

As the name implies, TPD was developed with the intention to study the amount of gas that a specific material, catalysts, or surface, can adsorb. The desired material was pretreated with an adsorbing gas, and when equilibrium had been reached, the desorption process took place under specific controlled conditions, allowing the user to determine the exact amount of gas that had been previously adsorbed. This technique can also be used to quantify the amount of basic, or acidic, sites in a heterogeneous catalysts by studying the interactions of a gas, or liquid, with the solid surface. As presented in [11], ammonia, dimethylamine, methylamine, pyridine, among others, can be used for acidic site titration; while carbon dioxide, sulfur dioxide and pyrrole, can be used to titrate basic sites.

Temperature programmed desorption of ammonia (NH_3-TPD) is one of the most widely used and flexible techniques for characterizing acid sites on surfaces. Determining the quantity and strength of the acid sites is crucial to understand and predict the performance of a catalyst. Upon contact with a surface, the probe molecule (ammonia) is absorbed onto the surface either by physisorption, chemisorption, or by the formation of chemical bonds, minimizing the energy of the species. Then, on increasing the sample temperature, the species is desorbed from the surface. The amount of ammonia desorbed is proportional to the acid concentration. However, when using ammonia as a probe it is not possible to differentiate the site-specific adsorption as the adsorption may occur at Brønsted sites, Lewis sites, or as a result of any combination of surface/vapor attractive forces. This problem can be circumvented by using other amines like alkyl amines. The main advantage of the use of alkyl amines as a probe for acid-site densities is that the temperature-programmed decomposition reaction occurs only at Brønsted sites, providing an efficient method for direct quantification of Brønsted acidity. The results from this technique are strongly affected by mass transfer limitations as well as heat transfer limitations. This allows characterization of the strength of active sites but does not identify their type, and it gives an average value of acid strength, but not a distribution.

TPR, and similarly TPO, typically involve monitoring the surface or bulk processes between the solid catalyst and its gaseous environment via continuous analysis of the gas phase composition as the temperature is raised linearly with time. Instrumentation for temperature programmed investigations consist of a reactor charged with the catalyst in a furnace that can be temperature programmed and a thermal conductivity detector (TCD) to measure the conductivity of the gas stream passing through the sample before and after interaction, to find out its change in concentration. Mass spectrometer can be coupled with the detector for identification of the various species formed or evolved during the desorption process.

In TPR, the catalyst is exposed to a flowing reducing gas while the temperature increases. Following the amount of reducing gas that is being consumed, like hydrogen or CO, it can determine the degree of reduction and therefore, the oxidation state of the catalysts, or material, under study. Figure 1.25 shows the TPR profiles for fresh catalyst EUROCAT [27].

Figure 1.25 shows six plots of TPR for a specific catalytic material, EUROCAT. Three of the plots have been carried out at conditions (D-1, CH-1, and F-3 set 3) different from the other two (F-3 set 1 and E-2). However, in all cases, two reduction peaks can be discerned [27].

In the case of TPO, an oxidation gas, generally oxygen, is used in order to oxidize the material. As with TPR, well-controlled and known conditions are required in order to carry out the measurement and determination. This technique has been used in several different applications, like coke deposition studies and their effect in catalyst deactivation within the petrochemical area [28], to carbonaceous deposit in jet fuel [29].

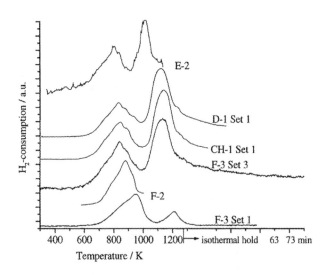

FIGURE 1.25 TPR for catalysts EUROCAT. (*Reproduced with permission* [27])

1.4.7 Raman Spectroscopy

Raman spectroscopy is based on the inelastic scattering of photons, which lose energy by exciting vibrations in the sample. Photons of the laser are absorbed by the sample and then re-emitted. Frequency of the re-emitted photons is shifted up or down in comparison with original monochromatic frequency, which is called the Raman effect. This shift provides information about vibrational, rotational, and other low frequency transitions in molecules. A vibration is Raman active if it changes the polarizability of the molecule. Raman and infrared spectroscopy complement each other, in particular for highly symmetrical molecules like CO_2. Raman spectroscopy is commonly used in chemistry, since vibrational information is specific to the chemical bonds and symmetry of molecules. Therefore, it provides a fingerprint by which the molecule can be identified. From Raman spectral data, we can determine the composition of material, crystal symmetry and orientation, quality of crystal, amount of material, and stress or strain.

The Raman scattering of light can be divided into two subareas, Stokes–Raman and Anti-Stokes–Raman scattering. This is related to how the frequencies are before and after the scattering. If both frequencies are identical, before and after the scattering, then we have Rayleigh scattering. The three scenarios presented in Figure 1.26 are Rayleigh scattering, $hv_0 = hv$; Stokes scattering, $hv_0 = hv + hv_1$; and anti-stokes, $hv_0 = hv - hv_1$. This means that in a Stokes–Raman situation, the adsorbed wavelength is higher than the emitted wavelength, while in the anti-Stokes–Raman situation, it is the opposite.

This technique has several advantages; it can be used in all states of matter (solids, liquids, or gases), and it does not require any pretreatment of the sample. The Raman spectra obtained is unique for each material, like a fingerprint which can allow identification of different materials. In addition, it is a nondestructive method and it takes a very short time to perform.

Nevertheless, it is a relatively expensive technique, even though it does not require vacuum equipment, it is almost impossible to use it in pure metals or alloys or for

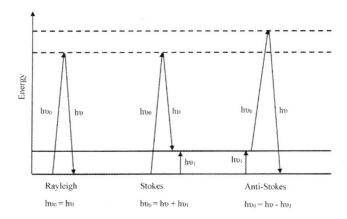

FIGURE 1.26 Representation of different the Raman effects.

materials that are present in low concentration. Dutta and Twu [30] have used Raman on a faujasitic zeolite, they have seen that as the Si/Al ratio increases, the single peak spectra is then changed into a double peak spectrum due to the increase in the amount of Si. This can be seen in Figure 1.27.

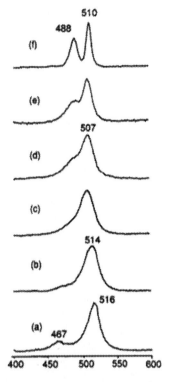

FIGURE 1.27 Raman spectra of faujasitic zeolites as a function of Si/Al ratio in the region between 400 and 600 cm^{-1}. The Si/Al ratios of the framework are (a) 1, (b) 1.3, (c) 2.6, (d), 3.3, (e) 4.5, and (f) ∞. (*Reproduced with permission* [30])

1.4.8 X-ray Photoelectron Spectroscopy (XPS)

X-ray photoelectron spectroscopy has come into the limelight as a result of the pioneering work by Kai Siegbahn (Nobel prize winner, 1981) and his colleagues at Uppsala university, Sweden. This technique is based on the photoelectric effect discovered by Heinrich Hertz and was later explained by Albert Einstein. It involves the bombing of a solid surface with X-rays and the measurement of the associated photo-emitted electrons. It is a widely used technique for obtaining chemical information of various material surfaces. The low kinetic energy ($0 \leq 1500$ eV) of emitted photoelectrons limits the depth from which it can emerge, making XPS a very surface-sensitive technique with sample depths in the range of few nanometers.

Catalysis Preparation and Characterization

Photoelectrons are collected and analyzed by the instrument to produce a spectrum of emission intensity versus electron binding (or kinetic) energy. XPS uses either monochromatic aluminum Kα or non-monochromatic magnesium Kα X-rays to eject a photoelectron from an atom at the sample's surface. Due to this vacancy, an electron from a higher energy level will then fall to a lower energy level to fill in this vacancy, and the energy produced in such process will permit the ejection of an Auger electron. Therefore, the XPS technique produces both photoelectrons and Auger electrons which can be seen in the spectrum. The electrons ejected are analyzed with the help of an XPS detector by measuring the kinetic energy. This information is then used to determine the type of elements present in the sample. Figure 1.28 illustrates a schematic representation of the X-ray photoelectron process which shows five steps: (i) the material before any interaction; (ii) the interaction of the X-ray photon with the surface and in particular with a 1s electron, (iii) the emission of the photoelectron with a certain kinetic energy described by Equation 1.10, (iv) the holes left behind in the lower energy levels are then filled with electrons that fall to fill those vacancies, and (v) the energy produced from the falling of electrons to lower levels is used to emit an Auger electron from the material.

The photoemitted electrons have discrete kinetic energy that is the characteristics of the emitting atoms and their bonding states. The kinetic energy, E_k, of these photoelectrons is determined from the energy of the incident X-ray radiation ($h\nu$) and the electron binding energy (E_b), is given by:

$$E_k = h\nu - E_b \tag{1.10}$$

The experimentally measured energies of the photoelectrons are given by:

$$E_k = h\nu - E_b - E_w \tag{1.11}$$

where E_w is the work function of the spectrometer.

Each element produces a characteristic set of XPS peaks at different binding energy values that directly identify each element present on the surface of the material being analyzed. These characteristic peaks correspond to the electronic configuration of the atoms, like 1s, 2s, 2p, or 3s. To account for the multiple splitting and satellites accompanying the photoemission peaks, the photoelectron spectra have to be interpreted in terms of many-electron states of the final ionized state of the sample, rather than the occupied one-electron states of the neutral species.

Bian et al. [31] studied the use of hydroiodic acid when producing $CsPbI_3$-based perovskite solar cells. They used this XPS technique in order to confirm the material composition and the interactions among different materials. They concluded that pri-PbI_2 and pri$CsPbI_3$ does not show the N 1s peak as presented by for example DMAI. However, syn-PbI_2 and syn-$CsPBI_3$, does show the N 1s peak. This can be seen in Figure 1.29 as an example of the use of the XPS technique.

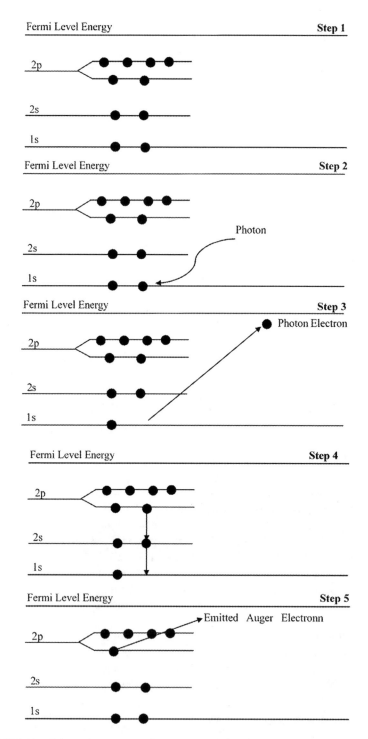

FIGURE 1.28 Schematic representation of a XPS technique.

Catalysis Preparation and Characterization 33

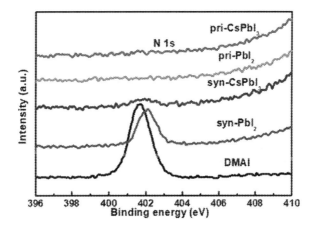

FIGURE 1.29 XPS spectra of N 1s of DMAI, Syn-PbI2, Pri-PbI2, Syn-CsPbI3, and Pri-CsPbI3 powders [31].

1.5 INDUSTRIAL CATALYTIC PROCESSES

The accomplishments of any chemical plant depend on the catalyst technology it uses. Therefore, catalysis became the backbone of the chemical industry contributing substantially to our societal needs and wealth. The continuous discovery of novel catalytic processes has led to major innovations in chemical processing. Both, homogeneous and heterogeneous catalysis have been utilized in numerous well-established industrial applications, such as chemical, food, pharmaceutical, automobiles and petrochemicals, and in emerging fields, such as greener production of fuels, biotechnology, and fuel cell technology. Furthermore, there has been an ever-increasing effort to make the processes environmentally benign.

Areas of the industry where homogeneous catalysts are used include hydroformylation (rhodium- and cobalt-based catalysts), hydrocyanation in DuPont (nickel-based catalyst), Metochlor in Novartis (sulfuric acid, iridium-based catalyst), Glycidol (diethyl tartrate), among other. While sectors of the industry where heterogeneous catalysts are used include hydrogenation (zinc- and copper-based catalysts, Raney Nickel), oxidation reaction (vanadium- and silver-based catalysts), alkylation (silica-supported phosphoric acid catalyst), olefin reactions (supported chromium oxide catalysts), among others.

Catalysis research and development of chemical industry go hand in hand. It is therefore, not surprising that many basic chemical processes like hydrogenations and oxidations, which are of utmost interest to both academia and industry, bring forth more and more effective catalysts, new technologies and in-depth understanding of fundamental catalytic principles. The important parameters to produce commodity chemicals are always linked to the feed costs, catalyst life, energy efficiency, process integration, environmental taxes, product selectivity and of course, payback times.

Due to the high number of different processes and an even greater amount of catalytic materials that are being used, have been used, and are being developed, it is impossible that all of them are included in the scope of this book. Therefore, some industrial process and some catalytic alternatives will be presented. However, other possibilities are equally relevant, and this selection is not based on performance of the catalysts or the process. We encourage the reader to read Jess and Wasserscheid book for more processes [32].

1.5.1 Synthesis Gas

Synthesis gas (CO and H_2, sometimes with CO_2) is being produced via the chemical transformation of petroleum, biomass, or coal, via steam reforming, or gasification reaction. The production of synthesis gas is highly related to the number of applications for instance, the production of hydrogen as an energy vector, to the use of hydrogen to produce ammonia as well as methanol, toluene, aromatic olefins, among several other process.

The production of synthesis gas can be done via gasification where different biomass and/or carbon sources are transformed into CO and H_2, and in some cases with CO_2. The gasification process is carried out under a control quantity of oxygen (steam) in order to avoid combustion, generally at temperatures higher than 700°C. Typically, the biomass is first dried, then fed into the gasifier together with some steam, and the products are then purified or cleaned to obtain the product of interest. The use of this final product could be into methanization for CH_4, and hydrogen could be used for fuel cells, for heat, and power.

The gasification reaction takes place in the presence of a catalysts, different types have been studied such as those based on dolomites, alkaline metals (sodium, potassium, magnesium), nickel, platinum, and palladium, are the most common ones among others. The reactions considered, when reacting with steam or CO_2 are:

$$C_nH_m + nH_2O \leftrightarrow nCO + \left(n + \frac{m}{2}\right)H_2$$

$$C_nH_m + nCO_2 \leftrightarrow 2CO + \left(\frac{m}{2}\right)H_2$$

Synthesis gas is regularly produced in a tubular reactor call reformer, therefore the name of steam reforming. The catalyst in the reformer tube must possess high mechanical stability because the reaction is performed at high temperatures. Furthermore, the catalyst particle size is of high importance, as very low particle size would result in pressure drop and less energy efficiency. The group VII metals are active for the steam reforming reactions, with Rh and Ru being the most active and Ni, the most favorable because of the cost. However, the Ni catalyst is the most sensitive to sulfur poisoning. The stability of heterogeneous catalyst in a high temperature atmosphere and reactant contact time is very critical for the catalytic gas phase reactions. A popular example for high-temperature gas-phase catalytic reaction, with residence time of microseconds, is the oxidation of ammonia to produce nitric oxide. Catalytic partial oxidation of natural gas over platinum group metal catalysts, to produce synthesis gas and olefins is another promising example of a fast chemical reaction. This reaction is:

$$CH_4 + H_2O \rightarrow CO + 3H_2$$

This reaction could be accompanied by the water gas shift reaction in order to improve the production of hydrogen; the latest reaction is:

$$CO + H_2O \rightarrow CO_2 + H_2$$

1.5.2 Ammonia Synthesis

Ammonia is a commodity chemical and its production rate might be directly proportional to the growth of human population as it has been widely used as a fertilizer. Natural gas is the major source for ammonia production, and the production plants are usually located where the feedstock is less expensive. The number of stages involved in the production of ammonia from natural gas with operating temperature, pressure, and catalyst are tabulated in Table 1.1 [33].

The different types of catalysts that are required and some operational conditions for each step for ammonia production are also presented in the table. The Haber–Bosch process associated with the iron catalyst developed by Mittasch was used for the commercial synthesis of ammonia. Ammonia synthesis is well understood and is used as a model reaction for fundamental studies of novel catalytic materials; but even after 100 years, a lucrative replacement for iron catalyst has not yet been found.

TABLE 1.1
Typical Catalysts and Process Parameters for a Natural Gas-Based Ammonia Plant

Reaction	Catalysts	P (bar)	T (°C)	Gas
Desulfurization				
$RHS + H_2 = RH + H_2S$	$CoMo/Al_2O_3$	45	365	10 ppb S
$H_2S + ZnO = ZnS + H_2O$	ZnO	45	365	
Steam Reforming				
Prereforming				
$C_nH_m + nH_2 = 0 =$	Ni, X/MgO	40	450/500	$H_2O/C = 3.0$
$nCO + ½ (n+m)H_2$	Al_2O_3			
Tubular reforming				
$CH_4 + H_2O = CO + 3H_2$	$Ni/MgAl_2O4$	37	520/817	–
Secondary reforming				
$CH_4 + ½ O_2 = CO + 2H_2$	$Ni/MgAl_2O_4$	37	550/1000	0.3% CH_4, 13% CO
Water gas shift				
$CO + H_2O = CO_2 + H_2$	–	–	–	–
HTS	Fe_3O_4 (Cr), Al	6	360/430	3% CO
LTS	Cu, ZnO, Al_2O_3	34	205/225	0.3% CO
Methanation				
$CO + 3H_2 = CH_4 + H_2O$	Ni/Al_2O_3	32	300/323	10 ppm (CO + CO_2)
Synthesis				
$N_2 + 3H_2 = 2NH_3$	Fe (K, Ca, Al)	190	233/456	20% NH_3

Source: (Reproduce with permission [33])

1.5.3 Selective Oxidation

Selective oxidation of hydrocarbons is of crucial importance in activating hydrocarbons to form intermediates and final products for applications in chemical and pharmaceutical sectors. Around the world, about one quarter of all the organic compounds produced are synthesized via selective oxidation of hydrocarbons. During the last two decades, a significant progress has been made within the area of catalytic oxidations, which has led to a range of selective and mild processes from both industrial and synthetic point of view. These reactions may be based on organocatalysis, metal catalysis, or biocatalysis. Some of the large scale industrial catalytic oxidation processes include oxidation of (a) p-Xylene to terephthalic acid (oxidant–oxygen, liquid phase process), (b) cyclohexane to adipic acid (oxygen, liquid phase), (c) cumene to cumyl hydroperoxide (oxygen, liquid phase), and (d) n-butane to maleic anhydride (oxygen, gas phase). In all these processes, reactions are conducted at high temperatures and high O_2 pressures. In this context, oxidation catalysis has been playing and will play a leading role. The reason for this is that oxidation is the tool for the production of huge quantities of intermediates and monomers for the polymer industry. The impact of manufacturing of these components on the environment might have been much greater, if considerable efforts are not put into continuous improvement of the technologies used for the production of these chemicals, including the replacement of toxic and dangerous reactants, better heat recovery and energy integration in the plant, recovery of waste streams and abatement of tail emissions, and downstream use of by-products. In the chemical industry, oxidation is one of the technologies that has great potential for improvement, and this has led to a series of better processes in recent years. Improvements to catalysts, reactions, and process technologies have moved in the direction of an improved sustainability and have been widely documented in books and research publications in recent years. Although the developments were driven by economics, they have also led to more sustainable processes. However, some processes still produce large amounts of waste or operate under conditions leading to non-optimal selectivity to the desired compound. A sustainable catalytic oxidation should present the following fundamental features: (i) capability to activate O_2 and H_2O_2, in aqueous phase, with solvent-free protocols, or by using environmental friendly solvents, including per fluorinated environment, ionic liquids and carbon dioxide, (ii) high selectivity, and (iii) oxidative, hydrolytic, and thermal stability in the reaction conditions. The contemporary presence of these features could provide the ideal oxidation catalyst. As an example we can look into adipic acid, which is produced by a two-step oxidation process, where the first one is (as prevalent today) the oxidation of cyclohexane with air, and the second one is the oxidation of cyclohexanone/cyclohexanol (called KA oil) with nitric acid.

1.5.4 Fischer–Tropsch Process

This process allows the transformation of CO and H_2 into hydrocarbons; this process regularly takes place in the presence of a metal catalyst and intermediate operational conditions. However, research is constantly being done for this process and therefore, new technologies and catalytic materials are constantly under development.

This process is also considered as the foundation of the gas to liquid (GTL) technology where short hydrocarbons like methane are combined to produce longer liquid hydrocarbons that can be used as fuel. It is not practical to present all possible scenarios and catalysts that could be used for the FT process. The catalyst first developed for this process was cobalt, which was used to change syn gas into liquid. Other alternatives are group VIII metals [34]. The main reaction is the production of hydrocarbons, as shown below:

$$nCO + (2n+1)H_2 \rightarrow nH_2O + C_nH_{2n+2}$$

where n is desired to be larger than 10.

It is important to mention that this technology has been expanded to other areas such as using biomass as raw materials which has introduced the concept of biomass to liquid (BTL), showing how this process is adaptable to other resources. In addition, this technology has also been studied in order to show the sustainability for the production of jet fuel [35].

As shown by Cavalett and Cherubini [35], the production of renewable jet fuel using FT was the best solution that fulfilled the UN goals under the evaluated parameters.

1.5.5 Biodiesel Production

Finally, it is of interest to show the different types of catalysts that can be used for biodiesel production via transesterification. There are many catalysts since this renewable fuel can be produced in homogeneous, heterogeneous, enzymatic or supercritical state, *i.e.*, with all possible types of catalysts [36–41]. There are many papers within this field, with several upcoming due to the relevance of the topic, which will be presented briefly in this book.

Basic homogeneous catalysts that can be used for high efficiency biodiesel production are generally based on potassium or sodium, and could be hydroxide or alkoxy, depending on the process [36–38]. In order to use basic catalysts, the raw material needs to be free of fatty acids; otherwise, they could react with the catalysts to produce soap and water [36,37]. These two components will decrease the productivity of the process and complicate the separation of the products. A solution to this problem is to use homogeneous acid catalysts, where the presence of free fatty acids and water are not a problem. However, the main problem of acid catalysts is the longer reaction time needed, and, like base catalysts, the products need to be separated from the catalyst which are in the same phase [37,42,43].

Due to the above-mentioned problems, solid heterogeneous catalysts such as metal oxide, Amberlyst resins, sulfonic resin, Nafion, zinc-based materials, zirconia-based catalysts, hydrotalcites, and metallic salts, have been extensively tested under different conditions [41,44–47]. In general terms, solid catalysts have shown to be a possible solution to produce biodiesel from cheap sources (high amount of impurities) with promising results that can make them competitive at the industrial scale. Other heterogeneous catalysts that are currently under study are enzymes [48–50].

These catalysts are of high selectivity and can be used in mild conditions; however, their high cost makes the process less attractive from an economic perspective.

Finally, non-catalytic supercritical processes have been developed in order to reduce the reaction time considerably. However, reaching these operational conditions, in terms of pressure and temperature, is expensive. In addition, the high pressure (50–70 bars) could lead to safety issues and safety precautions that are not present in other technological alternatives [51,52].

Besides the technologies and processes presented here, there are other process and other technologies that can be studied and presented.

REFERENCES

1. Singh, S.B., Tandon, P.K. "Catalysis: a brief review on nano-catalysts". *Journal of Energy and Chemical Engineering*. 2(3). (2014). 106–115.
2. Monnier, A., Schuth, F., Huo, Q., Kumar, D., Margolese, D., Maxwell, R.S., Stucky, G., Krishnamurty, M., Petroff, P., Firouzi, A., Janicke, M., Chmelka, N. "Cooperative formation of inorganic-organic interfaces in the synthesis of silicate mesostructures". *Science*. 261. (1993). 1299–1303.
3. Tanev, P.T., Pinnavaia, T.J. "Mesoporous silica molecular sieves prepared by ionic and neutral surfactant templating: a comparison of physical properties". *Chemistry of Materials*. 8. (1996), 2068–2079.
4. Zhou, X., Gan, Y., Du, J., Tian, D., Zhang, R., Yang, C., Dai, Z. "A review of hollow Pt-based nanocatalysts applied in proton exchange membrane fuel cells". *Journal of Power Sources*. 232. (2013). 310–322.
5. Zhang, L., Zhou, M., Wang, A., Zhang, T. "Selective hydrogenation over supported metal catalysts: from nanoparticles to single atoms". *Chemical Reviews*. 120. (2020). 683–733.
6. Lan, X., Wang, T. "Highly selective catalysts for the hydrogenation of unsaturated aldehydes: a review". *ACS Catalysis*. 10. (2020). 2764–2790.
7. Sasirekha, N., Sangeetha, P., Chen, Y.W. "Bimetallic Au-Ag/CeO$_2$ catalysts for preferential oxidation of CO in hydrogen-rich steam: effect of calcination temperature". *The Journal of Physical Chemistry C*. 118. (2014). 15226–15233.
8. Chandra, P., Enespa, Singh, R., Arora, P.K. "Microgial lipases and their industrial applications: a comprehensive review". *Microbial Cell Factories*. 19. (2020). 169–210.
9. Dimitrijevic, N.M., Vijayan, B.K., Poluektov, O.G., Rajh, T., Gray, K.A., He, H., Zapol, P. "Role of water and carbonated in photocatalytic transformation of CO$_2$ to CH$_4$ on titania". *Journal of the America Chemical Society*. 133. (2011). 3964–3971.
10. Sujatha, G., Shanthakumar, S., Chiampo, F. "UV light-irradiated photocatalytic degradation of coffee processing wastewater using TiO$_2$ as a catalyst". *Environments*. 7(6). (2020). 47–59.
11. Ertl, G., Knözinger, H., Schüth, F., Weitkamp, J. (ed). *Handbook of Heterogeneous Catalysis*, 2nd edition. (2008). Wiley-VCH Verlag GmbH & Co. Weinheim, Germany.
12. Shahinuzzaman, M., Yaakob, Z., Ahmed, Y. "Non-sulphide zeolite catalysts for bio-jet-fuel conversion". *Renewable and Sustainable Energy Reviews*. 77. (2017). 1375–1384.
13. Swastika, T., Ardy, A., Susanto, H. "Preparation of Catalysts Cu-ZnO-MgO-Al$_2$O$_3$ for direct synthesis of DME". *IOP Conference of Series: Materials Science and Engineering*. 543. (2019). 012063.

14. Schwarz, J., Contescu, C., Contescu, A. "Method for preparation of catalytic materials". *Chemical Reviews*. 95(3). (1995). 477–510.
15. Baerns, M. (ed). *Basic Principles in Applied Catalysis*. (2004). Springer, Springer-Verlag, Berlin.
16. Rashid, H.U., Yu, K., Umar, M.N., Anjum, M.N., Khan, K., Ahmad, N., Jan, M.T. "Catalysts role in chemical vapor deposition (CVD) process: a review". *Review on Advance Material Science*. 40(3). (2015). 235–248.
17. Johnson, R.W., Hultqvist, A., Bent, S.F. "A brief review of atomic layer deposition: from fundamentals to applications". *Materials Today*. 17(5). (2014). 236–246.
18. Ali, M.A., Rahman, M.A., Alam, A.M.S. "Use of EDTA-Grafted Anion-exchange resin for the separation of selective heavy metals ions". *Analytical Chemistry Letters*. 3(3). (2013). 199–207.
19. Avhad, M.R., Sanchez, M., Peña, E., Bouaid, A., Martinez, M., Aracil, J., Marchetti, J.M. "Renewable production of value-added jojobyl alcohols and biodiesel using a naturally-derived heterogeneous green catalysts". *Fuel*. 179. (2016). 332–338.
20. Avhad, M.R., Sanchez, M., Bouaid, A., Martinez, M., Aracil, J., Marchetti, J.M. "Modeling chemical kinetics of avocado oil ethanolysis catalyzed by solid glycerol-enriched calcium oxide". *Energy Conversion and Management*. 126. (2016). 1168–1177.
21. Sanchez, M., Avhad, M.R., Marchetti, J.M., Martinez, M., Aracil, J. "Enhancement of the jojobyl alcohols and biodiesel production using a renewable catalyst in a pressurized reactor". *Energy Conversion and Management*. 126. (2016). 1047–1053.
22. Avhad, M.R., Gangurde, L.S., Sanchez, M., Bouaid, A., Aracil, J., Martinez, M., Marchetti, J.M. "Enhancing biodiesel production using green glycerol-enriched calcium oxide catalysts: an optimization study". *Catalysis Letters*. 148(4). (2018). 1169–1180.
23. Sanchez, M., Marchetti, J.M., El-Boulifi, N., Aracil, J., Martinez, M. "Kinetics of jojoba oil methanolysis using a waste from fish industry as catalyst". *Chemical Engineering Journal*. 262. (2015). 640–647.
24. Mangesh, M.R. "Heterogeneous catalytic conversion of non-edible lipid biomass to biochemicals". Ph.D. Thesis. (2016). Norwegian University of Life Science. Thesis number 2016:84.
25. Alcaraz de la Osa, R., Iparragirre, I., Ortiz, D., Saiz, J.M. "The extended Kubelka-Munk theory and its applications to spectroscopy". *ChemTexts*. 6(2). (2020).1–14.
26. Uematsu, E., Itadani, A., Uematsu, K., Toda, K., Sato, M. "IR study on adsorption of carbon dioxide on alkaline earth metal ion exchanged titanate nanotubes at room temperature". *Vibrational Spectroscopy*. 109. (2020). 103088.
27. Reiche, M.A., Maciejewski, M., Baiker, A. "Characterization by temperature programmed reduction". *Catalysis Today*. 56. (2000). 347–355.
28. Minh, C.L., Jones, R.A., Craven, I.E., Brown, T.C. "Temperature-programed oxidation of coke deposited on cracking catalysts: combustion mechanism dependence". *Energy & Fuels*. 11. (1997). 463–469.
29. Eser, S., Venkataraman, R., Altin, O. "Utility of temperature-programed oxidation for characterization of carbonaceous deposits from heated jet fuel". *Industrial & Engineering Chemistry Research*. 45. (2006). 8956–8962.
30. Dutta, P.K., Twu, J. "Influence of framework Si/Al ratio on the Rama spectra of Faujasitic Zeolites". *Journal of Physical Chemistry*. 95. (1991). 2498–2501.
31. Bian, H., Wang, H., Li, Z., Zhou, F., Xu, Y., Zhang, H., Wang, Q., Ding, L., Liu, S., Jin, Z. "Unveiling the effects of hydrolysis-derived DMAI/DMAPbI$_x$ intermediate compound on the performance of CsPbI$_3$ solar cells". *Advance Science*. 7. (2020). 1902868.

32. Jess, A., Wasserscheid, P. *Chemical Technology: An Integrated Textbook.* (2013). Wiley-VCH Verlag GmbH & Co. Weinheim, Germany.
33. Belloer, M., Renken, A., van Santen, R.A. (ed). *Catalysis: From Principles to Applications.* (2012). Wiley-VCH Verlag GmbH & Co. Weinheim, Germany.
34. Vargas, D.X.M., Rangel, L.S., Calderon, O.C., Flores, M.R., Lozano, F.J., Nigam, K.D.P., Mendoza, A., Castellanos, A.M. "Recent advances in bifunctional catalysts for the Fischer-Tropsch process: one-stage production of liquid hydrocarbons from syngas". *Industrial & Engineering Chemistry Research.* 58. (2019). 15872–15901.
35. Cavalett, O., Cherubini, F. "Contribution of jet fuel from forest residues to multiple sustainable development goals". *Nature Sustainability.* 1. (2018). 799–807.
36. Marchetti, J.M. *Biodiesel Production Technologies.* (2010). Nova Science Publishers, New York.
37. Knothe, G., Krahl, J., Van Gerpen, J. (ed). *The Biodiesel Handbook.* 2nd edition. (2010). AOCS Press, Urbana, IL.
38. Avhad, M.R., Marchetti, J.M. "A review on recent advancement in catalytic materials for biodiesel production". *Renewable & Sustainable Energy Reviews.* 50. (2015). 696–718.
39. Singh, D., Sharma, D., Soni, S.L., Sharma, S., Sharma, P.K., Jhalani, A. "A review on feedstocks, production processes, and yield for different generations of biodiesel". *Fuel.* 262. (2020). 116553.
40. Chua, S.Y., Periasamy, L.A.P., Goh, C.M.H., Tan, Y.H., Mubarak, N.M., Kansedo, J., Khalid, M., Walvekar, W., Abdullah, E.C. "Biodiesel synthesis using natural solid catalysts derived from biomass waste – a review". *Journal of Industrial and Engineering Chemistry.* 81. (2020). 41–60.
41. Tang, Z.E., Lim, S., Pang, Y.L., Ong, H.C., Lee, K.T. "Synthesis of biomass as heterogeneous catalyst for application in biodiesel production: state of the art and fundament review". *Renewable & Sustainable Energy Reviews.* 92. (2018). 235–253.
42. Serrano, M., Marchetti, J.M., Martinez, M., Aracil, J. "Biodiesel production from waste salmon oil: kinetic modeling, properties of methyl esters, and economic feasibility of a low capacity plant". *Biofuel, Bioproducts and Biorefining.* 9. (2015). 516–529.
43. Bashir, M.J.K., Wong, L.P., Hilaire, D.S., Kim, J., Salako, O., Jean, M.J., Adeyemi, R., James, S., Foster, T., Pratt, L.W. "Biodiesel fuel production from brown grease produced by wastewater treatment plant: optimization of acid catalyzed reaction conditions". *Journal of Environmental Chemical Engineering.* 8. (2020). 103848.
44. Thangaraj, B., Solomon, P.R., Muniyandi, B., Ranganathan, S., Lin, L. "Catalysis in biodiesel production – a review". *Clean Energy.* 3(1). (2019). 2–23.
45. Mansir, N., Teo, S.H., Rabiu, I., Yap, Y.H.T. "Effective biodiesel synthesis form waste cooking oil and biomass residue sold green catalyst". *Chemical Engineering Journal.* 347. (2018). 137–144.
46. Abdullah, S.H.Y.S., Hanapi, N.H.M., Azid, A., Umar, R., Juahir, H., Khatoon, H., Endut, A. "A review of biomass-derived heterogeneous catalyst for a sustainable biodiesel production". *Renewable & Sustainable Energy Reviews.* 70. (2017). 1040–1051.
47. Gupta, J., Agarwal, M., Dalai, A.K. "An overview on the recent advancements of sustainable heterogeneous catalysts and prominent continuous reactor for biodiesel production". *Journal of Industrial and Engineering Chemistry.* 88. (2020). 58–77.
48. Dhawane, S.H., Kumar, T., Halder, G. "Recent advancement and prospective of heterogeneous carbonaceous catalysts in chemical and enzymatic transformation of biodiesel". *Energy Conversion and Management.* 167. (2018). 176–202.

49. Moazeni, F., Chen, Y.C., Zhang, G. "Enzymatic transesterification for biodiesel production form used cooking oil, a review". *Journal of Cleaner Production*. 216. (2019). 117–128.
50. Santos, S., Puna, J., Gomes, J. "A Review on bio-based catalysts (immobilized enzymes) used for biodiesel production". *Energies*. 132. (2020). 3013–3031.
51. Martinez, V.M.O., Martinez, P.A., Martinez, N.G., de los Ríos, A.P., Fernandez, F.J.H., Medina, J.Q. "Approach to biodiesel production from microalgae under supercritical conditions by the PRISMA method". *Fuel Processing Technology*. 191. (2019). 211–222.
52. Martinez, P.A., Martinez, V.M.O., Martinez, N.G., de los Ríos, A.P., Fernandez, F.J.H., Medina, J.Q. "Production of biodiesel under supercritical conditions: state of the art and bibliometric analysis". *Applied Energy*. 264. (2020). 114753.

2 Reactor Design
Mole and Energy Balance

2.1 INTRODUCTION

Chapter 1 discussed the production techniques of different types of catalysts tailored for various functionalities. Different methodologies to evaluate and measure desirable properties, before or after a reaction, were explored. It was shown that BET helps identify the surface area of the catalyst where a surface reaction takes place, the number of active sites, and other aspects, such as the extent of surface poisoning due to coke deposition. Important techniques were introduced to characterize catalysts, such as the use of SEM and TEM to understand material morphology and X-ray diffraction to study the crystallographic structure. All these aspects are crucial for selecting appropriate catalyst materials to be added to achieve a desirable reaction process.

However, in the previous chapter, the reaction processes themselves have not been discussed in detail. It is important to understand exactly where the reactions take place and how the progress of the reactions is experimentally measured and mathematically modeled. To address these aspects, this chapter will first focus on the equipment within which a reaction takes place, *i.e.*, a chemical reactor. In the following sections, the reaction, its kinetics, and the relevance of this information in different scenarios will be presented.

2.2 TYPES OF REACTORS

A reactor is mainly a vessel where a chemical reaction, such as that presented in Equation 2.1, takes place.[1]

$$\alpha A + \beta B \leftrightarrows \delta C + \Omega D \tag{2.1}$$

The reactor and its components can be assembled in various possible, built according to the purpose of the reaction, configurations. Reactors can be broadly categorized as ideal and real reactors. Since some industrial equipment behaves close to the ideal scenario, that will be our focus.

The most typical ideal reactors can be divided into two major categories, *i.e.*, batch and continuous. Among the continuous reactors, the common types include continuous stirred tank reactors (CSTRs), plug flow reactors (PFRs), membrane reactors (MRs), fluidized bed reactors (FBRs), and packed bed reactors (PBRs). Figures 2.1 to 2.6 show schematics of these reactors.

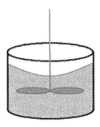

FIGURE 2.1 Batch reactor (BR).

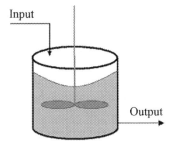

FIGURE 2.2 Continuous stirred tank reactor (CSTR).

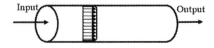

FIGURE 2.3 Plug flow reactor (PFR).

FIGURE 2.4 Packed bed reactor (PBR).

FIGURE 2.5 Fluidized bed reactor (FBR).

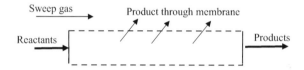

FIGURE 2.6 Membrane reactor (MR).

2.2.1 Batch Reactor

A batch reactor is a vessel that may or may not be pressurized, where there are no input or output flow streams. All the reactants are fed into the reactor at the start of the reaction, *i.e.*, at $t = 0$. When the desired time has passed, the process is stopped; the reactor emptied and cleaned, and is ready for a second batch.

The general balance can be written as follows:

$$\text{Input} - \text{Output} + \text{Generation} = \text{Accumulation}$$

which is expressed in terms of moles, by Equation 2.2, as follows:

$$F_A^0 - F_A + r_A V = \frac{dN_A}{dt} \quad (2.2)$$

In the case of a batch reactor, $F_A^0 = 0$ and $F_A = 0$. Thus, Equation 2.2 is reduced and the mole balance for the batch reactor is obtained as follows:

$$r_A V = \frac{dN_A}{dt} \quad (2.3)$$

A generic expression for the mole balance can be derived as follows:

$$r_i V = \frac{dN_i}{dt} \tag{2.4}$$

Here, V was used as the total volume of the vessel and in order to be rigorous, the term $r_A V$ must be calculated for a non-uniform control volume. Assuming that for a batch reactor, represented in Figure 2.7, differential size volumes should be considered where in each of them a reaction rate r_A takes place.

Then, the integration of all the differential terms will yield an accurate estimation of $r_A V$.

$$\int r_A \, dV = \frac{dN_A}{dt} \tag{2.5}$$

If the reaction rate, r_A, is constant in all the differential volumes (typical case for ideal scenarios), then it can be placed outside the integral. Equation 2.5 can be rewritten as:

$$r_A \int dV = \frac{dN_A}{dt} \tag{2.6}$$

If the integration is carried out between limits 0 and the volume V of the vessel, Equation 2.3 can be obtained.

If the volume of the reaction system is constant, which is typically the case for liquids and gas reactions when there is no change in pressure, temperature, or the number of moles between reactants and products, Equation 2.4 can be reorganized and written in terms of concentration:

$$r_A = \frac{dC_A}{dt} \tag{2.7}$$

In order to solve this expression, we need an initial time boundary condition, such as the concentration at $t = 0$ and the dependence of r_A^2 on the concentration of the

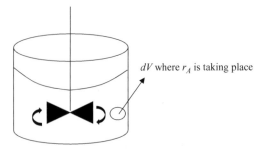

FIGURE 2.7 Scheme of a batch reactor.

Reactor Design

reactants and products. By solving this general expression, the time required to reach a desired final concentration, C_A, can be estimated as follows:

$$t = \int_{C_A^0}^{C_A} \frac{dC_A}{r_A} \tag{2.8}$$

Consider a first-order irreversible reaction, for reactant A, being carried isothermally, then $r_A = -k\, C_A$, where k is the reaction rate constant. Therefore, Equation 2.8 can be rewritten as:

$$t = \int_{C_A^0}^{C_A} \frac{dC_A}{-k\, C_A} \tag{2.9}$$

Now we can solve this expression to obtain the required reaction time:

$$t = -\frac{1}{k}\left(\ln C_A - \ln C_A^0\right) \tag{2.10}$$

This expression shows that by knowing the initial concentration of component A, we can estimate the final concentration after an arbitrary time, t, which can also be expressed as the time taken to reduce the reactant concentration to a certain value.

An even more generalized equation involves expressing the concentration as a function of time; therefore, Equation 2.10 can be rearranged as follows:

$$C_A = C_A^0\, e^{(-kt)} \tag{2.11}$$

On plotting Equation 2.11, as shown in Figure 2.8, the concentration of the reactant A is seen to be exponentially decreasing with the passage of time.[3] It is important to note that Figure 2.8 has been plotted under the assumption of an expression for r_A, in this case, a first-order reaction. This assumption will be reconsidered later on by studying how to estimate and verify r_A with experimental data.

Different reactions rates will have different tendencies when plotted. To show the impact of the type of selected reaction rate, Figures 2.9 and 2.10 show the plots of Equation 2.8 for an isothermal case where $r_A = -k$, i.e., a zero-order equation, and $r_A = -kC_A^2$, i.e., a second-order equation, respectively.

Solving Equation 2.8 using the zero- and second-order expressions, Equations 2.12 and 2.13, respectively, were obtained.

$$t = -\frac{1}{k}\left(C_A - C_A^0\right) \tag{2.12}$$

$$t = -\frac{1}{k}\left(\frac{1}{C_A^0} - \frac{1}{C_A}\right) \tag{2.13}$$

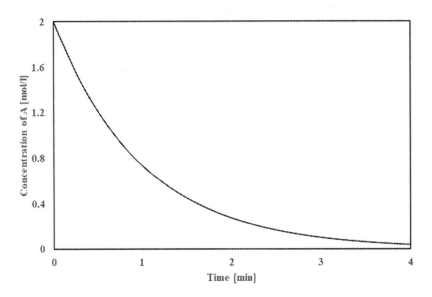

FIGURE 2.8 Variation of the concentration of reactant A as a function of time when a first-order irreversible reaction is being carried out isothermally in a batch reactor.

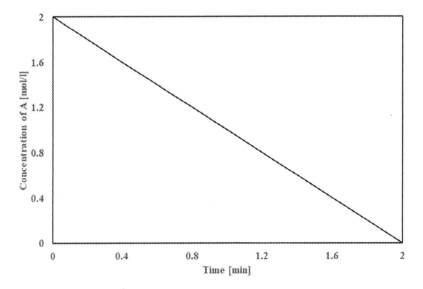

FIGURE 2.9 Variation of the concentration of reactant A as a function of time when a zero-order irreversible reaction is being carried out isothermally in a batch reactor.

Reactor Design

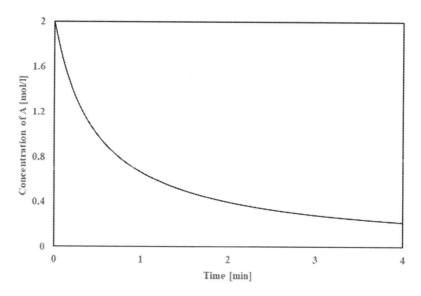

FIGURE 2.10 Variation of the concentration of reactant A as a function of time when a second-order irreversible reaction is being carried out isothermally in a batch reactor.

To obtain plots of concentration versus time, Equations 2.12 and 2.13 were rewritten as Equations 2.14 and 2.15, respectively:

$$C_A = C_A^0 - kt \tag{2.14}$$

$$C_A = \frac{C_A^0}{1 + ktC_A^0} \tag{2.15}$$

Based on Equations 2.14 and 2.15, Figures 2.9 and 2.10, respectively, can be plotted showing the variation of the concentration as a function of time for each scenario.

While Figures 2.8 and 2.10 may seem alike, it is important to note that they correspond to an exponential decrease and a hyperbolic decrease, respectively. Therefore, their rates of consumption of reactant A are different.

The previous analysis was done considering the reaction to be A ➔ B, *i.e.*, the reaction rate only depended on component A, which is the simplest case for a batch reactor. We will address a more complex scenario later in the book and recommend the reader to pay close attention to the problem and its solution presented in Chapter 4.

Complex reactions, *i.e.*, those with more than one raw material or reactant and the production of more than one product, are more common. This aspect will be addressed in Section 2.9 which covers multiple reactions.

2.2.2 Semi-batch Reactor

A semi-batch reactor consists of an input for reactants but not an output and it is a mix of a batch and continuous type reactor; the latter will be presented in the next section. This means that instead of mixing reactants A and B together at $t = 0$, it is preferable to have all the reactant B in the batch reactor while reactant A is added to it slowly, in small amounts. Figure 2.11 shows a schematic for the semi-batch process, which can be used, for instance, when the degree of reaction needs to be controlled or selectivity of products is important.

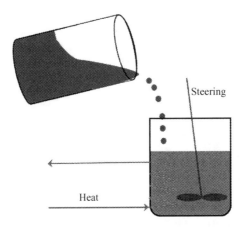

FIGURE 2.11 Schematic of the semi-batch process. One reactant is added in small amounts.

For a semi-batch reactor with only two reactants, the mole balance for reactants A and B is calculated separately, based on the premise that reactant A is constantly being fed to the reactor containing reactant B. Therefore, consider the following reaction:

$$A + B \rightarrow C + D$$

Before examining the concentration variations of the reactants and the products, it is important to see the global mass balance for the reaction system. In general terms, the accumulation of mass can be expressed in terms of the flow of mass into the system (m^0), the flow of mass out of the system (m), and the generation of mass, as follows:

$$\frac{dm}{dt} = m^0 - m + generation \tag{2.16}$$

According to Figure 2.11, there is no output flow, *i.e.*, $m = 0$ and there is no generation of new mass of any chemical component in this reaction, as the individual components are consumed and then proportionately produced, thus there is no generation of mass, *generation* = 0. Thus, Equation 2.16 can then be simplified as follows:

$$m^0 = \frac{dm}{dt} \tag{2.17}$$

Reactor Design

If the density of this liquid system (ρ) is considered to be constant, which is possible for certain types of chemicals being fed, then the mass, m, can be expressed as:

$$m = \rho * V \tag{2.18}$$

where V is the reactor's volume

The mass rate can be expressed as:

$$m^0 = \rho^0 * v^0 \tag{2.19}$$

where v^0 is the volumetric rate or volume per unit of time.

Consequently, Equation 2.19 can be rearranged as follows:

$$v^0 = \frac{dV}{dt} \tag{2.20}$$

This expression can easily be integrated between $t = 0$ and a generic time period, t, and the initial volume of the reactor, V^0 (mainly composed of component B) and the final volume, V. Equation 2.20 will then be written as:

$$V = v^0 * t + V^0 \tag{2.21}$$

When plotted, Equation 2.21 is a straight line with a slope, v^0, corresponding to the volumetric flow of the reactant that is added to the system. However, this expression, mathematically, will increase with time, and an upper limit corresponding to the physical restrictions, should be specified. The final volume, V, cannot be below zero and cannot be larger than the maximum volume of the reactor itself.

2.2.2.1 Balance for Reactant A

Considering the most generic scenario, where a reactant can flow into and out of the system as well as being involved in a chemical reactor, the mole balance for such system, including accumulation over time is given by Equation 2.22.

$$F_A^0 - F_A + r_A V = \frac{dN_A}{dt} \tag{2.22}$$

where F_A^0 is the flow into the reactor and F_A is the flow exiting the reactor.

For a semi-batch reactor, there is an input flow of moles of component A ($F_A^0 \neq 0$) while there is no output ($F_A = 0$); furthermore, we need to express F_A as a function of C_A it can be expressed as follows:

$$F_A = v_A * C_A \tag{2.23}$$

Here, v_A is the volumetric flow and $C_A = N_A/V$. Thus, given that component A flows into the system and there is no flow out of the system; Equations 2.22 can be rearranged as follows:

$$F_A^0 + r_A V = \frac{dN_A}{dt} \tag{2.24}$$

Since the volume is constantly changing, we can no longer assume it is constant and now the variations in the moles of component A need to be expressed as the product of volume and concentration, as given below:

$$\frac{dN_A}{dt} = \frac{d(V*C_A)}{dt} = V\frac{dC_A}{dt} + C_A\frac{dV}{dt} \tag{2.25}$$

Substituting Equation 2.20 into 2.25 and consequently into Equation 2.24, we obtain:

$$F_A^0 + r_A V = V\frac{dC_A}{dt} + CA*v^0 \tag{2.26}$$

Using the definition from Equation 2.23, applied to the input flow, and re-arranging Equation 2.26 we obtain:

$$\frac{dC_A}{dt} = r_A + \frac{v^0}{V}\left(C_A^0 - C_A\right) \tag{2.27}$$

In the case of the semi-batch reactor, there is a new term in the equation as the volume is not constant. It is also important to highlight that in Equation 2.27, C_A^0 is the initial concentration of A in the stream being fed to the reactor and not the concentration of A in the reactor. The concentration of A in the reactor is one of the initial conditions used to solve the differential equation. We will come back to the mole balance for different reaction systems where the volume is variable due to other factors.

2.2.2.2 Balance for Reactant B

Since there is no inlet and outlet for the component B which is present in the reactor, its mole balance is similar to that observed in the batch reactor, as described in the previous section. However, batch reactors have a constant total volume, unlike semi-batch reactors in which the total volume increases since reactant A is being added. Therefore, the mole balance for reactant B needs to be modified. Starting from Equation 2.22, without flow in and out of the reactor, for component B, the mole balance reduces to:

$$r_B V = \frac{dN_B}{dt} \tag{2.28}$$

Using Equation 2.26 for component B, and applying it to Equation 2.28, the following mole balance is obtained:

$$\frac{dC_B}{dt} = r_B - \frac{v^0}{V}C_B \quad (2.29)$$

In contrast to Equation 2.7, Equation 2.29 contains a new term which considers the changes in volume. Similar expressions can be obtained for products C and D, as follows:

$$\frac{dC_C}{dt} = r_C - \frac{v^0}{V}C_C \quad (2.30)$$

$$\frac{dC_D}{dt} = r_D - \frac{v^0}{V}C_D \quad (2.31)$$

In order to solve this problem, four differential equations need to be solved simultaneously and therefore, four initial conditions are needed. It is very common to have $C_A = 0$, $C_C = 0$, and $C_D = 0$, when only reactant B is present in the reactor, at the beginning of the reaction. Based on Equations 2.27, 2.29, 2.30, and 2.31, the change in the concentrations of each component can be plotted as a function of time.

Example 2.1

A third-order irreversible elementary reaction, 2A + B → C + D, is being carried out at a constant temperature of 350 [K] and at constant pressure. The reaction is being carried out in a semi-batch reactor where A is fed into the system with a volumetric flow of 0.2 [dm³/min] and with a concentration of 1.2 [mol/l]. Reactant B is in the reactor with an initial concentration of 0.8 [mol/l] and there is no product at the beginning of the process. The reaction constant is 2.35 [min*l²/mol²]. Solve the problem and plot the variation of all the concentrations and the volume as a function of time. Discuss the results.

Solution

First, all the relevant information, assumptions, and unknowns will be identified with respect to a semi-batch reactor schematic (Figure 2.12).

Since this is an elementary reaction, the reaction rate can be calculated from the stoichiometry of the reaction itself. For this case:

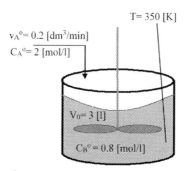

FIGURE 2.12 Scheme for example 2.1.

$$r_A = -k * C_A^2 C_B \qquad (2.32)$$

According to the rate law equivalence for all reactants and products, all the other reaction rates can be obtained as follows:

$$\frac{r_A}{-2} = \frac{r_B}{-1} = \frac{r_C}{1} = \frac{r_D}{1} \qquad (2.33)$$

Thus, reaction rate expressions for all the components can be obtained:

$$r_B = -\frac{1}{2} k * C_A^2 C_B \qquad (2.34)$$

$$r_C = \frac{1}{2} k * C_A^2 C_B \qquad (2.35)$$

$$r_D = \frac{1}{2} k * C_A^2 C_B \qquad (2.36)$$

Then, mass balance is applied to the whole process, taking into account that the reactor has not outlet flow, no production or consummation of mass occurs, and assuming that the density remains constant. The general mass balance can be summarized as:

$$\frac{dV}{dt} = v_A^0 \qquad (2.37)$$

The mole balances are equal to those form Equations 2.27, 2.29, 2.30, and 2.31. A solver, like Polymath, can be used to solve the following equations:

Balances

$$\frac{dV}{dt} = v_A^0$$

$$\frac{dC_A}{dt} = r_A + \frac{v^0}{V}\left(C_A^0 - C_A\right)$$

$$\frac{dC_B}{dt} = r_B - \frac{v^0}{V} C_B$$

$$\frac{dC_C}{dt} = r_C - \frac{v^0}{V} C_C$$

$$\frac{dC_D}{dt} = r_D - \frac{v^0}{V} C_D$$

Reactor Design

Reaction rates

$$r_A = -k * C_A^2 C_B$$

$$r_B = -\frac{1}{2}k * C_A^2 C_B$$

$$r_C = \frac{1}{2}k * C_A^2 C_B$$

$$r_D = \frac{1}{2}k * C_A^2 C_B$$

Constants

v_A^0 = 0.2 [l/min]; $C_A(0)$ = 0 [mol/l]; C_A^{04} = 1.2 [mol/l]; $C_B(0)$ = 0.8 [mol/l]; $C_C(0)$ = 0 [mol/l]; $C_D(0)$ = 0 [mol/l]; k = 2.35 [min*l²/mol²]

Time dependency

Initial time = 0 min; Final time = 50 min

Solving all the equations together, the concentrations profiles of all the products can be obtained as functions of time. The trends for the variation in volume and concentration of each component is shown in Figures 2.13 and 2.14, respectively. The following script can be used in Polymath:

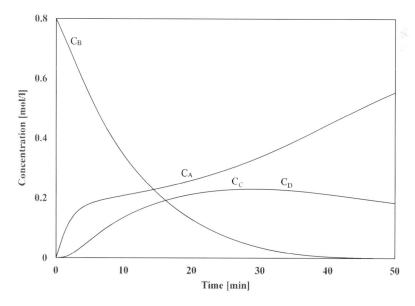

FIGURE 2.13 Concentration profiles of components A, B, C, and D as a function of time.

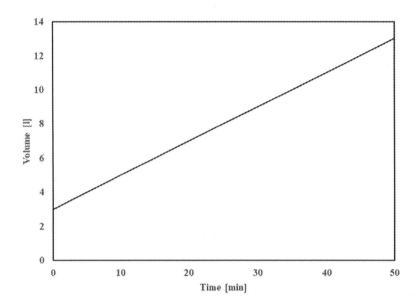

FIGURE 2.14 Variation of the total volume of the components in reactor, as a function of time.

Script:

#Mole balance
 d(Ca)/d(t) = ra+((vao/V)*(Ca0 − Ca))
 Ca(0) = 0
 d(Cb)/d(t) = rb − vao/V*Cb
 Cb(0) = 0.8
 d(Cc)/d(t) = rc − vao/V*Cc
 Cc(0) = 0
 d(Cd)/d(t) = rd − vao/V*Cd
 Cd(0) = 0
#Volume balance
 d(V)/d(t) = vao
 V(0) = 3
#Reaction rate
 ra = −k*Ca^2*Cb
 rd = 1*ra/−2
 rc = 1*ra/−2
 rb = −1*ra/−2
#Constants
 vao = 0.2
 Ca0 = 1.2

Reactor Design

```
    k = 2.35
#Time
    t(0) = 0
    t(f) = 50
```

It is important to note that in Figure 2.13, the concentration profiles of components C and D overlap perfectly since their initial conditions were identical.

Mathematically, the volume of the components in the reactor can increase to infinity, which is not physically possible. The reactor has a finite volume, and the time used in the simulations should not be larger than the time required to fill 90% of the reactor volume. Therefore, it is very important that the mathematical results obtained have physical relevance to the problem.

It is possible to have semi-batch reactors which have only an output stream and no inlet stream. If the reaction takes place in liquid phase, all the components will flow out of the reactor through the outlet. In this case, it will be much more complicated to solve the problem since the amount of each component remaining in the reactor should be known. Westerterp *et al.* [1] in their reactor engineering book, presented an interesting approach for a reaction that produced a liquid product and a gas product, and we strongly recommend the reader to study this specific example. Since both reactants were in liquid phase, a semi-batch reactor was achieved by removing the pure product gas phase. In this case, only one component flows out and the steps and method used to solve the Equations are analogous to that of the semi-batch reactor with only an inlet, as presented above.

2.2.3 Continuous Reactors

There are quite a few different types of ideal continuous reactors, such as CSTRs, PFRs, PBRs, MRs, FBRs, among others. In this section, the analysis for the CSTR, the PF, the PBR, and the MR, will be presented and the rest of the reactors will not be included as they are out of the scope of this book.

The main difference between the continuous and batch reactors is that in the continuous system, there is a flow of reactants and products via the inlet and outlet. However, since there is no accumulation of components with time, the system is in steady state. Therefore, starting from the general balance written as:

$$\text{Input} - \text{Output} + \text{Generation} = \text{Accumulation}$$

which is expressed as following in terms of moles:

$$F_A^0 - F_A + r_A V = \frac{dN_A}{dt} \tag{2.38}$$

Here, F_A^0 is the flow through the inlet and F_A is the flow from the outlet.

2.2.3.1 Continuous Stirred Tank Reactor (CSTR)

This type of system is the simplest in this class of reactors and is very useful for different types of reactions, making it very common in industry. The CSTR is a system with a constant volume inside the reactor, *i.e.*, the reactants are fed into the reactor and the product and unreacted reactants flow out of the reactor. The components are considered to be perfectly well mixed and for isothermal systems, there are no temperature profiles. Additionally, it is assumed that the concentration at the outlet is the same as the concentration in the reactor. Figure 2.15 shows a schematic diagram of a CSTR system.

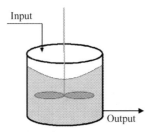

FIGURE 2.15 Scheme of a CSTR.

Since the system is in steady state, Equation 2.38 can be rewritten as:

$$F_A^0 - F_A + r_A V = 0 \qquad (2.39)$$

This can further be rearranged as follows:

$$V = \frac{F_A^0 - F_A}{-r_A} \qquad (2.40)$$

The generalized representation of Equation 2.40, for any component, is as follows:

$$V = \frac{F_i^0 - F_i}{-r_i} \qquad (2.41)$$

Equation 2.41 will be used to determine the appropriate volume, and therefore, the size of the CSTR reactor. It is important to notice that among all the reactor systems, this is the only system that is not represented by a differential equation.

Just like for the batch system, an expression for the reaction rate is needed here, for the CSTR. For the time being, the expression will be considered arbitrarily and later it will be demonstrated how to obtain it based on experimental data.

Therefore, if the reaction is a first-order irreversible reaction, then $r_A = -kC_A$, then it can be substituted in Equation 2.40 to obtain:

Reactor Design

$$V = \frac{F_A^0 - F_A}{-kC_A} \quad (2.42)$$

Now, in order to solve this expression, the relationship between concentration and molar flow: $F_A = C_A * v_A$, is used, which will lead to:

$$V = \frac{C_A^0 * v_A^0 - C_A v_A}{-kC_A} \quad (2.43)$$

For a liquid system, the metric flow remains constant, meaning that $v_A = v_A^0$, which can be referred to as v. Therefore, Equation 2.43 can be rearranged in two different ways, i.e., Equations 2.44 and 2.45.

$$V = \frac{v\left(C_A^0 - C_A\right)}{-kC_A} \quad (2.44)$$

This expression provides the size of the reactor as a function of the concentration of component A at the outlet.

Equation 2.45, shown below, is a more common expression that is found in several textbooks [1–4], where a new parameter, the space time, is defined. Space time, τ, represents the time that is required to process the fluids in the reactor volume in question. This space time is very useful to compare reactor times and to predict operational conditions.

$$\tau = \frac{V}{v} = \frac{\left(C_A^0 - C_A\right)}{-kC_A} \quad (2.45)$$

With Equation 2.44, for a first-order irreversible reaction, we can plot the variations of the volume as a function of C_A, as well as the variation of C_A as a function of the volume; the latter is plotted in Figure 2.16. Then, Equation 2.44 can be rewritten as:

$$C_A = \frac{C_A^0}{\left(1 + \left(V/v\right)k\right)} \quad (2.46)$$

where $\tau = V/v$ and the Damköhler number is given by $Da = \tau * k$. This number is very important, and a more general definition and different uses will be provided later. Therefore, Equation 2.46 can also be expressed as:

$$C_A = \frac{C_A^0}{\left(1 + Da\right)} \quad (2.47)$$

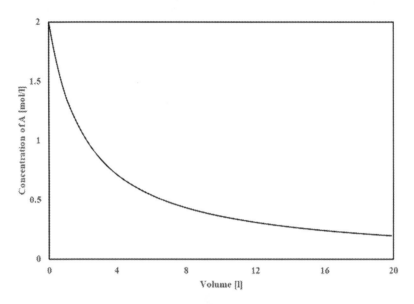

FIGURE 2.16 Variation of the concentration of A as a function of time.

Example 2.2

The reaction of 2A to give component B takes place in liquid phase, in an isothermal CSTR system. The reaction is an elementary second-order reaction and therefore, follows the reaction rate equation:

$$r_A = -kC_A^2 \tag{2.48}$$

The initial concentration of reactant A is 2 [mol/l] while B is zero. Find the space time required to achieve 65% conversion. If the volumetric flow is 3 [l/min], how big should the reactor be? The kinetics parameter is 0.5 [mol/(l*min)]. How big should the reactor be to achieve 99% conversion?

Solution

In order to solve this problem, we first need to obtain the new balance for the new reaction system required.

The mole balance for the CSTR remains:

$$V = \frac{F_A^0 - F_A}{-r_A} \tag{2.49}$$

By substituting Equation 2.48 and $F_A = C_A * v_A$, we obtain:

Reactor Design

$$V = \frac{v\left(C_A^0 - C_A\right)}{kC_A^2} \qquad (2.50)$$

Thus, the space time, according to Equation 2.45 is:

$$\tau = \frac{\left(C_A^0 - C_A\right)}{kC_A^2} \qquad (2.51)$$

Equation 2.51 needs to be solved to obtain the space time for the given conditions. In order to solve this problem, for a conversion of 65%, the value of $C_A = 0.65\%$ of the value of C_A^0, i.e., $C_A = 0.7$ [mol/l] which can be substituted into Equation 2.51 and obtain the value for the space time:

$$\tau = \frac{\left(C_A^0 - C_A\right)}{kC_A^2} = \frac{(2 - 0.7)}{0.5(0.7)^2} = 5.30 \, [\text{min}]$$

Since the volumetric flow is known and $\tau = V/v$, the volume of the reactor is:

$$v * \tau = V = 5.3 * 3 = 15.9 \, [\text{l}]$$

Now, we can compare this result with the size of the reactor if 99% conversion is to be achieved. In this case, $C_A = 0.99$ of C_A^0, i.e., $C_A = 0.02$ [mol/l]. Substituting these values into Equation 4.30, the value for the space time is:

$$\tau = \frac{\left(C_A^0 - C_A\right)}{kC_A^2} = \frac{(2 - 0.02)}{0.5(0.02)^2} = 9,900 \, [\text{min}]$$

Furthermore, the volume of the reactor is:

$$v * \tau = V = 9,900 * 3 = 29,700 \, [\text{l}]$$

This shows that an increase of conversion by 34% increases the reactor volume by a factor of 1,800 times.

Extra:

In order to see the effect of the volume better, we will now plot (Figure 2.17) the variation of the volume of the reactor as a function of the concentration of component A. As expected, the decrease in the concentration of component A that exits the reactor, leads to a higher volume.

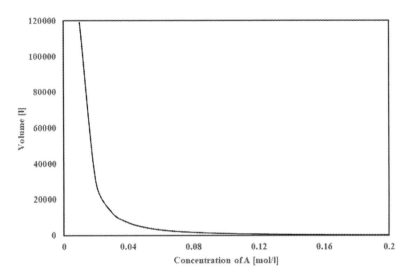

FIGURE 2.17 Variation of the volume as a function of C_A.

2.2.3.2 Plug Flow Reactor (PFR)

The second ideal type of flow reactor is the PFR, which is also known as the piston flow reactor since the fluid moves through the reactor as a piston. It is schematically represented in Figure 2.18. The system moves like a plunge; therefore, it is in a turbulent regime that brings uniformity and no gradients in the radial direction and there is no recirculation of fluid. The temperature profile is uniform in the radial direction, with no gradients inside the equipment, and the temperature of the system can only vary along the reactor.

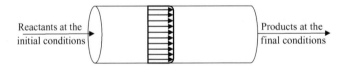

FIGURE 2.18 Scheme of a plug flow reactor.

Like the CSTR, the PFR also runs in a steady-state scenario and there is no accumulation of components within the reactor. So, Equation 2.37 can be rewritten as:

$$F_A^0 - F_A + r_A V = 0 \qquad (2.52)$$

In this case, since the reaction takes place along the system, a differential analysis will be used to obtain the desired differential equation that represents this problem. Figure 2.19 shows a differential cut of the reactor where the mole balance is conducted.

Reactor Design

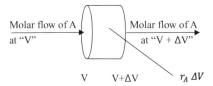

FIGURE 2.19 Differential section for mole balance.

Now that we have selected a differential area of the mole balance for that differential of volume, the general balance expression is:

$$\text{Input} - \text{Output} + \text{Generation} = \text{Accumulation}$$

This is still valid for the new volume defined above; however, a differential volume requires an input flow of component A at location V, and an output flow of component A at location $V+\Delta V$. The accumulation is zero as the system is in steady state and the generation rate is the uniform reaction rate, r_A, that takes place in the volume, $\Delta V = V + (\Delta V - V)$.

The flow expression for component A is given as:

$$F_V^A - F_{V+\Delta V}^A + r_A \Delta V = 0 \tag{2.53}$$

We can rearrange this expression as:

$$F_{V+\Delta V}^A - F_V^A = r_A \Delta V \tag{2.54}$$

Diving the expression by ΔV, we get:

$$\frac{F_{V+\Delta V}^A - F_V^A}{\Delta V} = r_A \tag{2.55}$$

Considering the limit $\Delta V \to 0$, the volume can be made as small as possible:

$$\lim_{\Delta V \to 0} \frac{F_{V+\Delta V}^A - F_V^A}{\Delta V} = \frac{dF_A}{dV} \tag{2.56}$$

Therefore, the mole balance for a plug flow for component A is:

$$\frac{dF_A}{dV} = r_A \tag{2.57}$$

This type of analysis can be done for all the components present in the PFR; therefore, we can generalize Equation 2.57 to:

$$\frac{dF_i}{dV} = r_i \tag{2.58}$$

In order to solve Equation 2.58, a reaction rate expression and the volume of the reactor are needed. It is important to see that in many cases, PFRs are also called tubular reactors. If that is the case and the area is constant, it is also typical to express the variations of the molar flow as a function of the length of the tube instead of the volume.

For this first demonstration of how to solve Equation 2.58, a reaction of component A that gives B (A➔B) irreversibly and in an elementary step, is considered, such that:

$$r_A = -kC_A \tag{2.59}$$

Substituting this into Equation 2.57, the following is obtained:

$$\frac{dF_A}{dV} = r_A \tag{2.60}$$

In order to solve this expression, one initial condition for the molar flow is needed, however, it is more common to know the initial concentration of component A and the volumetric flow. So, for this case, we know that:

$$v = 5\,[1/\min]$$

$$C_A^0 = 2\,[\text{mol}/1]$$

$$k = 3\,[1/\min]$$

With this information, the Polymath script to solve this differential equation is:

Script:
#Mole balance
   d(Fa)/d(V) = ra
   Fa(0) = 10
#Reaction rate
   ra = −k*Ca
#Auxiliary equations
   Ca = Fa/v
#Constants
   k = 3
   v = 5
#Independent variable
   V(0) = 0
   V(f) = 10

Based on the script and solving the differential equation we can obtain the profile of the molar flow of component A as a function of the volume. It is important to point out that a constant value of k indicates that the process is isothermal. Figure 2.20 shows the molar flow variation as a function of volume.

Reactor Design

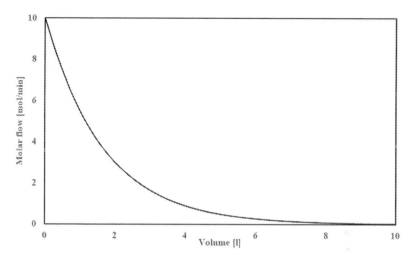

FIGURE 2.20 Molar flow variations for a plug flow reactor.

This problem is based on a one reactant and one product in an irreversible reaction; a more complex situation is presented in Chapter 4. The following example considers two reactants, in an irreversible reaction where the molar flow of the product is important to estimate.

Example 2.3
Consider a reaction with components A and B that form C, that follows the following kinetics:

$$2A + B \rightarrow C$$

This elementary reaction takes place in a plug flow that is operated at 350 [K]. The flow into the reactor is an equimolar flow of A and B with a volumetric flow of 13 [l/min]. Initial concentration of A and B was 2.5 [mol/l]. How long should the tubular reactor be in order to achieve a 95% conversion of component A? If the temperature is decreased to 300 [K], how much conversion can be achieved within the same length as the previous scenario? Plot the variations of F_A, F_B, and F_C.

Solution
Based on the data that it is an elementary reaction; the reaction rate is:

$$r_A = -k C_A^2 C_B \tag{2.61}$$

Based on the general expression (Equation 2.58), the mole balance for each component is:

$$\frac{dF_A}{dV} = r_A \tag{2.62}$$

$$\frac{dF_B}{dV} = r_B \qquad (2.63)$$

$$\frac{dF_C}{dV} = r_C \qquad (2.64)$$

In order to be able to solve this system of differential equations, we need the relationship between r_A, r_B, and r_C. This is given by a similar analysis as the one done in Equation 2.33 and for this case, it is:

$$\frac{r_A}{-2} = \frac{r_B}{-1} = \frac{r_C}{1} \qquad (2.65)$$

Substituting Equation 2.61 into 2.62, 2.63, and 2.64, using Equation 2.65, gives:

$$\frac{dF_A}{dV} = -kC_A^2 C_B \qquad (2.66)$$

$$\frac{dF_B}{dV} = -\frac{1}{2}kC_A^2 C_B \qquad (2.67)$$

$$\frac{dF_C}{dV} = \frac{1}{2}kC_A^2 C_B \qquad (2.68)$$

In order to solve this system, the expression of concentration should be changed to molar flow, with the relation, $F_i = C_i * v_i$. In this case, the reaction takes place in liquid phase and therefore, the volumetric flow can be considered equal for all components.

Additional data:

$$k = 13{,}500\, e^{\left(\frac{-Ea}{RT}\right)} \left[\frac{l^2}{mol^2 * min}\right]$$

here, Ea is the activation energy and R is the universal gas constant.
R = 8.314 [J/(mol*K)];   Ea = 33,000 [J/mol];
Diameter = 0.2 [dm]
Putting all this information into Polymath, we can get the following script:

Script:

    #Mole balance
        d(Fa)/d(V) = ra
        Fa(0) = 32.5
        d(Fb)/d(V) = rb

Reactor Design

```
    Fb(0) = 32.5
    d(Fc)/d(V) = rc
    Fc(0) = 0
#Reaction rate
    ra = -k*Ca^2*Cb
    rb = ra/2
    rc = -ra/2
#Auxiliary equations
    Ca = Fa/v
    Cb = Fb/v
    Cc = Fc/v
    x = (Fa0 - Fa)/Fa0
    k = koo*EXP(-Ea/(R*T))
    L = V/A
    A = 3.14159/4*D^2
#Constants
    Fa0 = 32.5
    v = 13
    koo = 135,000
    R = 8.314
    T = 350
    Ea = 33,000
    D = 2
#Independent variable
    V(0) = 0
    V(f) = 50
```

To achieve 95% conversion, we need to run the script several times (trial and error). A volume value is proposed, the script is run, and the final conversion is achieved. This procedure is done until the desire conversion is reached. For this case, a final volume of 50 [l] will provide a final conversion of 95%, knowing the area of the tube, the length of the reactor can be estimated, 1.591 [m]

For this scenario, Figure 2.21 shows the variations of the molar flow. It is evident that the molar flow of component A declines faster than B even though both of them have the same initial conditions, corresponding to the reaction stoichiometry. The molar flow of component C increases since it is the product of the reaction.

Figure 2.22 shows the variation of the conversion as a function of the length of the reactor. It can be seen that the conversion increases rapidly initially but flattens along the reactor length since it is reaching the highest conversion. This also produce that the reaction rate declines. The variation of conversion is directly linked to the profile for the limiting reactant of the process, chemical A in this example, which has decreased and eventually will decline to zero.

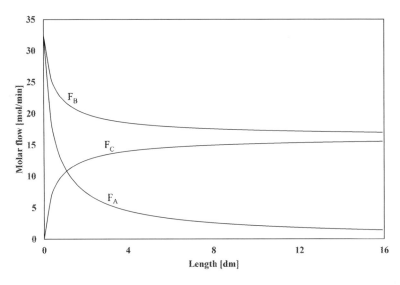

FIGURE 2.21 Molar flow profiles as a function of the reactor length.

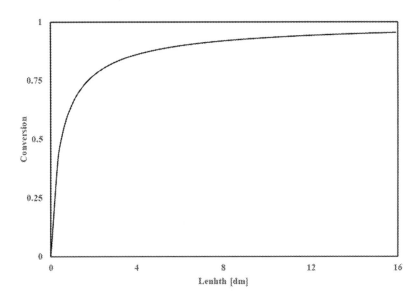

FIGURE 2.22 Conversion profile as a function of reactor length.

Finally, it is desired to know the conversion that can be achieved in this reactor when the temperature decreases to 300 [K].

So, for this, we need to run the script by changing only the temperature to 300 [K]. The corresponding conversion is plotted as a function of the length, as shown in Figure 2.23, and the final conversion obtained is 80%.

We can also plot the variations of the flow, for each component, as a function of the length of the reactor. As expected and presented in Figure 2.24, the flow of

Reactor Design

each component is less pronounced as the reaction rate is lower. When decreasing the temperature, less energy is available for the system, and therefore, the reaction takes place at a slower pace.

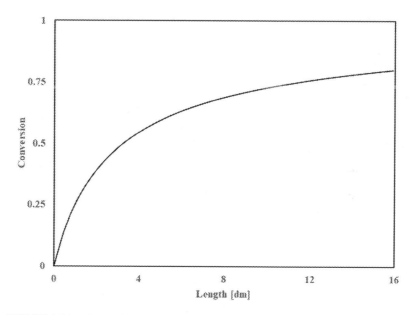

FIGURE 2.23 Conversion profile as a function of reactor length.

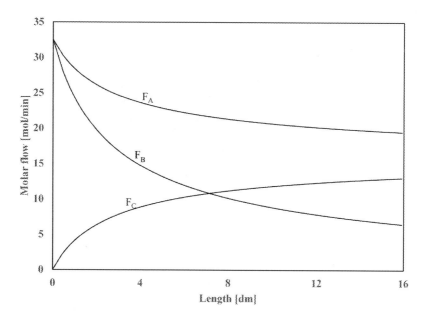

FIGURE 2.24 Molar flow profiles as a function of the length.

In the previous model, the flow profiles and conversion were described as a function of the length of the reactor. However, as we will see, many reactions can take place in a heterogeneous system, where the reactants are in liquid or gas phase and the catalyst is in solid phase. For those cases, it is quite common to use a PBR or FBR; the latter is out of the scope of this book.

2.2.3.3 Pack Bed Reactor (PBR)

The PBR is another ideal flow system, as shown schematically in Figure 2.25. In the PBR, the reactants pass through the catalyst, which is fixed and does not move, and the products are generated at the other end. The fluid is not as uniform, as in the previous case, since different patterns can be produced due to the distribution of the catalytic particles. However, it is assumed that there is no gradient in the radial direction and that there is no recirculation of fluid. The temperature profile is uniform in the radial direction, with no gradients inside the equipment, and the temperature of the system can vary along the reactor.

Like other flow systems discussed, this one is also in a steady-state regime; therefore, Equation 2.37 is rewritten as:

$$F_A^0 - F_A + r_A^w W = 0 \tag{2.69}$$

It is important to emphasize that the reaction does not take place in the whole volume of the reactor, but only over the catalytic surface. Therefore, a differential analysis is needed in order to obtain the desired differential equation that represents this problem. Figure 2.26 shows a differential cross section of the reactor where the mole balance is conducted.

Here, the term r_A^w is introduced, this term is the reaction rate per unit of mass of catalysts, with units of [mol/(kg*min)] as compared with the PF where for r_A the units where [mol/(l*min)]

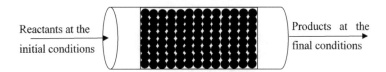

FIGURE 2.25 Schematic diagram of a pack bed reactor.

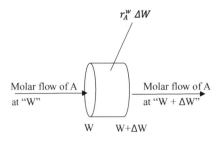

FIGURE 2.26 Differential section for mole balance.

Reactor Design

Based on Figure 2.26 and an analogue analysis for the PFR, we can obtain the expression for the mole flow as a function of the weight of the catalyst.

$$\frac{dF_A}{dW} = r_A^w \tag{2.70}$$

This type of analysis can be done for all the components present in the PFR; therefore, we can generalize this expression as:

$$\frac{dF_i}{dW} = r_i^w \tag{2.71}$$

The solution to this problem is analogous to the PFR and the results are similar, but in a PBR, the reaction occurring per mass (usually in kg) of catalyst is more important than per volume of reactor.

2.2.3.4 Membrane Reactor (MR)

Membrane reactors are a type of flow system where the reactants flow through the reactor but at least one of the products is constantly removed through the membrane for its separation and purification. Membrane reactors have been widely used for hydrogen production [5], reforming of ethanol [6], treatment of toxic chemicals [7], production of oligosaccharides [8], production of polymers for medical applications [9], to name a few.

There are different types of membranes, *i.e.*, reactive or inert. Inert membranes are only for separation of components while reactive membranes allow separation and can also support depositions of catalytic material. The functioning of a membrane reactor is summarized in Figure 2.27.

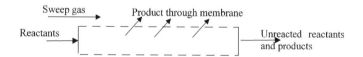

FIGURE 2.27 Schematic representation of a membrane reactor.

A sweep gas is used in order to remove the product that is passing through the membrane. The selection of the sweep gas is case dependent since it is crucial that it is inert toward the product, that will not destroy the membrane, and that it is easily separated from the product produced.

To show the balance for this equipment, we will solve the problem for the reaction system:

$$A + B \rightarrow C + D$$

In order to analyze a membrane reactor, we need to consider which of the desired products (or raw materials) have to pass through the membrane. In this case, component C is the one that is being produced in the process and removed simultaneously through the membrane.

For the remaining components, a membrane reactor functions as a PFR or a PBR. Consider a liquid reaction for this section, which will be designed like a plug flow system, with an elementary reaction taking place in liquid phase.

Therefore, for components A, B, and D, the balance is based on Equation 2.58 giving:

$$\frac{dF_A}{dV} = r_A \qquad (2.72)$$

$$\frac{dF_B}{dV} = r_B \qquad (2.73)$$

$$\frac{dF_D}{dV} = r_D \qquad (2.74)$$

The analysis for component C is more complex, for which the general balance for any volume should be considered:

$$\text{Input} - \text{Output} + \text{Generation} = \text{Accumulation}$$

In the case of a membrane reactor, a differential volume (Figure 2.28) is taken for analysis, similar to the PFR. It is crucial to notice that the output happens at $V+\Delta V$ due to the flow and also through the membrane.

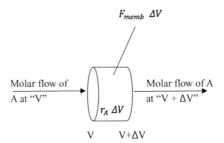

FIGURE 2.28 Differential section for mole balance.

Therefore, the mole balance obtained is analogous to the plug flow, as follows:

$$\frac{dF_C}{dV} = r_c - F_{memb} \qquad (2.75)$$

In order to solve this problem, we need to find an expression for the flow through the membrane, F_{memb}, as a function of component C. Usually component C is not expected to be a part of the sweep gas since the $\Delta C_C = (C_C - C_{CSW})$ is the driving

Reactor Design

force to move component C through the membrane. In mass transfer, the flow is proportional to a gradient of concentration. Therefore, if the concentration in the sweep gas is zero, $C_{CSW} = 0$, then $\Delta C_C = (C_C - 0) = C_C$ and the flow can be obtained as follows:

$$F_{memb} = k_{mem} C_C \tag{2.76}$$

Here k_{mem} is a constant that is related to the properties and characteristics of the membrane, such as permeability, size, and porosity.

In order to solve this problem, the concept of reaction in gas phase needs to be introduced and is provided in Section 2.7.

2.3 CONVERSION

So far, we have addressed the mole balances of different ideal reactors in order to obtain the concentration or the molar flow profiles. This section addresses the definition of conversion and how it can be used to solve the mole balance for the three ideal reactors, Batch, CSTR, Plug Flow, that have been presented in the previous sections.

Conversion is defined by the following expression:

$$x = \frac{N(t=0) - N(t=t_i)}{N(t=0)} = \frac{N_0 - N_t}{N_0} \tag{2.77}$$

where $N(t = 0)$ is the number of moles at the beginning of the reaction, *i.e.*, those that are being fed to the system and $N(t = t_i)$ are the number of moles at an arbitrary time, t. The numerator of Equation 2.77 represents the number of moles that have been converted into the product and the denominator is the number of moles that were fed into the reactor. Therefore, x should be a number between 0 and 1.

Conversion can only be calculated in the context of one of the reactants and it is preferable to use the reactant that is the most relevant (and most likely to be the limiting reactant) of the process in question. Assuming that reactant A is that chemical that is more expensive and therefore, important to monitor, Equation 2.77 can be rewritten as:

$$x = \frac{N_A(t=0) - N_A(t=t_i)}{N_A(t=0)} = \frac{N_A^0 - N_A}{N_A^0} \tag{2.78}$$

If the volume remains constant, then the conversion can be presented as a function of the concentration of component A. Thus, Equation 2.78 can be rewritten as:

$$x = \frac{C_A(t=0) - C_A(t=t_i)}{C_A(t=0)} = \frac{C_A^0 - C_A}{C_A^0} \tag{2.79}$$

If the volumetric flow is also constant, which will be the case if the volume is constant, then Equation 2.79 can be rewritten in terms of the molar flow of component A:

$$x = \frac{F_A(t=0) - F_A(t=t_i)}{F_A(t=0)} = \frac{F_A^0 - F_A}{F_A^0} \quad (2.80)$$

All the previous equations for calculating the conversion are equivalent and they can be used depending on the type of problem to be solved. Section 2.7 will address how to obtain these equations when there is volume change in a gas phase system.

With this new definition for conversion, the mole balances for all the reactors can be expressed in terms of conversion. Equation 2.79 can be rewritten as a function of conversion:

$$C_A = C_A^0 - C_A^0 x = C_A^0(1-x) \quad (2.81)$$

Considering Equations 2.7 and 2.81, the following is obtained:

$$r_A = \frac{d(C_A^0 - C_A^0 x)}{dt} \quad (2.82)$$

which can further be rewritten as:

$$r_A = -C_A^0 \frac{d(x)}{dt} \quad (2.83)$$

In order to solve this equation, we need an expression for the reaction rate, r_A. Assuming that the reaction under consideration is a first-order irreversible reaction, then:

$$r_A = -kC_A \quad (2.84)$$

We can substitute C_A from Equation 2.81 to obtain:

$$r_A = -kC_A^0(1-x) \quad (2.85)$$

Now we can substitute Equation 2.85 into Equation 2.83:

$$-kC_A^0(1-x) = -C_A^0 \frac{d(x)}{dt} \quad (2.86)$$

which on rearranging gives:

$$k(1-x) = \frac{d(x)}{dt} \quad (2.87)$$

Reactor Design

This differential equation can be solved by separating variables as follows:

$$kdt = \frac{d(x)}{(1-x)} \tag{2.88}$$

Both sides of Equation 2.88 can be integrated between limits, *i.e.*, zero conversion at $t = 0$, and an arbitrary amount of conversion, x, at time t, respectively. Then, the general expression for the variation of the conversion as a function of time, for a batch reactor with a first-order irreversible reaction, can be obtained as follows:

$$\int_0^t -kdt = \int_0^x \frac{dx}{(1-x)} \tag{2.89}$$

This will lead to:

$$x = 1 - e^{(-kt)} \tag{2.90}$$

Therefore, we can now estimate the time taken to reach a certain extent of conversion, and we can estimate the conversion achieved after different time periods. As presented in Figure 2.29, the conversion is seen to increase from zero to a maximum of 1.

Figure 2.29 shows the variation of conversion for a fist order elementary reaction. This methodology to establish the variation of conversion as a function of time, for reactions with different reaction order, is similar and analogous to the one presented above. We leave the plotting and comparison of the variations of conversion as a function of time, for a second and a third-order reaction, for the reader.

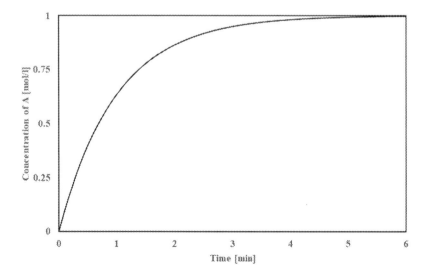

FIGURE 2.29 Conversion profile for an elementary first-order reaction in a batch reactor.

Mole balances, in terms of conversion, are useful not only for batch reactors, but also for all types of reactors such as CSTR, PF, or PBR.

For a CSTR and a batch reactor, it is crucial to know the reaction rate expression. As a first example, a simple first-order elementary reaction of component A that produces B, is considered. This leads to a reaction rate:

$$r_A = -kC_A \tag{2.91}$$

Using Equation 2.40:

$$V = \frac{F_A^0 - F_A}{-r_A} \tag{2.92}$$

The definition of the molar flow as a function of conversion, as shown in Equation 2.80, together with Equation 2.91 can be combined to rewrite Equation 2.92 as:

$$V = \frac{F_A^0 X}{kC_A} \tag{2.93}$$

In order to express Equation 2.93 as a function of conversion, C_A will be substituted using Equation 2.79, as follows:

$$V = \frac{F_A^0 x}{kC_A^0(1-x)} \tag{2.94}$$

Using the relation, $F_A^0 = C_A^0 * v_A$, we have:

$$V = \frac{C_A^0 v_A x}{kC_A^0(1-x)} \tag{2.95}$$

leading to:

$$V = \frac{v_A x}{k(1-x)} \tag{2.96}$$

In terms of the space time (with $v_A = v$):

$$\tau = \frac{V}{v} = \frac{x}{k(1-x)} \tag{2.97}$$

Here, the conversion at the outlet is related to the equipment size, i.e., volume of the reactor or to the space time. On plotting Equation 2.97, Figure 2.30 is obtained which shows the size of the reactor as a function of the output conversion of the reactor.

Reactor Design

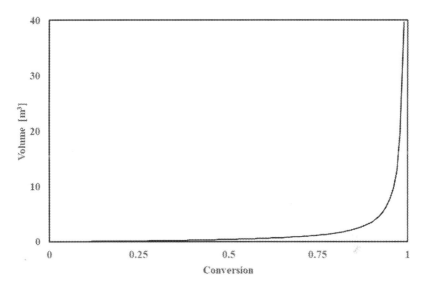

FIGURE 2.30 Volume variations for different conversions for a CSTR reactor.

Drawing an analogy for a first-order elementary irreversible reaction in a PFR, the mole balance is:

$$\frac{dx}{dV} = \frac{k}{v}(1-x) \tag{2.98}$$

Similarly, the expression for a PBR can be obtained:

$$\frac{dx}{dW} = \frac{k}{v}(1-x) \tag{2.99}$$

On solving and plotting Equation 2.98, the variation of the volume as a function of conversion is shown in Figure 2.31, which can also be presented as the variation of the space time. While Figure 2.31 is for a first-order irreversible reaction, any change in kinetics to a second-order reaction will impact the plot accordingly.

The Levenspiel plot [3] provides a useful way to estimate the volume of a reactor, from the area under the plotted curve. The variation of the initial molar flow of any component A normalized by the reaction rate $\left(F_A^0 / -r_A\right)$, in the y-axis, is plotted as a function of variation of conversion in the x-axis. As presented in Figure 2.32, the Levenspiel plot provides how much volume is needed for a CSTR and for a PFR. Since the mole balance for a CSTR is an algebraic equation, the volume in the case of a CSTR can be easily calculated as the $\Delta x * \Delta\left(F_A^0 / -r_A\right)$ from the plot (highlighted as the gray area (a)). For the PF, the volume is calculated by solving a differential equation; therefore, the volume is the area under the curve (highlighted by the lines and marked (b)). For these cases, it can easily be seen that the volume of a PFR is always smaller than that of a CSTR, for a given conversion and a given reaction. However, this may not be true for non-isothermal cases, which will be presented later on.

We have seen the equation for one ideal reactor; however, reactors can be put in series or in parallel to achieve different goals such as having intermediate products

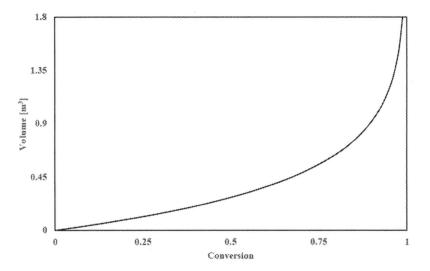

FIGURE 2.31 Volume profile for a PB reactor as a function of conversion.

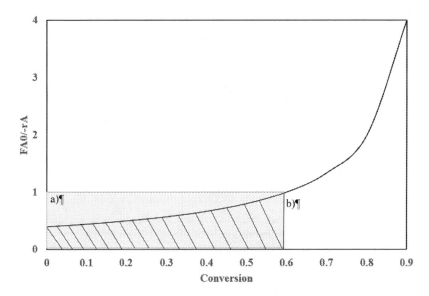

FIGURE 2.32 Comparison of volume calculations for (a) CSTR and (b) PFR using the Levenspiel plot.

Reactor Design

removed, or to avoid high temperature profiles. The following sections will discuss reactors, CSTR, and PF, in series and in parallel.

2.4 REACTORS IN SERIES

When reactors are put in series, the output of one reactor is the input of the next reactor. For the time being, it will be assumed that no heat is transferred between the reactors and no chemicals are added or removed, either in the reactors or in between reactors. Therefore, this system can be used in case a reactant can produce more than one product, for instance, at different temperatures. Thus, one reactor can be maintained at a lower temperature than the second reactor so that different products are obtained in the two reactors. These types of system are also useful for equilibrium reactions. Combinations of different types of reactors are possible, *i.e.*, two CSTRs, or one PFR and one CSTR. If there is no input flow to the process except to the first reactor, solving the equations requires a new definition of conversion.[5]

First, the solution for a problem with one CSTR and one PFR, as shown in Figure 2.33, will be presented. Here, component A is fed into a CSTR of known volume and a final conversion, x_1, is achieved.

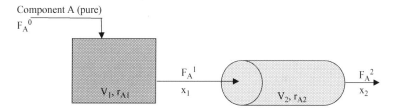

FIGURE 2.33 Schematic representation of a combination of a CSTR and PFR.

The molar flow, F_A^0, and the reaction rate, r_A, are known and the volume of the first reactor can be obtained from Equation 2.40, to give the following:

$$V = \frac{F_A^0 - F_A}{-r_A} \quad (2.100)$$

Then, from Equation 2.100, using, $F_A = F_A^0 (1 - x_1)$, V_1 is given by:

$$V_1 = \frac{F_A^0 x_1}{-r_{A1}} \quad (2.101)$$

Knowing the reaction rate and the desired conversion, the volume for the first reactor, *i.e.*, the CSTR, can be calculated using the Levenspiel plot [3].

For the second reactor, *i.e.*, the PFR, a differential equation needs to be solved and the mole balance or reactor design equation is:

$$\frac{dF_A}{dV} = r_A \quad (2.102)$$

For this case, we know that $F_A^2 = F_A^1 - F_A^1 x_2$ ⁶ and that $F_A^1 = F_A^0 - F_A^0 x_1$; therefore, Equation 2.102 can be solved based on the information from the first reactor.

Example 2.4
For a first-order irreversible reaction, based on the scheme presented in Figure 2.33, calculate the volume of a CSTR and a PFR when $x_1 = 35\%$ and the final molar flow is 2 [mol/min]. The initial flow is 11 [mol/min] and only reactant A is fed to the system. Calculate x_2.

Data: $k = 2$ [min⁻¹] and $v = 2$ [l/min]

Solution
For the first reactor, the volume can be estimated using Equation 2.101, substituting r_A, we get:

$$V_1 = \frac{F_A^0 x_1}{-r_{A1}} = \frac{F_A^0 x_1}{kC_A} \quad (2.103)$$

Re-writing C_A as a function of the conversion:

$$V_1 = \frac{F_A^0 x_1}{kC_A^0 (1-x_1)} \quad (2.104)$$

Using the relation between concentration and molar flow, we have:

$$V_1 = \frac{vF_A^0 x_1}{kF_A^0 (1-x_1)} \quad (2.105)$$

Simplifying Equation 2.105:

$$V_1 = \frac{vx_1}{k(1-x_1)} \quad (2.106)$$

Equation 2.106, where v, x_1, and k are provided, needs to be solved. Substituting the values for volume, and molar flow and concentration at the outlet, we have the following:

For $x_1 = 0.35$, $k = 2$ [min⁻¹], $v = 2$ [l/min] and $F_A^0 = 11$ [mol/min] we obtain:

$$V_1 = 0.53 \text{ [l]}, \, F_A^1 = 7.15 \text{ [mol/min]}, \, C_A^1 = 3.57 \text{ [mol/l]}$$

For the second reactor, the differential equation presented in Equation 2.102, will be solved. Substituting the reaction rate for the irreversible first-order reaction, we obtain:

$$\frac{dF_A^2}{dV_2} = -kC_A^2 \quad (2.107)$$

Reactor Design

Using the relationship, $F_A^2 = C_A^2 v$, we have:

$$\frac{dF_A^2}{dV_2} = -\frac{kF_A^2}{v} \qquad (2.108)$$

This differential equation obtained:

$$\int_{F_A^1}^{F_A^2} \frac{dF_A^2}{F_A^2} = \int_0^{V_2} \frac{-k}{v} dV_2 \qquad (2.109)$$

The solution to Equation 2.109 is as follows:

$$V_2 = \frac{-v}{k}\left(\left(\ln(F_A^2)\right) - \left(\ln(F_A^1)\right)\right) \qquad (2.110)$$

Substituting the data from the previous reactor, we obtain:

$$V_2 = 1.27 \text{ [l]}, \; x_2 = 0.7203$$

As presented in Figure 2.32, the volume of a CSTR is calculated as the area of a rectangle with the base equal to the conversion value and the high equal to the initial flow divided by the reaction rate. If instead of one CSTR, an infinite number of CSTR are put in series, each of them allowing an infinitesimal increment in the conversion, the sum of all these infinite number of reactors will be equal, in performance and result to have one PFR. Therefore, a PFR can be considered as the sum of an infinite number of CSTRs, with an infinitesimal small volume.

2.5 REACTORS IN PARALLEL

Besides having different or the same type of reactors in series, it is possible to have reactors in parallel, as shown in Figure 2.34. If all the reactors have the same volume and reactions conditions, it can be said that the same reaction takes place in each reactor. The purpose of having equipment in parallel is to distribute the flow in more than one unit and the final conversion obtained will be equal in each reactor.

The initial flow, F_A^0, will be divided among the total number of reactors. This generic case is assumed to have n number of reactors and a generic evaluation will be done for reactor i.

For reactor i, the flow at the inlet is, $F_i^0 = F_A^0/n$, the reaction rate is r_i and the volume V_i is V/n where $V = \sum_{i=1}^{n} V_i = n*V_i$, since all the volumes are equal.

Therefore, the expression for the CSTR reactor i is:

$$V_i = \frac{F_i^0 x_i}{-r_i} \qquad (2.111)$$

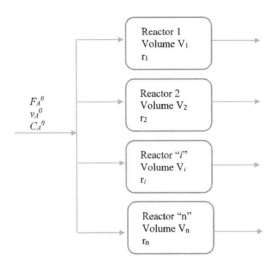

FIGURE 2.34 Scheme of CSTRs in parallel.

As mentioned, the conversion in each equipment is the same since the reaction rate is the same and the volume and operational conditions are the same, this means that $x = x_1 = x_2 = \ldots = x_i = x_n$

Also, the reaction rate is the same; $r_A = r_1 = r_2 = \ldots = r_i = r_n$

Substituting all these variables, we get:

$$\frac{V}{n} = V_i = \frac{F_i^0 x_i}{-r_i} = \frac{F_A^0}{n}\frac{x}{-r_A} \tag{2.112}$$

Simplifying it:

$$V = \frac{F_A^0 x}{-r_A} \tag{2.113}$$

Equation 2.113 is also valid for a larger CSTR, with a total volume equal to the sum of each individual reactor.

2.6 MOLES BALANCE CALCULATIONS

Before we analyze cases that consider volume changes due to variations in the number of moles, temperature, or pressure in a gas phase reaction, it is important to have a clear procedure to calculate the variations of moles of each component involved.[7] Consider the reaction:

$$\alpha A + \beta B \rightarrow \gamma C + \Omega D \tag{2.114}$$

Here, α and β moles of A and B, respectively, react to give γ and Ω moles of C and D, respectively. One of the reactants is chosen as the main component for the

Reactor Design

following calculations, *i.e.*, component A, in this case. Generally, this is done by selecting the limiting component to assure proper consumption of the more expensive raw material. For this reaction, we feed N_A^0, N_B^0, N_C^0, and N_D^0, for each component, respectively.[8]

After a certain reaction time, the amount of components A and B reduces and the amount of the products increases. Conversion based on component A is defined as:

$$x_A = \frac{N_A^0 - N_A}{N_A^0} \qquad (2.115)$$

which can be rearranged to obtain the mole quantity of component A at any given time:

$$N_A = N_A^0 - N_A^0 x_A = N_A^0 (1 - x_A) \qquad (2.116)$$

Now a similar analysis for all the remaining components involved, will be presented, based on Equation 2.114.

For each α moles of A that are consumed, β moles of component B are consumed as well. Given the component B is already present in the reactor at $t = 0$, the total amount of moles of B at any given time, t, is the initial amount of moles of B minus the moles of B that have reacted:

$$N_B = N_B^0 - \frac{\beta}{\alpha} N_A^0 x_A \qquad (2.117)$$

Similarly, the variation of the products, *i.e.*, components C and D are given by:

$$N_C = N_C^0 + \frac{\gamma}{\alpha} N_A^0 x_A \qquad (2.118)$$

$$N_D = N_D^0 + \frac{\Omega}{\alpha} N_A^0 x_A \qquad (2.119)$$

Therefore, we can express the total amount of moles as the sum of the individual contributions:

$$N_{Total} = \sum_{i=1}^{n} N_i \qquad (2.120)$$

For this specific case:

$$N_{Total} = N_A + N_B + N_C + N_D \qquad (2.121)$$

Substituting Equations 2.116, 2.117, 2.118, and 2.119 into 2.121, we obtain:

$$N_T = N_A^0 - N_A^0 x_A + N_B^0 - \frac{\beta}{\alpha} N_A^0 x_A + N_C^0 + \frac{\gamma}{\alpha} N_A^0 x_A + N_D^0 + \frac{\Omega}{\alpha} N_A^0 x_A \quad (2.122)$$

This can be rearranged as:

$$N_T = N_A^0 + N_B^0 + N_C^0 + N_D^0 + \frac{\gamma}{\alpha} N_A^0 x_A + \frac{\Omega}{\alpha} N_A^0 x_A - \frac{\beta}{\alpha} N_A^0 x_A - N_A^0 x_A \quad (2.123)$$

Taking into consideration that the initial amount can be calculated as:

$$N_T^0 = \sum_{i=1}^{n} N_i^0 \quad (2.124)$$

Then, Equation 2.123 can be rearranged as:

$$N_T = N_T^0 + \left(\frac{\gamma}{\alpha} + \frac{\Omega}{\alpha} - \frac{\beta}{\alpha} - 1 \right) N_A^0 x_A \quad (2.125)$$

This expression represents the total change in moles at any arbitrary point/time during the reaction.

While there are different methods to handle the calculations, we strongly recommend the strategy presented by Fogler [2], where he uses a table to organize all the information, as shown in Table 2.1 for this case.

Table 2.2 can be extrapolated to a concentration-based or a flow-based situation. In both cases, it is important that the volume remain constant. For the concentration-based condition, the moles need to be divided by the volume. In the following section, the impact of volume changes on this problem will also be explored.

TABLE 2.1
Fogler's Methodology for Mole Balance [2]

Component	Initial Amount	Consumed/Produced	Final
A	N_A^0	$-N_A^0 x_A$	$N_A^0 - N_A^0 x_A$
B	N_B^0	$-\frac{\beta}{\alpha} N_A^0 x_A$	$N_B^0 - \frac{\beta}{\alpha} N_A^0 x_A$
C	N_C^0	$+\frac{\gamma}{\alpha} N_A^0 x_A$	$N_C^0 + \frac{\gamma}{\alpha} N_A^0 x_A$
D	N_D^0	$+\frac{\Omega}{\alpha} N_A^0 x_A$	$N_D^0 + \frac{\Omega}{\alpha} N_A^0 x_A$
Inert[0]	N_I^0	0	$N_I = N_I^0$
TOTAL	N_T^0	$\left(\frac{\gamma}{\alpha} + \frac{\Omega}{\alpha} - \frac{\beta}{\alpha} - 1 \right) N_A^0 x_A$	$N_T^0 + \sigma N_A^0 x_A$

where $\sigma = \left(\frac{\gamma}{\alpha} + \frac{\Omega}{\alpha} - \frac{\beta}{\alpha} - 1 \right)$

Reactor Design

TABLE 2.2
Fogler's Methodology for a Flow Scenario [2]

Component	Initial Amount	Consumed/Produced	Final
A	F_A^0	$-F_A^0 x_A$	$F_A^0 - F_A^0 x_A$
B	F_B^0	$-\dfrac{\beta}{\alpha} F_A^0 x_A$	$F_B^0 - \dfrac{\beta}{\alpha} F_A^0 x_A$
C	F_C^0	$+\dfrac{\gamma}{\alpha} F_A^0 x_A$	$F_C^0 + \dfrac{\gamma}{\alpha} F_A^0 x_A$
D	F_D^0	$+\dfrac{\Omega}{\alpha} F_A^0 x_A$	$F_D^0 + \dfrac{\Omega}{\alpha} F_A^0 x_A$
Inert	F_I^0	0	$F_I = F_I^0$
TOTAL	F_T^0	$\left(\dfrac{\gamma}{\alpha} + \dfrac{\Omega}{\alpha} - \dfrac{\beta}{\alpha} - 1\right) F_A^0 x_A$	$F_T^0 + \sigma F_A^0 x_A$

where $\sigma = \left(\dfrac{\gamma}{\alpha} + \dfrac{\Omega}{\alpha} - \dfrac{\beta}{\alpha} - 1\right)$

The flow-based situation is presented as follows, as summarized in Table 2.2 for a PFR.

Example 2.5
For the following reaction, 1 [mol/l] of A and 3 [mol/l] of B are fed into the system in the presence of 5 [mol/l] of an inert component to produce C. For an 82% conversion of A, calculate the final concentration of A, B, and C, and the inert component, as well as the final total concentration. The reaction takes place in liquid form and therefore, the volume can be considered constant:

Reaction: 1A + 2B → 2C

We can therefore produce Table 2.3 based on the information from the problem description:

TABLE 2.3
Solution to the Problem based on Table 2.2. Problem Solved in Concentration [mol/l]

Comp.	Initial Amount	Consumed/Produced	Final
A	1	$-1(0.82) = -0.82$	$1 - 1(0.82) = 0.18$
B	3	$-\dfrac{2}{1} 1(0.82) = -1.64$	$3 - \dfrac{2}{1} 1(0.82) = 1.36$
C	0	$\dfrac{2}{1} 1(0.82) = 1.64$	$0 + \dfrac{2}{1} 1(0.82) = 1.64$
Inert	5	0	$5 = 5$
TOTAL	9	-0.82	$9 - 0.82 = 8.18$

2.7 REACTION IN GAS PHASE

Reactions in liquid and gas phases were shown in the previous sections. However, no change in pressure, temperature, or number of moles was considered, while it is very likely that in a gas phase system, one or many of these parameters will simultaneously vary. Here, a typical reaction in gas phase, *i.e.*, the production of ammonia from nitrogen and hydrogen, in an ideal reactor, will be considered. As the following reaction shows, the total number of moles varies from 4 moles (reactants) at the beginning of the reaction, to 2 moles (product) at the end.

$$N_2 + 3H_2 \rightarrow 2NH_3$$

The variation in the number of moles is expected to influence the volume of the system. The concentration of a generic component A (for the previous example is N_2) is given by the number of moles of A divided by the volume occupied:

$$C_A = \frac{N_A}{V} \tag{2.126}$$

The volume can also be expressed according the ideal gas law:

$$V = \frac{N_T RTZ}{P} \tag{2.127}$$

Based on which, the initial volume is:

$$V^0 = \frac{N_T^0 RT^0 Z^0}{P^0} \tag{2.128}$$

On dividing Equation 2.127 with 2.128, and rearranging:

$$\frac{V}{V^0} = \frac{N_T}{N_T^0} \frac{P^0}{P} \frac{Z}{Z^0} \frac{T}{T^0} \frac{R}{R} \tag{2.129}$$

Assuming that the compressibility factor does not change and since R is a constant, Equation 2.129 can be rewritten as:

$$V = V^0 \frac{N_T}{N_T^0} \frac{P^0}{P} \frac{T}{T^0} \tag{2.130}$$

Reactor Design

Substituting Equations 2.130 and 2.116 into 2.126, we have:

$$C_A = \frac{N_A}{V} = \frac{N_A^0(1-x_A)}{V^0} \frac{N_T^0}{N_T} \frac{P}{P^0} \frac{T^0}{T} \tag{2.131}$$

Based on Equation 2.125 and $C_A^0 = \frac{N_A^0}{V^0}$, we can obtain:

$$C_A = \frac{N_A}{V} = C_A^0(1-x_A) \frac{N_T^0}{N_T} \frac{P}{P^0} \frac{T^0}{T} \tag{2.132}$$

which can be rearranged to give:

$$C_A = \frac{N_A}{V} = \frac{C_A^0(1-x_A)}{(1+\varepsilon x_A)} \frac{P}{P^0} \frac{T^0}{T} \tag{2.133}$$

where $\varepsilon = \frac{N_A^0}{N_T^0} \sigma$.

Equation 2.133 can be simplified if the process is assumed to be isothermal and isobaric:

$$C_A = \frac{C_A^0(1-x_A)}{(1+\varepsilon x_A)} \tag{2.134}$$

This expression can be generalized for a generic component, i, as follows:

$$C_i = \frac{C_i^0 - \alpha_i/\alpha_A \, C_A^0 x_A}{(1+\varepsilon x_A)} \frac{P}{P^0} \frac{T^0}{T} \tag{2.135}$$

where α_i and α_A are the stoichiometric numbers for component i and component A, respectively.

By analogy, we can obtain the expression for the volumetric flow as follows:

$$v = v^0 \frac{F_T}{F_T^0} \frac{P^0}{P} \frac{T}{T^0} \tag{2.136}$$

Example 2.6
Consider a gas phase, elementary, and irreversible reaction being carried out in a plug flow system, for the formation of 2 moles of product C:

$$3A + B \rightarrow 2C$$

The initial concentrations for A and B are 3 and 2 [mol/l], respectively, while component C is not present at the beginning of the reaction. The reaction temperature is set to 350 [K] and the initial molar flow of component A is set to 1 [mol/min]. Since the reaction is an elementary reaction, the reaction rate can be derived as follows:

$$r_A = -k * C_A^3 C_B \qquad (2.137)$$

While k follows an Arrhenius equation where:

$k = k_{00} \exp\left(\frac{-E_a}{RT}\right)$ with $k_{00} = 0.05$ [l³/(mol³*min)] and $E_A = 1{,}500$ [J/(mol*K)].

This reaction is carried out under isothermal and isobaric conditions; therefore, there is no change in the pressure or temperature. However, since the total number of moles changes in the course of the reaction and the reaction is in gas phase, this has to be considered and is elaborated as follows.

Solution
Equation 2.135, can be rewritten it for A, B, and C, as follows:

$$C_A = \frac{C_A^0 - C_A^0 x_A}{(1 + \varepsilon x_A)} \qquad (2.138)$$

$$C_B = \frac{C_B^0 - \frac{1}{3} C_A^0 x_A}{(1 + \varepsilon x_A)} \qquad (2.139)$$

$$C_C = \frac{C_C^0 + \frac{2}{3} C_A^0 x_A}{(1 + \varepsilon x_A)} \qquad (2.140)$$

where $\varepsilon = \sigma * \frac{N_A^0}{N_T^0}$ and for this case, $\sigma = 2/3 - 1/3 - 1 = -2/3$ and the ratio of moles at $t = 0$ is $3/5 = 0.6$.

With all this information, we can write a code in Polymath to plot the variations of the concentration of each component A, B, and C, as well as the conversion as a function of the volume.

Script:
```
#Mole balance
    d(x)/d(V) = -ra/Fa0
    x(0) = 0
#Reaction kinetics
    ra = -k1*Ca^3*Cb
```

Reactor Design

```
#Auxiliary equations
    k1 = k00*EXP(-Ea/(R*T))
    Ca = Ca0*(1-x)/(1+Epsilon*x)
    Cb = (Cb0-((1/3)*Ca0*x))/(1+Epsilon*x)
    Cc = (Cc0+((2/3)*Ca0*x))/(1+Epsilon*x)
#Constants
    k00 = 0.05
    Ea = 1,500
    R = 8.3143
    T = 350
    Fa0 = 1
    Ca0 = 3
    Cb0 = 2
    Cc0=0
    Epsilon = Sigma*Ya0
    Ya0=Ca0/(Ca0+Cb0)
    Sigma=(2/3-1/3-1)
#Independent variable
    V(0) = 0
    V(f) = 3
```

Based on the script, the variation of conversion as a function of volume can be plotted, as shown in Figure 2.35 which shows that a final conversion of around 78% was obtained.

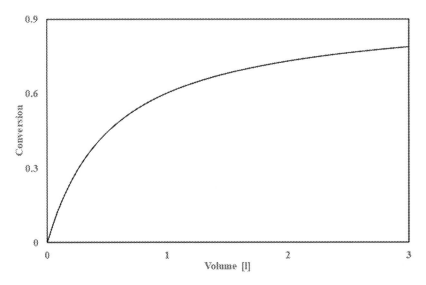

FIGURE 2.35 Variation of conversion as a function of the volume for a gas phase reaction.

Figure 2.36 shows the variation of the concentration of components A, B, and C, for the reaction under study. Due to the stoichiometric balance, component A is rapidly consumed in comparison to component B, while component C is produced rapidly. Figure 2.37 provides a comparison between the variation of component C in a gas and liquid phase reaction and shows that by not taking into consideration the variation of moles, the concentration of the product will have been underestimated by around 40%.

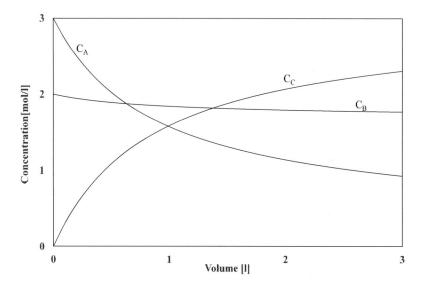

FIGURE 2.36 Variation of Concentration C as a function of the volume for a gas phase reaction.

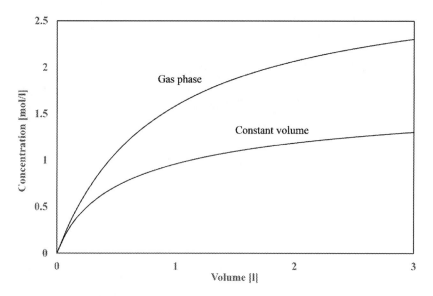

FIGURE 2.37 Comparison of the variation of the concentration of C.

Reactor Design

Example 2.7 (Membrane Reactor Continuation)

As the impact of the gas phase reaction has been presented, we can obtain a more accurate solution for a membrane reactor. To recall, a schematic of the membrane reactor is shown in Figure 2.38.

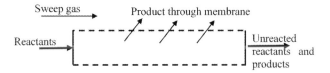

FIGURE 2.38 Membrane reactor representation.

Solution

The reaction taking place in a gas phase is given as follows:

$$2A + B \rightarrow C + D$$

Now, we will consider the balance of a PFR with variation of the concentration due to the change in the number of moles, and possibly temperature and pressure. Therefore, for components A, B, and D, the balance is based on Equation 2.58, giving:

$$\frac{dF_A}{dV} = r_A \qquad (2.141)$$

$$\frac{dF_B}{dV} = r_B \qquad (2.142)$$

$$\frac{dF_D}{dV} = r_D \qquad (2.143)$$

For component C,

$$\frac{dF_C}{dV} = r_C - F_{memb} \qquad (2.144)$$

In order to solve this problem, we need to find an expression for F_{memb} as a function of component C. Usually, component C is move through the membrane and immediately removed; this is because the ΔC_C is the driving force to move component C through the membrane and therefore if there is none in the sweep gas, the driving force will be maximum. From mass transfer courses, you have seen that the flow is proportional to the change in the concentration. Therefore, if the concentration in the sweep gas is zero, then:

$$F_{memb} = k_{mem} C_C \qquad (2.145)$$

where k_{mem} is a constant that is related to the properties and characteristics of the membrane, such as permeability, size, and porosity.

Therefore, we can solve this problem by knowing some more information about the reaction. The reaction under consideration is an elementary reaction, taking place under isothermal and isobaric conditions in gas phase. The value of k_{mem} has been established as well as other experimental parameters. The reaction rate is expressed as:

$$r_A = -k_1 C_A^2 C_B \qquad (2.146)$$

The reaction rates for the rest of the variables was obtained with the following relationship:

$$\frac{r_A}{-2} = \frac{r_B}{-1} = \frac{r_C}{1} = \frac{r_D}{1} \qquad (2.147)$$

The variations of the concentration for each of the components can be obtained by the relation, $C_i = F_i/v$, where v is defined according to Equation 2.136.

Now, taking into account all that information and assuming that $k_{oo} = 0.5$ [l²/(mol²*min)], $E_A = 1,500$ [J/(mol*K)], $T = 350$ [K], $k_{mem} = v_0 = 3$ [l/min], $F_A^0 = 10$ [mol/min], $F_B^0 = 10$ [mol/min], and $F_C^0 = F_D^0 = 0$, the variation of the flow of each component as a function of the volume can be obtained. The script used for this example is provided as follows:

Script:

```
#Mole balance
    d(Fa)/d(V) = ra
    Fa(0) = 10
    d(Fb)/d(V) = rb
    Fb(0) = 10
    d(Fc)/d(V) = rc-Fmem
    Fc(0) = 0
    d(Fd)/d(V) = rd
    Fd(0) = 0
#Reactions kinetics
    ra = -k1*Ca^2*Cb
#Auxiliary equations
    k1 = k00*EXP(-Ea/(R*T))
    rb = 1/2*ra
    rc = -1/2*ra
    rd = -1/2*ra
    Fmem = kmem*Cc
    Ca = Fa/v
    Cb = Fb/v
    Cc = Fc/v
```

Reactor Design

```
    Cd = Fd/v
    v = v0*(Ft/Ft0)
    Ft = Fa+Fb+Fc+Fd
    Ft0 = Fa0+Fb0+Fc0+Fd0
#Constants
    k00 = 0.5
    Ea = 1,500
    R = 8.3143
    T = 350
    kmem = 5
    v0 = 3
    Fa0 = 10
    Fb0 = 10
    Fc0 = 0
    Fd0 = 0
#Independent variable
    V(0) = 0
    V(f) = 3
```

Using this script in Polymath gives the plot shown in Figure 2.39, where the variations of the molar flow of all components are presented. It is clear that components A and B are being consumed while D is being produced. It is interesting to see that while component C is being produced, its molar flow is seen to decrease, and this is not due to the consumption of component C in a reaction but due to mass/flow being removed through the membrane.

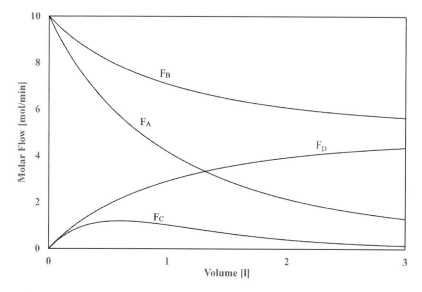

FIGURE 2.39 Variations of the molar flow in a membrane reactor.

2.8 PRESSURE DROP

The above section shows that the equation of variation of concentration of component A for a gas phase reaction can be modified due to conversion, the variation in the total number of moles, the change in temperature and the change in pressure. So far, we have dealt with isothermal and isobaric systems, and now we will add the variation of pressure in a gas phase reaction to obtain a more complete situation. We strongly recommend reading Chapter 4 where an entire problem, with increasing complexity, has been solved from start to end, considering all possible steps.

For the reaction:

$$A \rightarrow B + C$$

This elementary irreversible reaction takes place in gas phase, isothermally, in a PBR; therefore, the reaction rate is:

$$r_A^w = -k_1 C_A \tag{2.148}$$

The mole balance of the reactor can be written as:

$$F_A^0 \frac{dx_A}{dW} = -r_A^w \tag{2.149}$$

We need to write the concentration of A as a function of the conversion, the variation of the number of moles and the changes in pressure; therefore:

$$C_A = \frac{N_A}{V} = \frac{C_A^0 (1-x_A)}{(1+\varepsilon x_A)} \frac{P}{P^0} \tag{2.150}$$

From the reaction equation, we can see that $\varepsilon = 1$.

On substituting Equation 2.150 into 2.148 and consequently into Equation 2.149, an expression for the variation of conversion as a function of the mass of catalyst is obtained. However, this differential equation will be a function of conversion and pressure. Therefore, one more differential equation is needed, which considers the variation of pressure as a function of the mass of catalyst.

The Ergun equation [10] is the one that is most commonly used to estimate the pressure drop in a PBR, as shown as follows and the relevant parameters definitions and units are provided in Table 2.4.

$$\frac{dP}{dz} = \frac{U}{\rho d_p} \left(\frac{1-\theta}{\theta^3} \right) \left(\frac{150(1-\theta)\mu}{d_p} - 1.75U \right) \tag{2.151}$$

However, there is one parameter in this expression that requires a little more attention, i.e., density (ρ). Here, the pressure and temperature are changing in the reactor; therefore, density also changes accordingly. Density will be rewritten as a function of

TABLE 2.4
Definition and Units of the Terms in Equation 2.151

Abbreviation	Definition	Units
U	Superficial mass velocity	$\dfrac{kg}{m^2 s}$
ρ	Fluid density	$\dfrac{kg}{m^3}$
d_p	Catalyst diameter	m
θ	Porosity = empty volume	
μ	Fluid viscosity	$\dfrac{kg}{ms}$

pressure and temperature, and eventually the initial density at the entrance of the reactor. As the system is in a steady state, the mass flow at the inlet has to be the same as the mass flow at the outlet, and considering the definition of mass as density * volume, we can express density as follows:

$$\rho = \rho_0 \frac{P}{P_0} \frac{T_0}{T} \frac{F_{T0}}{F_T} \tag{2.152}$$

We can now substitute Equation 2.152 into 2.151, to obtain:

$$\frac{dP}{dz} = -\frac{U}{\rho_0 d_p}\left(\frac{1-\theta}{\theta^3}\right)\left(\frac{150(1-\theta)\mu}{d_p} + 1.75U\right)\frac{T}{T_0}\frac{P_0}{P}\frac{F_T}{F_{T0}} \tag{2.153}$$

Equation 2.153 is now a function of pressure, temperature, and indirectly, of conversion (*i.e.*, reactants and products flows). Different versions of this equation can be found in literature, with certain parameters assumed to be constant and based on a problem in particular. Also, further simplifications can be done if the flow is laminar or turbulent.

Now, a mathematical equation in terms of the weight (W) of catalyst is needed to solve this expression together with the mole balance, and eventually the energy balance. The weight of the catalyst can be expressed in terms of the volume occupied:

$$W = V * \rho_c \tag{2.154}$$

Considering that W is the weight of the catalyst and ρ_c is the density, we need to express the volume of the catalyst in terms of reactor length (z). The volume of the catalyst is the total volume of the pipe (cross-sectional area (A) times the length (z)) multiplied by the fraction of that volume being occupied by the solid. The volume that is being used by the solid is related to the porosity (θ), which was a way to quantify the empty volume. Therefore, the volume occupied will be the total volume minis the empty part. This can be written as:

$$V_{solid} = V_{total} - V_{empty} \tag{2.155}$$

Substituting the definitions mentioned above:

$$V_{solid} = A*z - A*z*\theta \qquad (2.156)$$

Rearranging:

$$V_{solid} = A*z(1-\theta) \qquad (2.157)$$

Substituting this Equation 2.157 into 2.154 and then into Equation 2.153, we obtain:

$$\frac{dP}{dW} = -\frac{1}{A\rho_c(1-\theta)}\frac{U}{\rho_0 d_p}\left(\frac{1-\theta}{\theta^3}\right)\left(\frac{150(1-\theta)\mu}{d_p}+1.75U\right)\frac{T}{T_0}\frac{P_0}{P}\frac{F_T}{F_{T0}} \qquad (2.158)$$

Equation 2.158 represents the variation of pressure in its general form when a reaction takes place in a PBR. This expression is a function of pressure (P), temperature (T), and conversion (F_T/F_{t0}). In addition, this expression can be used for all regimes of flow since no simplifications have been added. Variations of this expression will be found in literature and books on reaction engineering.

Example 2.8
The following gas phase, irreversible, and elementary reaction takes place in a PBR, which is being operated isothermally at $T = 555$ [K].

$$A + 3B \rightarrow C$$

Here, the molar flow of components A and B, at the input of the reactor, are of 5 and 15 [mol/min] with a volumetric flow of 15 [l/min]. Component C is not present at the input of the reactor. The total reactor diameter is 0.2 [meters]. Plot the variations of flow for each component as well as the pressure profile as a function of the catalyst mass.

Additional information:

$$P_0 = 2,500\,[\text{kPa}]; k = 40\left[\frac{l^3}{\text{mol}^3}\right]\exp\left(\frac{-3,000\left[J/\text{molK}\right]}{8.3143\left[\frac{J}{\text{mol}}\right]\cdot655[K]}\right); \theta = 0.6;$$

$$dp = 0.03\,[\text{m}]; \mu = 0.01\left[\frac{\text{kg}}{\text{ms}}\right]; \rho_0 = 1.2\left[\frac{\text{kg}}{\text{m}^3}\right]; U = 5.5\left[\frac{\text{kg}}{\text{m}^2*\text{s}}\right];$$

$$\rho_c = 2,500\left[\frac{\text{kg}}{\text{m}^3}\right]; W = 10\,[\text{kg}].$$

Reactor Design

Solution

For an irreversible elementary reaction, we can write the reaction rate for component A as:

$$r_A^w = -k_1 C_A C_B^3 \qquad (2.159)$$

While the expression for B and C can be obtained using the relationship:

$$\frac{r_A^w}{-1} = \frac{r_B^w}{-3} = \frac{r_C^w}{1} \qquad (2.160)$$

The mole balance for each component is:

$$\frac{dF_A}{dW} = r_A^w; \quad \frac{dF_B}{dW} = r_B^w; \quad \frac{dF_C}{dW} = r_C^w$$

The pressure balance follows Equation 2.158:

$$\frac{dP}{dW} = -\frac{1}{A\rho_c (1-\theta)} \frac{U}{\rho_0 d_p} \left(\frac{1-\theta}{\theta^3}\right) \left(\frac{150(1-\theta)\mu}{d_p} + 1.75U\right) \frac{P_0}{P} \frac{F_T}{F_{T0}} \qquad (2.161)$$

Note that:

$$C_i = \frac{F_i}{v}; \quad v = v_0 \left(\frac{F_T}{F_{T0}}\right)\left(\frac{P^0}{P}\right); \quad F_T = F_A + F_B + F_C; \quad x_A = \frac{F_A^0 - F_A}{F_A^0}$$

$$F_{T0} = F_A^0 + F_B^0 + F_C^0$$

Putting all these expressions together in Polymath, we have the following script.

Script:

```
#Mole balance
    d(Fa)/d(w) = ra
    Fa(0) = 5
    d(Fb)/d(w) = rb
    Fb(0) = 15
    d(Fc)/d(w) = rc
    Fc(0) = 0
#Pressure drop
    d(P)/d(w) = −1*Q*O*J*Y*Z
    P(0) = 2,500
    Z = (((150*mu*(1−tita))/Dp)+(1.75*U))
    Y = (P0/P)*(FT/FT0)
```

```
J = ((1−tita)/(tita^3))
O = (U/(Dp*Rho0))
Q = (1/(Ac*Rhoc*(1−tita)))
#Reaction rate
    ra = −k1*Ca*Cb^3
    rb = 3*ra
    rc = −ra
#Constants and initial values
    k00 = 40
    Ea = 3,000
    R = 8.3143
    T = 555
    Fa0 = 5
    Fb0 = 15
    Fc0 = 0
    P0 = 2,500
    v0 = 15
    Ac = 3.14159/4*D^2
    D = 0.2
    tita = 0.6
    Dp = 0.03
    mu = 0.01
    Rho0 = 1.2
    U = 5.5
    Rhoc = 2,500
#Auxiliary equations
    Ca = Fa/v
    Cb = Fb/v
    Cc = Fc/v
    k1 = k00*EXP(−Ea/(R*T))
    v = v0*(FT/FT0)*(P0/P)
    FT = Fa+Fb+Fc
    FT0 = Fa0+Fb0+Fc0
    x = (Fa0−Fa)/Fa0
#Independent variable
    w(0) = 0
    w(f) = 10
```

When solving, the evolution of the different flows as a function of the amount of catalyst used, the variation of pressure, and conversion of A, can be plotted.

Reactor Design

Figure 2.40 presents the variation of conversion and pressure as a function of the amount of catalyst. It can be seen that the pressure decreases considerably from 20 bars to 16 bars while the conversion increases to 84%.

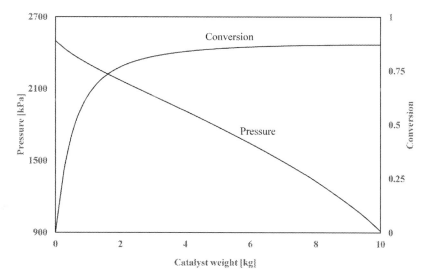

FIGURE 2.40 Variations of pressure and conversion.

Figure 2.41 shows the variation of the flow of all the reactants and products as a function of the catalyst amount. The flow of component A reduces considerably till the reaction is completed. We recommended the reader to plot the variation of the reaction rate of as a function of the catalyst amount and discuss its tendency.

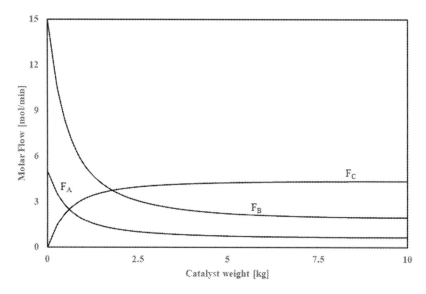

FIGURE 2.41 Variations of flows in a packed bed reactor.

Figure 2.42 represents the variation of conversion for this problem, at constant pressure and the case with a pressure drop. It is evident that when no pressure drop is considered, the final conversion is slightly higher; implying that assuming constant pressure for a PBR would overestimate the results. The extent of error depends on the conditions which need to be taken into account to decide whether the pressure contributions are substantial or can be neglected, such as, the viscosity of the fluid, the amount of catalyst or the length of the reactor, or the type and size of the catalyst. To be rigorous, pressure variations should always be considered but it may not always be practical, especially if the final results vary very little (~1%) by considering the variation in pressure.

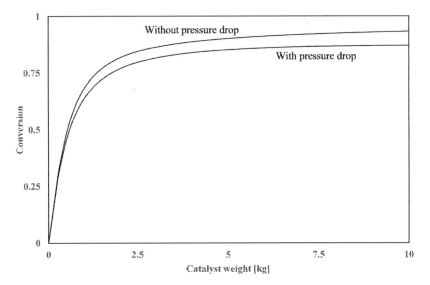

FIGURE 2.42 Variations of flows in a packed bed reactor.

2.9 MULTIPLE REACTIONS

So far, the variation of conversion, flow, and pressure for one single elementary reaction were studied. Now, we will consider a system with multiple reactions which can take place in parallel, in series, or a combination of both types, as shown:

$A \rightarrow B \rightarrow C$
Series

$A \begin{smallmatrix} \nearrow B \\ \searrow C \end{smallmatrix}$
Parallel

$A + B \rightarrow C$
$A + C \rightarrow D$
Combined

Reactor Design

When dealing with multiple reactions, generally one of those reactions is of special interest, *i.e.*, either an intermediate component in a series reaction, or one of the outputs of reactions in parallel. For instance, we can say that we are interested in component B in the parallel and series reaction, while the main interest is the product or intermediate chemical C, for the combined scenario. A typical case for a series reaction is a partial oxidation, for instance, when we want to produce CO from carbon and oxygen, instead of CO_2. We know that if we add too much O_2 or let the reaction to continue to the end, we will produce CO_2 which is very stable and will not revert to reactants or CO. Thus, selectivity is needed, and it can be defined in different ways; here are three possible definitions that are equally valid:

$$S = \frac{\text{molar flow of desired product}}{\text{molar flow of undesired products}} \quad (2.162)$$

$$S = \frac{\text{moles of desired product}}{\text{moles of undersired products}} \quad (2.163)$$

$$S = \frac{\text{rate of formation for the desired product}}{\text{rate of formation for the undersired products}} \quad (2.164)$$

It is important to point out that Equation 2.162 is mainly applied for flow systems while Equation 2.163 is generally used for batch systems. Nevertheless, Equation 2.164 is generally used since it is a more global expression that allows the user to understand the effects of different variables over the selectivity of the system better.

2.9.1 Series Reactions

We will take a look at the following reaction:

$$A \xrightarrow{k_D} D \xrightarrow{k_U} U \quad (2.165)$$

where A is the reactant, D is the desired product, and U is the undesired product. The main goal is to obtain the maximum amount of component D. To solve this problem and to have the highest concentration of D, and therefore, the highest selectivity, we need to know the reactions involved. We will be assuming that all reactions involved are first-order irreversible reactions. Therefore, we can obtain the reaction rate for each component, as follows. Component A is only being consumed via the first reaction, therefore $r_A = -k_D C_A$, component D is being produce via the first reaction but consumed via the second reaction, giving $r_D = k_D C_A - k_U C_D$ and finally, component U is being produced following $r_U = k_U C_A$. We can then solve the mole balance expressions analytically for this system and see the profiles of concentration as a function of time to see if component D has a maximum value. The analytical solution has been done by Levenspiel [3] and we recommend its reading. We will use Polymath for the solution and provide a discussion of the data. Besides the reaction rates that we have derived, above we need to solve the variation of concentration with respect to the volume for each chemical. Therefore, we need to solve the mole balance for each component simultaneously. The equations to be solved are:

$$\frac{dC_A}{dt} = r_A = -k_D C_A \tag{2.166}$$

$$\frac{dC_D}{dt} = r_D = k_D C_A - k_U C_D \tag{2.167}$$

$$\frac{dC_U}{dt} = r_U = -k_U C_D \tag{2.168}$$

We also need the values for k_D and k_U which have constant values, i.e., 2 and 1 [1/min], respectively, for this case. No effects of temperature are being considered at this stage. Initial concentration of A = 5 [mol/l] while there is no presence of D and U at the beginning of the reaction.

Script:
#Mole balance
 d(Ca)/d(t) = ra
 Ca(0) = 5
 d(Cd)/d(t) = rd
 Cd(0) = 0
 d(Cu)/d(t) = ru
 Cu(0) = 0
#Reaction rate
 ra = −kd*Ca
 rd = kd*Ca−ku*Cd
 ru = ku*Cd
#Constants
 kd = 2
 ku = 1
#Independent variable
 t(0) = 0
 t(f) = 6

By solving this system, we can obtain the variation of the concentration of each chemical as a function of time, as shown in Figure 2.43. It is evident that component D has a maximum value of production, and it also represents the maximum selectivity. The reader should try to obtain the optimal value of selectivity, and therefore, the optimal value of C_D, mathematically. This optimal value for component D, which takes place at around 0.7 min of reaction suggests that the reaction would have run for too long if we let it run for 6 min and all the desired products will convert into undesired chemicals.

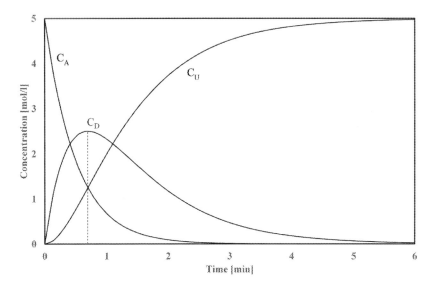

FIGURE 2.43 Concentration profiles of a reaction in serie.

It is possible to mathematically obtain the optimal time to reach the maximum concentration of the desired component. This is a very useful tool that can be used in the design to avoid over-dimensioning (reactor volume larger than required) or under-dimensioning (reactor volume smaller than optimal) of the reaction systems. To calculate this optimal time, consider $dC_D/dt = 0$, and solve[10] to obtain the following:

$$t_{optimal} = \left(\frac{1}{k_U - k_D}\right) \ln\left(\frac{k_U}{k_D}\right) \qquad (2.169)$$

This approach can be used for more than one intermediate component and for reversible reactions as well. We have worked on the production of biodiesel using different catalysts. This process is a set of second-order reactions in both forward and the reverse reactions, with two intermediate components. The system under study is presented in Figure 2.44 where triglycerides (TG), diglycerides (DG), and mono glycerides (MG) react with ethanol (E) to give DG, MG, and glycerol (G), respectively, along with fatty acid ethyl esters (FAEE).

$$TG + E \xleftrightarrow{Catalyst} FAEE + DG$$

$$DG + E \xleftrightarrow{Catalyst} FAEE + MG$$

$$MG + E \xleftrightarrow{Catalyst} FAEE + G$$

FIGURE 2.44 Multistep transesterification reaction.

The solution to this problem has been published [11,12] and here, we will show the variation of the concentration of TG, DG, MG, and FAEE as a function of time. Figure 2.45 is a representation of this functionality and is not the solution to any particular problem. As expected, the concentration of DG increases rapidly as the concentration of TG decreases. The concentration of MG is initially low as sufficient concentration of DG is needed for its production. The maximum concentration of DG is obtained before the maximum for MG, which is expected, and in the meantime, TG decreases constantly. All reactants and intermediates are fully consumed and therefore, the concentration of FAEE increases constantly until reaching the maximum value after sufficient time has passed.

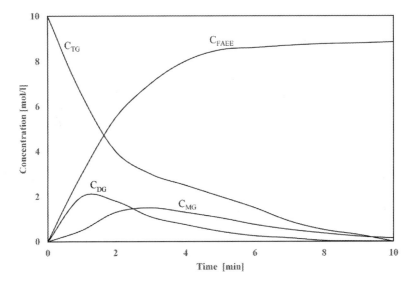

FIGURE 2.45 Concentration profile of the multistep transesterification reaction.

2.9.2 Parallel Reactions

We will analyze the situation with parallel reactions, *i.e.*, when one reactant, A, is used to produce a desired (D) and an undesired (U) product, shown as follows:

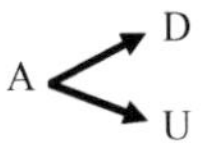

$$A \begin{smallmatrix} \nearrow D \\ \searrow U \end{smallmatrix}$$

Before showing some examples on this case, different possible scenarios will be examined.

2.9.2.1 Scenario 1

Both reactions are elementary first-order irreversible reactions, the only difference is that they have different activation energies. This means that the selectivity can be written as:

$$S = \frac{r_D}{r_U} = \frac{k_D C_A}{k_U C_A} \tag{2.170}$$

where $k_i = k_{00i} \exp(-E_{Ai}/(R*T))$, therefore we can rewrite this expression as:

$$S = \frac{k_{D00}}{k_{U00}} \frac{\exp\left(\frac{-E_{aD}}{RT}\right)}{\exp\left(\frac{-E_{aU}}{RT}\right)} \tag{2.171}$$

If both k_{00i} are similar, then the selectivity depends only on the temperature and can be rearranged based on the values of activation energy. Therefore, Equation 2.171 can be expressed as:

$$S = \exp\left(\frac{-E_{aD} + E_{aU}}{RT}\right) \tag{2.172}$$

This expression demonstrates the effect that the temperature has over the selectivity. However, it is important to find the relationships between the activation energies to be sure of the effect of increasing temperature. This will lead to case (i) $E_{aD} > E_{aU}$ and case (ii) $E_{aD} < E_{aU}$, as shown in Figure 2.46.

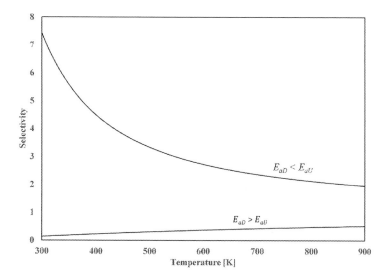

FIGURE 2.46 Variation of the selectivity as a function of temperature.

If $E_{aD} > E_{aU}$, then an increase in the temperature tends to increase the selectivity toward a desired product. Therefore, this process is an endothermic process. While, if $E_{aD} < E_{aU}$, then the increment in temperature produces an increase in the concentration of the undesired product, which has a higher activation energy, leading to a decrease in the selectivity.

2.9.2.2 Scenario 2

Similar to scenario 1, now the reaction to produce the desired product, D, is a first-order irreversible reaction while the reaction to produce the undesired product, U, is a zero-order reaction. Assuming that both kinetic constants are equal, we can write the selectivity equation as follows:

$$S = \frac{r_D}{r_U} = \frac{k_D C_A}{k_U} = \alpha * C_A \qquad (2.173)$$

Equation 2.173 is a simplification since it is not possible that $k_D = k_U$. However, this will lead to C_A being multiply by a number, α. We can analyze Equation 2.173 and based on the expression; we can see that temperature does not play an important a role as it did in the previous scenario. However, variations in the concentration of component A do play a role. An increase in C_A will lead to a linear increase in the selectivity. This indicates that we should run this process with as high concentration of reactant A, as possible since that will improve the production of our desired product.

Other scenarios can be proposed to be analyzed; we recommend the reader to try them with similar values of activation energy and similar reaction kinetics, with the undesired product having higher reaction order than the desired product and to plot the selectivity as a function of the different independent variables.

As we used one reactant, the number of possible scenarios would expand considerably if we have two reactants, since there is a larger possible combination of concentrations. We will now take a look at two reactants, which give two products (desired and undesired):

$$A+B \begin{array}{c} \nearrow D \\ \searrow U \end{array}$$

In generic terms, we can propose a reaction rate equation for the production of D and one for the production of U, these equations are:

$$r_D = k_{D_0} \exp\left(\frac{-E_{aD}}{RT}\right) C_A^\alpha C_B^\beta \qquad (2.174)$$

$$r_U = k_{U_0} \exp\left(\frac{-E_{aU}}{RT}\right) C_A^\gamma C_B^\varphi \qquad (2.175)$$

Reactor Design

With these reaction rate equations, we can write the selectivity equation in terms of all the concentrations and activations energies:

$$S = \frac{r_D}{r_U} = \frac{k_{D_0} \exp\left(\frac{-E_{aD}}{RT}\right) C_A^\alpha C_B^\beta}{k_{U_0} \exp\left(\frac{-E_{aU}}{RT}\right) C_A^\gamma C_B^\varphi} \tag{2.176}$$

We can then analyze Equation 2.176. If we first assumed that all temperature effects are of similar magnitude and the activation energies are close to each other, we can reduce Equation 2.176 to:

$$S = \frac{C_A^\alpha C_B^\beta}{C_A^\gamma C_B^\varphi} \tag{2.177}$$

where we can predict how the different concentrations will affect the selectivity. We will present this discussion in Table 2.5.

TABLE 2.5
Analysis of the Selectivity for Different Orders of Reaction

Reaction Orders	Selectivity Equation[a]	Conclusions
$\alpha = \beta = \gamma = \varphi$	$S = constant$	The concentration of the reactants has no role in the selectivity.
$\alpha = \gamma$ and $\beta = \varphi$	$S = constant$	The concentration of the reactants plays no role in the selectivity.
$\alpha > \gamma$ and $\beta > \varphi$	$S = \Theta * C_A^n C_B^m$	As $n > 0$ and $m > 0$, higher concentration of reactants will give higher selectivity.
$\alpha > \gamma$ and $\beta < \varphi$	$S = \Theta * \dfrac{C_A^n}{C_B^m}$	The selectivity increases when reactant A is in high concentration while reactant B is in low concentration.
$\alpha < \gamma$ and $\beta < \varphi$	$S = \dfrac{\Theta}{C_A^n C_B^m}$	Both reactants should be in as low concentration as possible to improve the selectivity.
$\alpha < \gamma$ and $\beta > \varphi$	$S = \Theta * \dfrac{C_B^m}{C_A^n}$	The concentration of B should be as high as possible while A is kept in a low concentration.

[a] Θ is a constant number (proportionality value), n and m are values > 0 in all cases.

In all the cases presented in Table 2.5, it has been considered that the temperature does not play any role. We can analyze the effect of temperature for two or more reactants in a similar way, as for one reactant. If the concentration of all reactants does not play any role for the selectivity, then the temperature effects will follow the same trends as for the case with one reactant, depending on the values of the activation energies.

The most typical case in real life is neither of the simplifications we have presented here. It is always some combination of the effects of the concentration and temperature and as we will see, concentrations can also be temperature dependent, so the selectivity should be calculated for each problem separately. Nevertheless, having some general idea and intuition of the different effects can help predict behaviors without the need of high computational cost.

Example 2.9
A two-reactant, parallel reaction to produce a desired product (and an undesired product) are being carried out in a batch reactor, at a constant temperature. The temperature effects on both reactions are similar and can be considered equivalent. Calculate the variation of the selectivity when the reactor is run for 5 min. Discuss the effect of changing the concentration of B. C_A is constant and equal to 10 [mol/l], while k_D and k_U are equal in value to 0.01 [l/(mol*min)] and 0.01 [l²/(mol²*min)], respectively. The reaction rates involved are:

$$r_D = k_{D_0} \exp\left(\frac{-E_{aD}}{RT}\right) C_A C_B \tag{2.178}$$

$$r_U = k_{U_0} \exp\left(\frac{-E_{aU}}{RT}\right) C_A C_B^2 \tag{2.179}$$

Solution
With these reaction rate equations, we can write the selectivity equation:

$$S = \frac{r_D}{r_U} = \frac{\Theta}{C_B} \tag{2.180}$$

Then, a table is made with the values of selectivity and concentration of the product as a function of C_B.

The script used to solve this problem is presented below, with the consideration that C_B^0 changes according to the values in Table 2.6.

Script:
```
#Mole balance
    d(Ca)/d(t) = ra
    Ca(0) = 10
    d(Cb)/d(t) = rb
    Cb(0) = 10
    d(Cu)/d(t) = ru
    Cu(0) = 0
    d(Cd)/d(t) = rd
    Cd(0) = 0
#Reaction rates
    ra = -(r1+r2)
    rb = -(r1+r2)
    rd = r1
    ru = r2
    r1 = k1*Ca*Cb
```

Reactor Design

```
    r2 = k2*Ca*Cb^2
#Constants
    k1 = 0.01
    k2 = 0.01
#Auxiliary equations
    S = rd/ru
#Independent variable
    t(0) = 0
    t(f) = 5
```

Based on the information of Table 2.6, we can plot the variation of the product and the variation of the selectivity as function of the concentration of B. Figure 2.47 shows the changes on C_D and it can easily be seen that the increase in the concentration of reactant B produces a decrease in the selectivity toward the product P. This tendency will tell us that we need to keep C_B as low as possible to have a good selectivity. However, by plotting the variations of the concentration of D, we can see that even a very high selectivity does not produce a significant amount of product. This implies that a balance between a good selectivity while producing a significant amount of the desired product needs to be found. In this case, we can see a maximum production when $C_B^0 = 10$ [mol/l] which is a stoichiometric fed.

TABLE 2.6
Sensitivity Variations for a Parallel System

C_B [mol/l]	C_D [mol/l]	S
0.01	0.003926	165.503
0.1	0.038526	17.11528
1	0.3245	2.243271
10	1.085166	0.368147
100	0.104261	0.011111

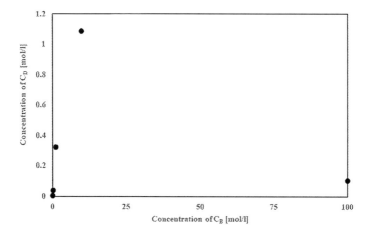

FIGURE 2.47 Variation of the C_D as a function of C_B.

2.9.3 Combined Reactions

We have look into parallel and series reactions separately. However, practically, these processes occur simultaneously, sometimes together with other that have not been presented here such as autocatalytic process. In this case, we face a reaction that can be written as follows:

1. A + B → C + D
2. A + C → F + G
3. D + G → E
4. B + E → H

where reactants A and B are being consumed in parallel, while reactants C and D are being produced and consumed in series, similarly, for reactant E and G, while H is just being produced.

In order to establish the selectivity for this system of equations, we need to first identify the desired product (D) and the undesired products (C, F, G, E, and H). With this in mind, we can then propose for each reaction step a reaction expression and then we can establish the reaction rate for each chemical. Based on the numbering each step has, we can obtain the following reaction rates expressions, assuming all the step are elementary steps.

$$r_1 = k_1 C_A C_B \tag{2.181}$$

$$r_2 = k_2 C_A C_C \tag{2.182}$$

$$r_3 = k_3 C_D C_G \tag{2.183}$$

$$r_4 = k_4 C_B C_E \tag{2.184}$$

Based on Equations 2.181 to 2.184, we can express the reaction rate for each chemical as presented in Table 2.7.

Based on all the equations presented in Table 2.7, we can write the selectivity equation for component D, as follows:

$$S = \frac{r_D}{r_C + r_E + r_F + r_G + r_H} \tag{2.185}$$

TABLE 2.7
Reaction Rate for all Chemicals Involved

Component	Step Reaction Rate Involved	Final Reaction Rate
A	$r_A = -r_1 - r_2$	$r_A = -k_1 C_A C_B - k_2 C_A C_C$
B	$r_B = -r_1 - r_4$	$r_B = -k_1 C_A C_B - k_4 C_B C_E$
C	$r_C = r_1 - r_2$	$r_C = k_1 C_A C_B - k_2 C_A C_C$
D	$r_D = r_1 - r_3$	$r_D = k_1 C_A C_B - k_3 C_D C_G$
E	$r_E = r_3 - r_4$	$r_E = k_3 C_D C_G - k_4 C_B C_E$
F	$r_F = r_2$	$r_F = k_2 C_A C_C$
G	$r_G = r_2 - r_3$	$r_G = k_2 C_A C_C - k_3 C_D C_G$
H	$r_H = r_4$	$r_H = k_4 C_B C_E$

Reactor Design

Equation 2.185 shows that when the system starts to have combined parallel and series reactions, the definition, at least by hand, for the selectivity, is not as straightforward as it was in the previous cases. Nevertheless, is still possible to find the selectivity and how the different factors affect it by solving the same problem several times, by making variations, as it was done in the previous example. For a particular case, we can present the script to be solved. Here, it can be seen that the selectivity can be determined at every given time and we will present the variations of the different components in Figure 2.48.

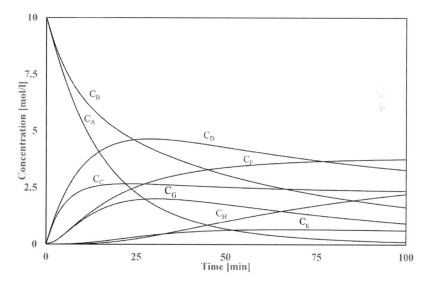

FIGURE 2.48 Variation of all components for a complex reaction system.

Script:

#Mole balance
d(Can)/d(t) = ra
Ca(0) = 10
d(Cb)/d(t) = rb
Cb(0) = 10
d(Cc)/d(t) = rc
Cc(0) = 0
d(Cd)/d(t) = rd
Cd(0) = 0
d(Ce)/d(t) = re
Ce(0) = 0
d(Cf)/d(t) = rf
Cf(0) = 0
d(Cg)/d(t) = rg

    Cg(0) = 0
    d(Ch)/d(t) = rh
    Ch(0) = 0
#Kinetics expressions
    ra = −r1−r2
    rb = −r1−r4
    rc = r1−r2
    rd = r1−r3
    re = r3−r4
    rf = r2
    rg = r2−r3
    rh = r4
    r1 = k1*Ca*Cb
    r2 = k2*Ca*Cc
    r3 = k3*Cd*Cg
    r4 = k4*Cb*Ce
#Auxiliary equations
    k1=k001*exp(−Ea1/(R*T))
    k2=k002*exp(−Ea2/(R*T))
    k3=k003*exp(−Ea3/(R*T))
    k4=k004*exp(−Ea4/(R*T))
#Selectivity
    S = rd/(rc+rf+rg+rh+re)
#Constant
    T = 700
    R = 8.3143
    k001 = 0.01
    k002 = 0.02
    k003 = 0.01
    k004 = 0.04
    Ea1 = 3,200
    Ea2 = 3,900
    Ea3 = 4,200
    Ea4 = 5,100
#Independent variable
    t(0) = 0
    t(f) = 100

Reactor Design

As shown in Figure 2.48, the main reactants are being consumed, some intermediates are being produced and consumed, and some products are only being produced; those that are being produced and consumed have some maximum in their concentration that is dependent on the kinetic parameters used in the script. But it is important to notice that they do have a maximum and therefore, in order to produce as much of the desired product as possible, the reactor should be run at the maximum production of component D.

2.10 EQUILIBRIUM REACTIONS

So far, we have been focusing on reactions that take place mainly following a non-reversible reaction, some examples of reversible systems have been introduced but here, we will introduce the concept of equilibrium reactions. An equilibrium reaction is a process in which there is a forward and a backward reaction, and in which for a given temperature, it reaches a final conversion (equilibrium conversion) that is not necessarily 100%.

Typically, we will describe the reactions with a reaction order for the forward reaction and another for the backward reaction. For example, the equilibrium reaction of A to B can be represented as:

$$A \underset{k_b}{\overset{k_f}{\rightleftarrows}} B \tag{2.186}$$

where k_f and k_b are the reaction rates for the forward and backward reactions, respectively.

For this case, we can write the reaction rate for component A, assuming that both are elementary reaction, as:

$$r_A = k_f C_A - k_b C_B \tag{2.187}$$

This equation can be written for any reaction order for the forward and backward reaction, they do not need to be the same, and a more general term will be:

$$r_A = k_f C_A^\alpha - k_b C_B^\beta \tag{2.188}$$

If the reaction has more than one reactant and more than one product, we can get a slightly more generic expression in Equation 2.189 as:

$$r_A = k_f \prod_{i=1}^{R} C_i^{\gamma_i} - k_b \prod_{j=1}^{P} C_j^{\gamma_j} \tag{2.189}$$

where associated with the forward reaction, we have the concentration of all reactants to the power of their reaction order, and for the backward reaction, we have a similar scenario.

This reaction, Equation 2.187, can be expressed as a function of the conversion, bearing in mind that $C_A = C_A^0 (1 - x)$ and $C_B = C_B^0 + C_A^0 x$, then we obtain:

$$r_A = k_f C_A^0 (1-x) - k_b \left(C_B^0 + C_A^0 x \right) \tag{2.190}$$

When the reaction has reached the equilibrium, the speed of the forward reaction is equal to the speed of the backward reaction; therefore, the reaction rate $r_A = 0$ and Equation 2.191 and be rewritten as:

$$k_f C_A^0 (1-x) = k_b \left(C_B^0 + C_A^0 x \right) \tag{2.191}$$

where we can define the equilibrium constant $K = k_f/k_b$ therefore, Equation 2.191 can be expressed as:

$$K = \frac{k_f}{k_b} = \frac{\left(C_B^0 + C_A^0 x \right)}{C_A^0 (1-x)} \tag{2.192}$$

where the conversion is now the equilibrium conversion since the system is at equilibrium. Based on the fact that both forward and backward reaction rate constants are temperature dependent, the equilibrium constant is only temperature dependent (technically an activity-based equilibrium constant is only temperature dependent, but in most of the cases, this is equal to the equilibrium constant based on concentration).

In order to see the variations of conversion with changes in the equilibrium constant (and therefore in temperature) we need to assume initial concentrations for both chemicals A and B; for this case, we will assume 1 [mol/l] for component A and zero for B, this reduce the Equation 2.193 to:

$$K = \frac{x}{(1-x)} \tag{2.193}$$

In order to plot Equation 2.193, it is important to know the dependence of K on temperature; this dependence follows the van Hoff expression, which allows us to calculate the value of the equilibrium constant for a given temperature, by knowing its value for another temperature plus the change of reaction enthalpy. The expression is:

$$\ln \left[K_2(T_2) \right] = \ln \left[K_1(T_1) \right] * \left[\left(\frac{-\Delta H}{R} \right) \left(\frac{1}{T_2} - \frac{1}{T_1} \right) \right] \tag{2.194}$$

Reactor Design

Therefore, we can rewrite it as:

$$K_2(T_2) = K_1(T_1) * \exp\left[\left(\frac{-\Delta H}{R}\right)\left(\frac{1}{T_2} - \frac{1}{T_1}\right)\right] \qquad (2.195)$$

By substituting Equation 2.195 into 2.193, we can obtain one expression that correlates temperature and equilibrium conversion for a first-order forward and first-order reverse reaction, with one reactant in each case:

$$\frac{x}{(1-x)} = K_2(T_2) = K_1(T_1) * \exp\left[\left(\frac{-\Delta H}{R}\right)\left(\frac{1}{T_2} - \frac{1}{T_1}\right)\right] \qquad (2.196)$$

In order to plot Equation 2.196, we need to know the value of ΔH, which will tell us if the reaction is endothermic or exothermic. Figure 2.49 shows the variation of the equilibrium conversion as a function of temperature for both endothermic and exothermic reactions. If $\Delta H < 0$, then the reaction is exothermic, while if $\Delta H > 0$ then, it is endothermic. The reader might find out that Equation 2.195 can be expressed in different ways mathematically; however, the physical meaning should be the same, independently.

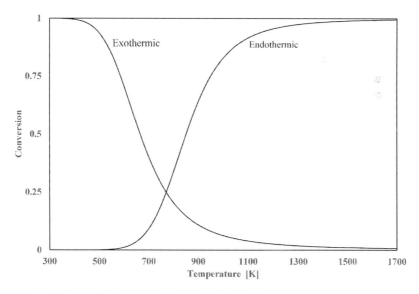

FIGURE 2.49 Equilibrium conversion as a function of temperature.

Similar plots can be produced for reactions of different orders of the forward and backward reaction and their temperature effects can also be analyzed.

From the previous information, we can obtain a generalized expression for the concentration-based equilibrium constant for a generic reaction, assuming the following reaction:

Reaction Engineering, Catalyst Preparation, and Kinetics

$$\alpha A + \beta B \leftrightarrow \gamma C + \phi D \tag{2.197}$$

Based on Equation 2.197, we can obtain the equilibrium constant as a function of the concentrations of each chemical to the power of their stoichiometric coefficient.

$$K = \frac{C_C^\gamma C_D^\phi}{C_A^\alpha C_B^\beta} \tag{2.198}$$

Generalizing:

$$K = \frac{\prod_{i=1}^{P} C_i^{\alpha_i}}{\prod_{j=1}^{R} C_j^{\alpha_j}} \tag{2.199}$$

Strictly speaking, the equilibrium constant that is only temperature dependent is the activity-based equilibrium constant that can be generalized as:

$$K = \frac{\prod_{i=1}^{P} a_i^{\alpha_i}}{\prod_{j=1}^{R} a_j^{\alpha_j}} \tag{2.200}$$

However, there are different relationships for the concentration-based, the pressure-based, and the activity-based constants. We will introduce them when required, however, for the purposes of this book, the concentration-based equilibrium constant is only temperature dependent.

Example 2.10

For the reaction where 3 moles of reactant A reacts with 2 moles of reactant B to produce 5 moles of reactant C, calculate the equilibrium conversion for a temperature of 750 [K]. The reaction is an elementary reaction in the forward and backward reaction. The initial concentration of A is 15 [moles/l] while for B is 25 [moles/l] and there is no C at the beginning.

Additional information:

$H_A^0 = -25,000\,[\text{J/mol}]; \quad H_B^0 = 15,000\,[\text{J/mol}]; \quad H_C^0 = -3,5000\,[\text{J}/(\text{mol} \ast \text{K})];$

$K_1(T = 298\,[\text{K}]) = 110,000.$

Solution

In order to solve this problem, we need to find out at what conversion our reaction will have when the reaction rate of zero. So, first we should write our reaction rate; since the reaction under consideration is elementary and reversible, then, based on the information provided, we know that the reaction is:

$$3A + 2B \overset{K}{\Leftrightarrow} 5C \tag{2.201}$$

Reactor Design

Based on Equation 2.201, we can obtain the reaction rate for this problem, this is:

$$r_A = k_f C_A^3 C_B^2 - k_b C_C^5 \tag{2.202}$$

When in equilibrium, we know that $r_A = 0$ and therefore, Equation 2.202 can be rewritten as:

$$K = \frac{C_C^5}{C_A^3 C_B^2} \tag{2.203}$$

where $K = k_f / k_b$

From the information provided, we know that we can rewrite the information on each side of the equations to have Equation 2.203 expressed as a function of temperature and of conversion. To do that we need to rewrite K as Equation 2.195, and for this case, it will be:

$$K_2(T_2) = 110{,}000 * \exp\left[\left(\frac{25{,}000 \left[\tfrac{J}{mol}\right]}{8.3143 \left[\tfrac{J}{molK}\right]}\right)\left(\frac{1}{T_2} - \frac{1}{298[K]}\right)\right] \tag{2.204}$$

If we substitute $T_2 = 750$ [K], we can then obtain the value of $K_2 = 2{,}514.47$.

In order to finally solve this problem, we need to express the concentration as function of conversion and then solve the problem to obtain the desired result. We, therefore, express $C_A = C_A^0 (1 - x)$, $C_B = C_B^0 - C_A^0 x$, and $C_C = C_C^0 + C_A^0 x$. Substituting and rearranging it, we obtain:

$$2{,}514.47 = \frac{(5/3)x^5}{(1-x)^3 \left(C_B^0 / C_A^0 - 2/3 x\right)^2} \tag{2.205}$$

Different software can be used to solve this problem; the way that this is solved is irrelevant; however, be aware that since it is a power of 5 you might get more than one answer and only those between 0 and 1 have physical relevance for this case. By solving this equation, we can obtain an equilibrium conversion of 0.87

2.11 NON-ISOTHERMAL REACTORS

So far, we have been dealing with the reactors working, or the reaction taking place, under isothermal conditions. However, this is not the general case for all the processes. Some reactions need to be isothermal; some are been carried out adiabatically, while others need a good control of the temperature to reach a high final conversion.

In order to develop a mathematical equation to be used in our balance, we need to see and understand where that dependency is. If we take a simple example of a first-order irreversible reaction (A ➔ B), with a reaction rate, $r_A = -kC_A$, considering

Arrhenius equation and expression C_A as a function of the conversion, the mole balance can be re-written into:

$$r_A = \frac{dC_A}{dt} = -C_A^0 \frac{dx}{dt} = -kC_A = -k_0 \exp\left(\frac{-E_A}{RT}\right) C_A = -k_0 \exp\left(\frac{-E_A}{RT}\right) C_A^0 (1-x) \quad (2.206)$$

This can be simplified to:

$$\frac{dx}{dt} = -k_0 \exp\left(\frac{-E_A}{RT}\right)(1-x) \quad (2.207)$$

where it can be seen that conversion and temperature are interdependent. Therefore, we have one differential equation with two unknowns. To solve this problem, we need to either establish a value for the temperature (isothermal process) or we need to find an equation that considers the variations of temperature with conversion and/ or with time. It is important to notice here that we have not considered the variations of pressure, which could have been added as well to the problem as it is presented in Chapter 4.

In order to obtain that temperature–conversion relationship, we need the first law of thermodynamics that considers the energy of the system and the variations due to heat and work, as follows:

$$\frac{dE_{sys}}{dt} = \dot{Q} - \dot{W} \quad (2.208)$$

where the left term is the rate of accumulation/loss of energy of the system, while the term on the right is the rate of heat into the system and the rate of work done by the system to the environment around it. This expression is useful for a closed system, since there is no flow of material in/out and therefore, no energy is in/out of the system.

For an open system, the work from Equation 2.208 needs to be divided into the work of the shaft, due to external mechanical movement, and the work due to the fluid moving. Furthermore, the expression needs to have two extra terms due to the energy that goes into the system and the energy that goes out of the system, due to the flow of material. In many thermodynamics books, this is done per unit of mass (kg) so that we can obtain the specific values of heat and work. Most reaction engineering books carry out this balance in specific mole basis, since that is what we need for unit compatibility with the rest of the equations.

Considering all this information, we can write the energy balance for an open system as:

$$\frac{dE_{sys}}{dt} = \dot{Q} - W_s + \sum_{i=1}^{P} F_i H_i \bigg|_{in} - \sum_{i=1}^{P} F_i H_i \bigg|_{out} \quad (2.209)$$

Reactor Design

where the work of the shaft, W_s, can be neglected for most cases and the work of the fluid as well as the energy in and out has been combined into the enthalpy (as done for thermodynamics first law for open systems). If the reader does not remember how to develop this equation, we recommend a revision of the first law of thermodynamics for closed and open systems [13,14].

It is important to develop an equation that can be used simultaneously with the mole balance; this means that we need to solve Equation 2.209 for different reactor scenarios (adiabatic, external cooling/warming fluid, etc.) in order to see the final expression we need to solve.

First of all, we will start with two major simplifications: (i) the work of the shaft can be neglected, this value is not zero but can be omitted in the presence of the enthalpy values; (ii) the process in consideration, at least for now, is in steady state. This means that $W_S = 0$ and that $dE_{sys}/dt = 0$.

With these two simplifications, we can reduce Equation 2.209 into:

$$0 = \dot{Q} + \sum_{i=1}^{P} F_i H_i \Big|_{in} - \sum_{i=1}^{P} F_i H_i \Big|_{out} \tag{2.210}$$

Now we can rewrite this expression for a generic example. Assuming that we have the reaction:

$$\alpha A + \beta B \rightarrow \gamma C + \varphi D \tag{2.211}$$

We can then write the Equation 2.210, substituting from Equation 2.211, as:

$$0 = \dot{Q} + F_A^{in} H_A^{in} + F_B^{in} H_B^{in} + F_C^{in} H_C^{in} + F_D^{in} H_D^{in} + F_I^{in} H_I^{in} - F_A^{out} H_A^{out} \\ - F_B^{out} H_B^{out} - F_C^{out} H_C^{out} - F_D^{out} H_D^{out} - F_I^{out} H_I^{out} \tag{2.212}$$

We rename, for convenient purposes F_i^{in} as F_i^0 and F_i^{out} as F so that we can express Equation 2.212 as:

$$0 = \dot{Q} + F_A^0 H_A^0 + F_B^0 H_B^0 + F_C^0 H_C^0 + F_D^0 H_D^0 + F_I^0 H_I^0 \\ - F_A H_A - F_B H_B - F_C H_C - F_D H_D - F_I H_I \tag{2.213}$$

We have added the F_I to represent the possibility of having an inert component in the system. This is not the case in all scenarios, but it is important to have a full description of the process.

We can now express each component as a function of conversion:

$$F_A = F_A^0 (1 - x)$$

$$F_B = F_B^0 - \frac{\beta}{\alpha} F_A^0 x$$

$$F_C = F_C^0 + \frac{\gamma}{\alpha} F_A^0 x$$

$$F_D = F_D^0 + \frac{\varphi}{\alpha} F_A^0 x$$

$$F_I = F_I^0$$

We can substitute all these definitions in our expression:

$$0 = \dot{Q} + F_A^0 H_A^0 + F_B^0 H_B^0 + F_C^0 H_C^0 + F_D^0 H_D^0 + F_I^0 H_I^0 - F_A^0 (1-x) H_A$$
$$-\left(F_B^0 - \frac{\beta}{\alpha} F_A^0 x\right) H_B - \left(F_C^0 + \frac{\gamma}{\alpha} F_A^0 x\right) H_C - \left(F_D^0 + \frac{\varphi}{\alpha} F_A^0 x\right) H_D - F_I^0 H_I \quad (2.214)$$

Rearranging:

$$0 = \dot{Q} + F_A^0 H_A^0 - F_A^0 H_A + F_A^0 x H_A + F_B^0 H_B^0 - F_B^0 H_B + \frac{\beta}{\alpha} F_A^0 x H_B + F_C^0 H_C^0$$
$$-F_C^0 H_C - \frac{\gamma}{\alpha} F_A^0 x H_C + F_D^0 H_D^0 - F_D^0 H_D - \frac{\varphi}{\alpha} F_A^0 x H_D + F_I^0 H_I^0 - F_I^0 H_I \quad (2.215)$$

This can be rearranged one more time as:

$$0 = \dot{Q} + F_A^0 \left(H_A^0 - H_A\right) + F_B^0 \left(H_B^0 - H_B\right) + F_C^0 \left(H_C^0 - H_C\right) + F_D^0 \left(H_D^0 - H_D\right)$$
$$+ F_I^0 \left(H_I^0 - H_I\right) - F_A^0 x \left(\frac{\varphi}{\alpha} H_D + \frac{\gamma}{\alpha} H_C - \frac{\beta}{\alpha} H_B - H_A\right) \quad (2.216)$$

where the term $\left(\frac{\varphi}{\alpha} H_D + \frac{\gamma}{\alpha} H_C - \frac{\beta}{\alpha} H_B - H_A\right) = \Delta H_{RX}$ is the heat of the reaction taking place at a temperature, T. Therefore, Equation 2.216 can be rewritten as:

$$0 = \dot{Q} + F_A^0 \left(H_A^0 - H_A\right) + F_B^0 \left(H_B^0 - H_B\right) + F_C^0 \left(H_C^0 - H_C\right)$$
$$+ F_D^0 \left(H_D^0 - H_D\right) + F_I^0 \left(H_I^0 - H_I\right) - F_A^0 x \Delta H_{RX} \quad (2.217)$$

It is important to notice that the heat of reaction is calculated at a temperature T; however, the tabulated values are at a reference temperature $T_R = 298$ [K]. Therefore, we need to find a way to calculate the enthalpy of reaction at the desired temperature based on its value at 298 [K].

As we know from thermodynamics, enthalpy is a function of state, meaning that we can go from point A to point B and the variation is determined by the values at the initial and final state but not based on the path followed. Based on this, we can express the value of H_i as follows:

$$H_i(T) = H_i(T_R) + \int_{T_R}^{T} C_{Pi}(T) dT \quad (2.218)$$

Reactor Design

The variation of the Cp_i as a function of temperature can be found in a thermodynamic book; it is a function of temperature and different expressions for this term can be found. Generally, this value is considered constant and while this is not entirely correct, however, for the purpose of this book, it will be assumed constant unless otherwise mentioned.

Therefore, we can write Equation 2.218 as:

$$H_i(T) = H_i(T_R) + C_{Pi}(T - T_R) \tag{2.219}$$

Using Equation 2.219, we can express the heat of reaction as:

$$\Delta H_{RX} = \frac{\varphi}{\alpha}\left(H_D(T_R) + C_{PD}(T - T_R)\right) + \frac{\gamma}{\alpha}\left(H_C(T_R) + C_{PC}(T - T_R)\right) \\ - \frac{\beta}{\alpha}\left(H_B(T_R) + C_{PB}(T - T_R)\right) - \left(H_A(T_R) + C_{PA}(T - T_R)\right) \tag{2.220}$$

Reorganizing:

$$\Delta H_{RX} = \left(\Delta H_{RX}(T_R)\right) + \left(\frac{\varphi}{\alpha}C_{PD} + \frac{\gamma}{\alpha}C_{PC} - \frac{\beta}{\alpha}C_{PB} - C_{PA}\right)(T - T_R) \tag{2.221}$$

So now, we can substitute Equation 2.221 into 2.217:

$$0 = \dot{Q} + F_A^0\left(H_A^0 - H_A\right) + F_B^0\left(H_B^0 - H_B\right) + F_C^0\left(H_C^0 - H_C\right) \\ + F_D^0\left(H_D^0 - H_D\right) + F_I^0\left(H_I^0 - H_I\right) \\ - F_A^0 x\left[\left(\Delta H_{RX}(T_R)\right) + \left(\frac{\varphi}{\alpha}C_{PD} + \frac{\gamma}{\alpha}C_{PC} - \frac{\beta}{\alpha}C_{PB} - C_{PA}\right)(T - T_R)\right] \tag{2.222}$$

We need to find some way to transform the values of H_i^0 and H_i into more manageable equations. From thermodynamics, we also know that when there is no phase change:

$$\Delta H = \int_{T_{i0}}^{T} C_{Pi}(T)\,dT = C_{Pi}(T - T_{i0}) \tag{2.223}$$

where T_{i0} represents the temperature in the inlet for component "i"; this means that each component could have a different initial temperature and therefore, all of them have to be taken into consideration. So, for our cases, then we can substitute Equation 2.223 into 2.222:

$$0 = \dot{Q} - F_A^0 C_{PA}(T - T_{A0}) - F_B^0 C_{PB}(T - T_{B0}) - F_C^0 C_{PC}(T - T_{C0}) \\ - F_D^0 C_{PD}(T - T_{D0}) - F_I^0 C_{PI}(T - T_{I0}) \\ - F_A^0 x\left[\left(\Delta H_{RX}(T_R)\right) + \left(\frac{\varphi}{\alpha}C_{PD} + \frac{\gamma}{\alpha}C_{PC} - \frac{\beta}{\alpha}C_{PB} - C_{PA}\right)(T - T_R)\right] \tag{2.224}$$

This is the energy balance for the generic reaction presented when there is no work on the shaft and there is steady-state scenario. We will now expand this to obtain a generic expression considering p number of components.

$$0 = \dot{Q} - \sum_{i=1}^{P} F_i^0 C_{P_i}(T - T_{i0}) - F_A^0 x \left[(\Delta H_{RX}(T_R)) + \sum_{i=1}^{P} \frac{\upsilon_i}{|\upsilon_A|} C_{P_i}(T - T_R) \right] \quad (2.225)$$

This is the generic energy balance to be used under the assumptions mentioned above. From this generic balance, we will now move into specific scenarios.

2.11.1 Adiabatic

Now we have managed to obtain an expression for the energy balance that is a function of temperature and conversion, this means that Equation 2.225 is the second equation (not a differential equation, in this case) to solve together with the mole balance, as previously explained.

In order to solve these two equations, we need some information regarding the value of heat involved (Q), to identify the scenario and dependence on other variables. As the title of this section implies, this is an adiabatic process ($Q = 0$).

Considering this new assumption, Equation 2.225 takes the form:

$$0 = -\sum_{i=1}^{P} F_i^0 C_{P_i}(T - T_{i0}) - F_A^0 x \left[(\Delta H_{RX}(T_R)) + \sum_{i=1}^{P} \frac{\upsilon_i}{|\upsilon_A|} C_{P_i}(T - T_R) \right] \quad (2.226)$$

This expression can be rearranged to get a conversion as a function of the temperature:

$$x = \frac{\sum_{i=1}^{P} F_i^0 C_{P_i}(T - T_{i0})}{-F_A^0 \left[(\Delta H_{RX}(T_R)) + \sum_{i=1}^{P} \frac{\upsilon_i}{|\upsilon_A|} C_{P_i}(T - T_R) \right]} \quad (2.227)$$

where Equation 2.227 is the general expression to be used when solving a non-isothermal process. This expression can be simplified if the initial temperature of all the chemicals is the same and $T_{i0} = T_0$. However, this simplification is not crucial to solve the problem.

One important result is that the heat of reaction contribution is considerably larger than the energy involved to heat the chemicals from the reference temperature to the reaction temperature. In that case, Equation 2.227 simplifies to:

$$x = \frac{\sum_{i=1}^{P} F_i^0 C_{P_i}(T - T_{i0})}{-F_A^0 \left[(\Delta H_{RX}(T_R)) \right]} \quad (2.228)$$

Reactor Design

Thus, the only dependency is a linear relationship between the conversion and temperature. This is a very useful result that could lead to simpler calculations but as it involves assumptions, the predictions of the system could have errors. However, the result from Equation 2.228 is accurate when the contributions of the Cp cancel each other out, as they are equal.

Example 2.11
Consider a liquid phase reaction where chemicals A and B transform via an elementary reversible reaction into chemical C. The process is carried out adiabatically, in steady state and with no work from the shaft involved. The reaction in consideration is:

$$A + B \rightarrow C$$

The reaction is carried out in a PFR where A and B are fed stoichiometrically and component C is not present at the beginning of the process. Components A and B are fed at a temperature of $T_0 = 300$ [K], with a molar flow of 5 [mol/min]. The volumetric flow of the process is 1 [l/min]. Find the variation of components A and C (B is identical to A, comment on this), along with the conversion and temperature profiles.

The forward reaction constant follows an Arrhenius equation $k_f = k_{00} \exp\left(\frac{-E_A}{RT}\right)$, while the equilibrium constant follows $K_2(T_2) = K_1(T_1) * \exp\left[\left(\frac{-\Delta H}{R}\right)\left(\frac{1}{T_2} - \frac{1}{T_1}\right)\right]$. This expression has the assumption that ΔH does not vary with temperature; this is a reasonable assumption that will be used throughout this book.

Additional information:

$H_A^0 = -43{,}000$ [J/mol]; $\quad H_B^0 = 23{,}000$ [J/mol]; $\quad H_C^0 = -33{,}000$ [J/mol];
$Cp_A = 160$ [J/(mol*K)]; $\quad Cp_B = 150$ [J/(mol*K)]; $\quad Cp_C = 180$ [J/(mol*K)];
$k_{00} = 0.01$ [l/(mol*min)]; $\quad T_1 = T_r = 298$ [K]; $\quad K_1(T_1) = 52$; $\quad R = 8{,}314$ [J/(mol*K)];
$E_A = 4{,}500$ [J/mol]; reactor volume = 600 [dm³].

Solution
Based on the information provided, we know that we have a reversible elementary reaction, the reaction rate for component A, is:

$$r_A = -k_1\left(C_A C_B - \frac{C_C}{K}\right)$$

(2.229)

To calculate the reaction rates for components B and C, we can use the following relationship: $\frac{r_A}{-1} = \frac{r_B}{-1} = \frac{r_C}{1}$; using the definition for k_1 and K as function of the temperature. Equation 2.229 can be written as a function of C_A (and therefore conversion) and temperature.

$$r_A = -k_{00} \exp\left(\frac{-E_A}{RT}\right)\left(C_A C_B - \frac{C_C}{K_1 \exp\left(\frac{-\Delta H}{R}\left(\frac{1}{T}-\frac{1}{T_1}\right)\right)}\right) \qquad (2.230)$$

The mole balance for each component is $\frac{dF_A}{dV} = r_A$; $\frac{dF_B}{dV} = r_B$; $\frac{dF_C}{dV} = r_C$; we also need the relationships that $C_i^* v = F_i$ and that $x = \frac{C_A^0 - C_A}{C_A^0}$. Finally, in order to solve this problem, we need to add the energy balance for this adiabatic scenario, this means solving Equation 2.227. This equation is a conversion as a function of temperature and in order to add it to Polymath, we need to have temperature as a function of conversion. This equation can be simplified if the contribution of the Cp to heat up to the reaction temperature can be neglected in comparison to the heat of the reaction (this is done in the following example for a gas phase case). However, if this is not possible, we need to rearrange Equation 2.227, so that all components enter the reactor at the same temperature ($T_{i0} = T_0$), we get:

$$-xF_A^0\left[\left(\Delta H_{RX}(T_R)\right) + \sum_{i=1}^{P}\frac{\upsilon_i}{|\upsilon_A|}C_{P_i}(T-T_R)\right] = \sum_{i=1}^{P}F_i^0 C_{P_i}(T-T_0) \qquad (2.231)$$

Reorganizing the terms and solving for temperature, the generic expression of temperature as a function of conversion is as follows:

$$T = \frac{-xF_A^0 \Delta H_{RX}(T_R) + xF_A^0 \sum_{i=1}^{P}\frac{\upsilon_i}{|\upsilon_A|}C_{P_i}T_R + \sum_{i=1}^{P}F_i^0 C_{P_i}T_0}{\sum_{i=1}^{P}F_i^0 C_{P_i} + xF_A^0 \sum_{i=1}^{P}\frac{\upsilon_i}{|\upsilon_A|}C_{P_i}} \qquad (2.232)$$

For our case, Equation 2.232 is reduced to:

$$T = \frac{-xF_A^0 \Delta H_{RX}(T_R) + xF_A^0 T_R (C_{P_C} - C_{P_A} - C_{P_B}) + (F_A^0 C_{P_A} + F_B^0 C_{P_B})T_0}{(F_A^0 C_{P_A} + F_B^0 C_{P_B}) + xF_A^0 (C_{P_C} - C_{P_A} - C_{P_B})} \qquad (2.233)$$

Now we can put Equation 2.233, together with 2.230, with all the additional data and supplementary equations into the Polymath script.

Script:

```
#Mole balance
    d(Fa)/d(V) = ra
    Fa(0) = 5
    d(Fb)/d(V) = rb
    Fb(0) = 5
    d(Fc)/d(V) = rc
```

Reactor Design

```
    Fc(0) = 0
#Energy balance
    T = ((-DH*x*Fa0)+((Fa0*Cpa+Fb0*Cpb)*T0)+(Fa0*Tr*x*(Cpc-Cpa-
        Cpb)))/((Fa0*Cpa+Fb0*Cpb)+(Fa0*x*(Cpc-Cpa-Cpb)))
#Kinetics expression
    ra = -k1*(Ca*Cb-(Cc/K))
    rb = ra
    rc = -ra
#Auxiliary expressions
    k1 = k00*EXP((-Ea/(R*T)))
    K = K1*EXP((-DH/R)*((1/T)-(1/T1)))
    DH = Hc-(Ha+Hb)
    Ca = Fa/v
    Cb = Fb/v
    Cc = Fc/v
    x = (Ca0-Ca)/Ca0
    Ca0 = Fa0/v
#Constants and initial values
    Tr = T1
    T1 = 298
    R = 8.3143
    K1 = 52
    Ea = 4,500
    T0 = 300
    Fa0 = 5
    Fb0 = 5
    Fc0 = 0
    Cpa = 160
    Cpb = 150
    Cpc = 180
    Ha = -43,000
    Hb = 23,000
    Hc = -33,000
    k00 = 0.01
    v = 1
#Independent variable
    V(0) = 0
    V(f) = 600
```

We can now run this script to find the desired profiles. The variation of the concentration of A and C can be seen in Figure 2.50, while the changes in conversion and temperature can be seen in Figure 2.51.

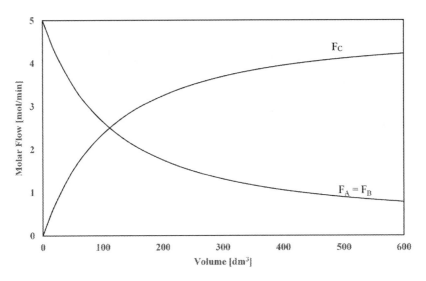

FIGURE 2.50 Molar flow profile.

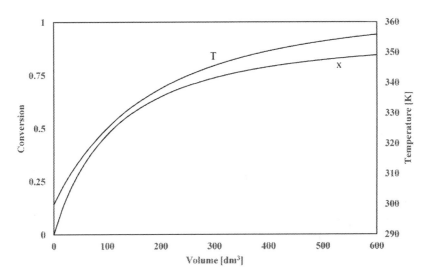

FIGURE 2.51 Conversion and temperature profiles.

Figure 2.50 shows that the amount of component A decreases but does not reach the value of zero. This is expected because the reaction is an equilibrium process and therefore, an equilibrium conversion will be achieved, and the reaction will not be able to continue forward. The same result can be seen in Figure 2.51, for the conversion tendency. Figure 2.51 also shows the changes in temperature and it can be seen that the temperature increases along the reactor. This is because the reaction under consideration is an exothermic reaction; therefore, energy is being produced along the system and the temperature increases. However, the equilibrium conversion for this case decreases with increasing temperature since this is an exothermic reaction. Additionally, here is the variation of the equilibrium conversion and conversion with temperature, in Figure 2.52. The variation of the conversion is almost a straight line starting from 300 [K], *i.e.*, the initial temperature and when it reaches the equilibrium plot, no more conversion can be achieved. In this example, that point has not been attained, but it is close. This means that after that, independent of the length of the reactor, no more product C will be produced. This will be discussed in detail in other examples; in the meantime, this should be considered as an introduction to this topic.

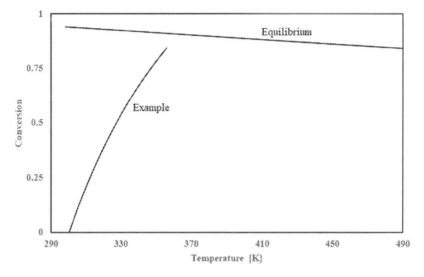

FIGURE 2.52   Conversion and equilibrium conversion profiles.

Example 2.12

The previous example was on a liquid phase reaction; therefore, there was no change in volume. For this case, we will solve the same problem as in the previous example with the difference that now the reaction takes place in gas phase. This modification introduce variations to the problem due to the change in the number of moles as well as to the temperature profile.

The major difference in this example, in comparison with the previous scenario, is that the volumetric flow, v, is not constant. This means that since it is a flow system, we need to use Equation 2.136 for the volumetric flow. With this modification, we can also modify the script accordingly.

Script:

```
#Mole balance
    d(Fa)/d(V) = ra
    Fa(0) = 5
    d(Fb)/d(V) = rb
    Fb(0) = 5
    d(Fc)/d(V) = rc
```

```
    Fc(0) = 0
#Energy balance
    T = ((-DH*x*Fa0)+((Fa0*Cpa+Fb0*Cpb)*T0)+(Fa0*Tr*x*(Cpc-Cpa-
        Cpb)))/((Fa0*Cpa+Fb0*Cpb)+(Fa0*x*(Cpc-Cpa-Cpb)))
#Kinetics expression
    ra = -k1*(Ca*Cb-(Cc/K))
    rb = ra
    rc = -ra
#Auxiliary expressions
    k1 = k00*EXP((-Ea/(R*T)))
    K = K1*EXP((-DH/R)*((1/T)-(1/T1)))
    DH = Hc-(Ha+Hb)
    Ca = Fa/v
    Cb = Fb/v
    Cc = Fc/v
    x = (Fa0-Fa)/Fa0
    Ca0 = Fa0/v
    v = v0*Ft*T/(Ft0*T0)
    Ft = Fa+Fb+Fc
    Ft0 = Fa0+Fb0+Fc0
#Constants and initial values
    Tr = T1
    T1 = 298
    R = 8.3143
    K1 = 52
    Ea = 4,500
    T0 = 300
    Fa0 = 5
    Fb0 = 5
    Fc0 = 0
    Cpa = 160
    Cpb = 150
    Cpc = 180
    Ha = -43,000
    Hb = 23,000
    Hc = -33,000
    k00 = 0.01
    v0 = 1
#Independent variables
    V(0) = 0
    V(f) = 600
```

Reactor Design

Solving this problem, we can obtain the new profiles for components A and C, as well as for the conversion and the temperature of the reactor. As shown in Figure 2.53, the molar flows for components A and C are similar to the previous case when the volume was constant as well as there was no influence from the change in the number of moles. However, there are small differences in the final values that can be obtained. This can be easily seen in the final conversion that is achieved in this case; Figure 2.54 shows this clear tendency. Furthermore, the temperature profile reaches a higher final temperature, *i.e.*, around 5 °C higher. This also has an impact on the final equilibrium conversion that is achieved, since the equilibrium constant is temperature dependent. The variation of conversion with temperature and the equilibrium limit can be seen in Figure 2.55.

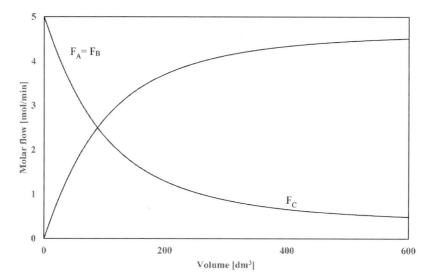

FIGURE 2.53 Molar flow profiles.

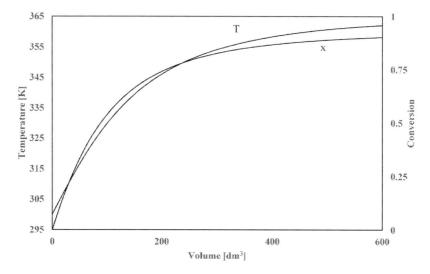

FIGURE 2.54 Conversion and temperature profiles.

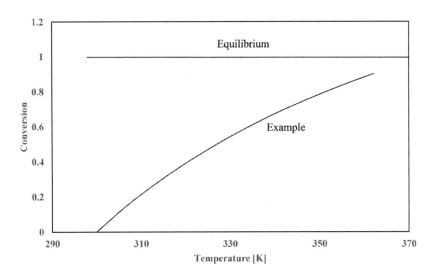

FIGURE 2.55 Conversion and equilibrium conversion profiles.

As presented in the previous examples and for some of the upcoming cases, the solution to the adiabatic process needs to be modified due to the phases the system is in (gas vs. liquid), if there is a pressure drop or a change in the number of moles. All these scenarios are presented in Chapter 4 where a full, extended example is solved for several possible scenarios.

As shown before, in Figures 2.52 and 2.55, the final highest conversion that can be achieved is the one that will take place at equilibrium. Therefore, we will present a little more information about this scenario, on how to deal with an equilibrium reaction, and on how to obtain the higher possible conversions (equilibrium value) for a given initial temperature. We will also show how to increase the conversion if one adiabatic process does not fulfill the desired requirements.

2.11.1.1 Adiabatic Process for a Reaction at Equilibrium

In the previous part of this chapter, we studied the reaction taking place in an adiabatic process. However, we put no restriction to the reaction and as it was shown in Figures 2.52 and 2.55 the conversion increases with temperature, but it has a limit that is given by the equilibrium conversion. In this section, we will solve problems for reactions at equilibrium and how to improve the process when the desired conversion is higher than the one that can be possibly achieved.

Let's developed the methodology for the simplest case, *i.e.*, an irreversible first-order reaction, for both the forward and the backward reaction, for example:

$$A \underset{k_b}{\overset{k_f}{\rightleftarrows}} B \tag{2.234}$$

Reactor Design

For the previous reaction, we want to calculate the equilibrium conversion for a given temperature. That means that we need to solve the reaction rate equation when it has reached equilibrium:

$$K = \frac{C_B}{C_A} \quad (2.235)$$

Knowing that $C_A = C_A^0 (1 - x)$ and that $C_B = C_B^0 + C_A^0 x$, and assuming that there is no B at the beginning, and knowing that $K(T) = K(T_1) * \exp\left[\left(\frac{-\Delta H}{R}\right)\left(\frac{1}{T} - \frac{1}{T_1}\right)\right]$ we can rewrite Equation 2.235:

$$K(T) = K(T_1) * \exp\left[\left(\frac{-\Delta H}{R}\right)\left(\frac{1}{T} - \frac{1}{T_1}\right)\right] = \frac{C_B}{C_A} = \frac{x_{eq}}{(1 - x_{eq})} \quad (2.236)$$

We have added the subscript *eq* to represent equilibrium conversion. By knowing the reaction itself, and therefore, the value of a given equilibrium constant at a given temperature and the heat of formation, we can obtain a plot of the equilibrium conversion as a function of temperature. This was also previously presented; however, it is presented here one more time since we will now use it. For this case, we want to know the equilibrium conversion when our reactor runs adiabatically. This means that we have a temperature change and therefore, that besides the mole balance (Equation 2.236), the energy balance is also required. Thus, Equation 2.236 and Equation 2.227, need to be jointly solved.

To solve this problem, we can do a trial and error approach that can be done using a simple Excel table. We need information for the different parameters. So, we can assign numbers that will help solve the desired problem.

The equations to solve this problem are:

$$K(T_1) * \exp\left[\left(\frac{-\Delta H}{R}\right)\left(\frac{1}{T} - \frac{1}{T_1}\right)\right] = \frac{x_{eq}}{(1 - x_{eq})} \quad (2.237)$$

$$x = \frac{\sum_{i=1}^{P} F_i^0 C_{P_i}(T - T_{i0})}{-F_A^0 \left[(\Delta H_{RX}(T_R)) + \sum_{i=1}^{P} \frac{\upsilon_i}{|\upsilon_A|} C_{P_i}(T - T_R)\right]} \quad (2.238)$$

If we have data for Cp_A and Cp_B (they are equal for example), and values for $T_R = T_1 = T_{i0} = 298$ [K], and we know the heat of reaction; then, we can solve Equations 2.237 and 2.238. When both conversions are equal, then we have found the solution to the problem. Table 2.8 presents this approach for some given temperatures.

As shown in Table 2.8, the solution to this problem can be done by trial and error. Another approach is to solve both equations simultaneously; in this case, it is possible by hand since we can insert one in the other and obtain the exact solution to the problem, *i.e.*, around 480.5 [K].

TABLE 2.8
Solution to the Equilibrium Problem

T[K]	K	x_{eq}	x
298	125	0.992	0.000
300	116.86	0.991	0.008
400	9.538	0.905	0.408
500	2.121	0.679	0.808
475	3.113	0.856	0.688
482	2.655	0.726	0.736
480	2.725	0.731	0.728
480.5	2.707	0.730	0.730

A third and quite useful method that we will use further on, is to represent both expressions graphically. We know that the mole balance cannot physically have a conversion higher than the equilibrium conversion; therefore, the plot of the mole balance is the limit of the higher conversion that can be achieved for any given temperature. Plotting the energy balance, only within the range of validity of this equation (mathematically can have any number we want but that does not mean it has a physical meaning), we can then see when both plots cross each other. This will give us the equilibrium conversion and the adiabatic temperature of the reactor. Figure 2.56 shows this solution, where only the physically relevant regions of the problem have been plotted.

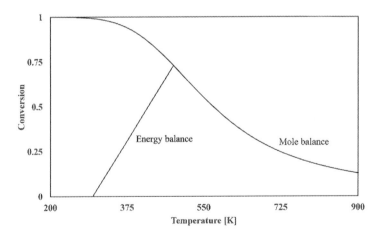

FIGURE 2.56 Graphical solution.

In practice, we cannot work exactly at equilibrium since that might require a driving force that will become larger and large the closer we are to the equilibrium, with

Reactor Design

a tendency to get close to infinity. Therefore, we can work for example at 98% of the equilibrium conversion.

However, in some cases, this is not high enough, as seen in Figure 2.56. If we want to have a conversion that is higher than 95%, we will not be able to achieve this value with the system we are working on. Some new solutions will be presented later on, like using a cooling/heating fluid; this can be co-current or counter-current. We can use a temperature profile so that we are always close to the equilibrium but with the possibility to increase the conversion.

A very typical practical approach is to have intermediate cooling (in this case, since this is an exothermic reaction), that will allow several series reactors with cooling. Figure 2.57 shows a schematic representation of the reaction set up with intermediate cooling.

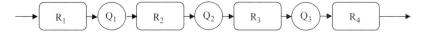

FIGURE 2.57 Schematic representation of the reactor with intermediate heat exchanger.

Figure 2.57 schematically shows four reactors and three intermediate heat exchangers. This system allows the reaction medium to cool down (as in the previous example) or heat up, when required, in order to allow higher conversions after the fourth reactor. As shown in Figure 2.56, schematically, this can be represented as in Figure 2.58 where these four reactors and three cooling systems are shown.

Figure 2.58 shows all the four reactors ($R_1 - R_4$) and in order to achieve the desired conversion, *i.e.*, above 95%, as well as the 3 heat exchangers ($Q_1 - Q_3$) that are required. In this case, after the reaction, each of the heat exchanger will decrease the temperature, to the same initial, T_0. This temperature should be as low as possible since it will decrease the number of reactors required, but not too low, since this will not generate enough energy to overcome the activation energy and the reaction will

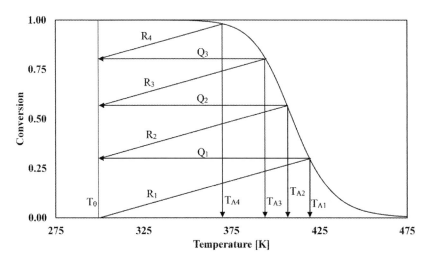

FIGURE 2.58 Graphical solution for schematic problem in Figure 2.57.

not take place. Figure 2.58 also shows $T_{A1} - T_{A4}$ which are the adiabatic temperatures for each of the four reactors involved in the process.

The solution presented in the previous case is for a simple first-order irreversible exothermic reaction; however, the same can be done for an n-order reversible reaction ($n > 1$) and/or endothermic process. For an endothermic reaction, with the first-order irreversible reaction, as presented before, Figure 2.58 will appear mirrored. This effect can be seen in Figure 2.59, where the similar schematic process of four reactors and three intermediate heat exchangers, was employed.

Figure 2.59 shows the results for an endothermic system, where energy is required to be given to the reaction in order to carry it out and to increase the final conversion of the process.

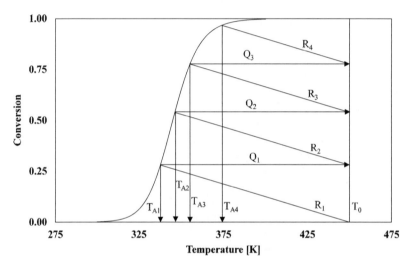

FIGURE 2.59 Graphical solution for the schematic problem in Figure 2.57 with an endothermic reaction.

Other strategies can be used in order to solve a non-isothermal problem; such as a non-adiabatic system which take place in a flow system, where the temperature of the reactor changes along the reactor for a PFR or a PBR. For a batch system, temperature will not be changing in a space variable but will be a time-dependent variable; this case will be presented in a latter section.

2.11.2 Plug Flow with Heat Added/Removed at a Constant Outside Temperature

We will now consider a steady-state scenario (not in a batch or semi-batch system) with a heat input/output to it but with the external temperature being constant. This could be the case for a CSTR, PFR, or PBR.

The solutions for a PFR and a PBR are analogous, so we will present only the plug flow and leave the derivation of the PBR for the student to attempt.

Reactor Design

Since now we have a tubular or piston reactor, we need to do a similar analysis as we did for the mole balance; we will, therefore, look at a differential part of that volume, as represented in Figure 2.60.

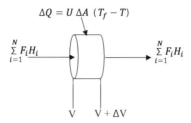

FIGURE 2.60 Differential plug flow representation for energy calculations.

Now we can try to present and solve the energy balance for the volume (ΔV) under consideration; we will assume, as previously done, that there is no W_s involved.

Therefore, the energy balance is the energy transferred into the system by flow, plus the energy added by the heat exchanger, minus the energy extracted by the flow. This is the case presented in Figure 2.60; if the heat is then removed from the reactor, the term $(T_f - T)$ will be written as $(T - T_f)$.

In our case then:

$$\Delta Q + \sum_{i=1}^{N} F_i H_i \bigg|_V - \sum_{i=1}^{N} F_i H_i \bigg|_{V+\Delta V} = 0 \tag{2.239}$$

We can substitute the definition of heat and on reorganizing this expression, we have:

$$U \Delta A (T_f - T) = \sum_{i=1}^{N} F_i H_i \bigg|_{V+\Delta V} - \sum_{i=1}^{N} F_i H_i \bigg|_V \tag{2.240}$$

The area involved in this expression is related to the heat transfer area. So, in the case of a cylinder, it is the perimeter of the base times the length of the tube. Since we want to express this in the terms of the volume, we can define $\Delta A = \varpi \Delta V$ giving:

$$\varpi = \frac{\Delta A}{\Delta V} = \frac{\pi D L}{\pi L D^2 / 4} = \frac{4}{D} \tag{2.241}$$

We can then substitute $\Delta A = \varpi \Delta V$ into Equation 2.240, divide the right hand side of the equation by ΔV and take the limit:

$$U\varpi(T_f - T) = \lim_{\Delta V \to 0} \left(\frac{\sum_{i=1}^{N} F_i H_i \Big|_{V+\Delta V} - \sum_{i=1}^{N} F_i H_i \Big|_{V}}{\Delta V} \right) \quad (2.242)$$

The term on the right hand side is the definition of the derivative; so Equation 2.242 will become:

$$U\varpi(T_f - T) = \frac{d\sum_{i=1}^{N} F_i H_i}{dV} \quad (2.243)$$

Mathematically, we obtain:

$$U\varpi(T_f - T) = \sum_{i=1}^{n} H_i \frac{dF_i}{dV} + \sum_{i=1}^{n} F_i \frac{dH_i}{dV} \quad (2.244)$$

In order to solve this expression, we need to solve the two differential parts in the right hand side of Equation 2.244. From the mole balance for a PFR, we have:

$$\frac{dF_i}{dV} = r_i \quad (2.245)$$

where $\dfrac{r_i}{\upsilon_i} = \dfrac{r_A}{-1}$

We also know from thermodynamics that the variations of enthalpy are a function of temperature, $dH = Cp \, dT$, therefore:

$$\frac{dH_i}{dV} = C_{Pi} \frac{dT}{dV} \quad (2.246)$$

We can now substitute Equations 2.245 and 2.246 into Equation 2.244:

$$U\varpi(T_f - T) = \sum_{i=1}^{n} H_i \upsilon_i (-r_A) + \sum_{i=1}^{n} F_i C_{Pi} \frac{dT}{dV} \quad (2.247)$$

Rearranging, we have:

$$U\varpi(T_f - T) = (-r_A)\sum_{i=1}^{n} H_i \upsilon_i + \frac{dT}{dV}\sum_{i=1}^{n} F_i C_{Pi} \quad (2.248)$$

Reactor Design

where the term $\sum_{i=1}^{n} H_i \upsilon_i$ is the heat of reaction (ΔH_{RX}). Therefore, we can reorganize Equation 2.248 into the energy balance that needs to be solved in order to acquire the temperature profiles inside the reactor:

$$\frac{dT}{dV} = \frac{U\varpi(T_f - T) + (r_A)\Delta H_{RX}}{\sum_{i=1}^{n} F_i C_{Pi}} \quad (2.249)$$

Here, the first term of the numerator is the heat to be transferred into the reactor and the second term is the heat produced due to the reaction. The first term is presented to show heat being added to the system; if the heat is removed, then the temperature term should be changed, as explained above.

Equation 2.249 is the energy balance that needs to be simultaneously solved with the mole balance in order to obtain the conversion and the temperature variation inside the reactor as a function of the volume.

Example 2.13

The elementary gas phase reaction to produce the desired product is being carried out in a tubular reactor where heat is removed from the process due to a constant external temperature. The heat is removed because the reaction is exothermic and because the catalyst can get deactivated at high temperatures. Obtain the conversion and temperature profile, as well as the flow profile for all the involved components. The process happens at constant pressure.

The reaction under study is:

$$2A + B \leftrightarrow C$$

In order to solve this problem, we will need some additional information such as:

Cp_A = 120 [J/(mol*K)]; Cp_B = 90 [J/(mol*K)]; Cp_C = 150 [J/(mol*K)];
$U\varpi$ = 1,300 [J/(min*dm³*K)]; H_A = 4,500 [J/mol]; H_B = −3,500 [J/mol];
H_C = 9,500 [J/mol]; T_f = 400 [K]; T_0 = 315 [K];

$T_1 = 298$ [K]; $k = 0.0052 \exp\left(\dfrac{-3,000}{8.3143 T}\right)$ [l²/(min*mol²)];

$K = 0.2 \exp\left(\dfrac{4,000}{8.3143} * \left(\dfrac{1}{T} - \dfrac{1}{T_1}\right)\right)$; $F_{A0} = F_{B0} = 15$ [mol/l] while $v_0 = 1.1$ [l/min].

Solution

For this problem, we know that $\frac{dF_A}{dV} = r_A$, $\frac{dF_B}{dV} = r_B$, and $\frac{dF_C}{dV} = r_C$, and based on the reaction, we can then obtain r_A:

$$-r_A = k_f \left(C_A^2 C_B - \frac{C_C}{K} \right) \quad (2.250)$$

Knowing that $\frac{r_A}{-2} = \frac{r_B}{-1} = \frac{r_C}{1}$, we can then express the mole balances related to the reaction rate of chemical A. Since the reaction is exothermic, constant heat removal from the process is required to not allow the reaction temperature to raise uncontrollably. Therefore, we need to add the energy balance to the problem. This general equation is shown in Equation 2.249, and for our case, it can be expressed as:

$$\frac{dT}{dV} = \frac{U\varpi(T_f - T) + (r_A)\Delta H_{RX}}{F_A C_{PA} + F_B C_{PB} + F_C C_{PC}} \quad (2.251)$$

Due to the system being in gas phase, with change in the number of moles and temperature, we need to write $C_A = F_A/v$ where $v = v_0 * (F_T/F_T^0) * (T/T_0)$ with $F_T = \sum_{i=1}^{n} F_i$ and $F_T^0 = \sum_{i=1}^{n} F_i^0$ and the total volume is 25 [dm³]. With all this additional data, we can write the script for Polymath:

Script:

```
#Mole balance
    d(Fa)/d(V) = ra
    Fa(0) = 15
    d(Fb)/d(V) = rb
    Fb(0) = 15
    d(Fc)/d(V) = rc
    Fc(0) = 0
#Energy balance
    d(T)/d(V) = (Uw*(Tf−T)+ra*DH)/((Fa*Cpa)+(Fb*Cpb)+(Fc*Cpc))
    T(0) = 315
#Auxiliary equations
    ra = −kf*(Ca^2*Cb−(Cc/K))
    rb = ra/2
    rc = ra/−2
    kf = koo*exp(−Ea/(R*T))
    K=K1*EXP((−DH/R)*((1/T)−(1/T1)))
    Ca = Fa/v
```

Reactor Design

```
        Cb = Fb/v
        Cc = Fc/v
        v = v0*(Ft/Ft0)*(T/T0)
        Ft = Fa+Fb+Fc
        Ft0 = Fa0+Fb0+Fc0
        DH = Hc-(2*Ha+Hb)
        x = (Fa0-Fa)/Fa0
    #Constants and initial values
        Fa0 = 15
        Fb0 = 15
        Fc0 = 0
        T0 = 315
        T1 = 298
        Tf = 400
        Uw = 1,300
        R = 8.3143
        v0 = 1.1
        Ea = 3,000
        Ha = 4,500
        Hb = -3500
        Hc = 9,500
        koo = 0.0052
        K1 = 0.2
        Cpa = 120
        Cpb = 90
        Cpc = 150
    #Independent variable
        V(0) = 0
        V(f) = 25
```

We can then solve this system of two differential equations with the same number of unknowns, and then plot the variation of the molar flow for each chemical. Figure 2.61 shows the variation of the different flows. As expected, chemical A decreases two times as fast as chemical B, since A is consumed at a double rate. Figure 2.62 represents the variation in the conversion of the process as well as the variation on the temperature profile along the reactor. This plot shows that the conversion has reached a maximum amount and cannot be overtaken under the operational conditions hereby described.

Figure 2.61 shows that the flow of the reactants decays considerably and that A falls faster than B. It is also important to see that since this is an equilibrium reaction, the final concentration or molar flow for chemical A is not zero. This can also be seen in Figure 2.62 where the conversion is presented as a function of the volume, it can be easily seen that the final value reached is around 0.915.

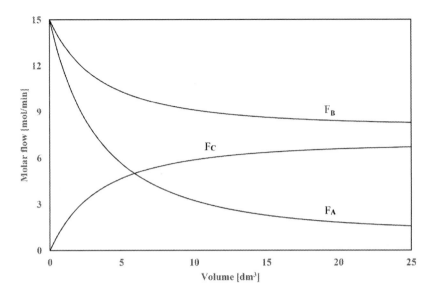

FIGURE 2.61 Profiles of the mole flow.

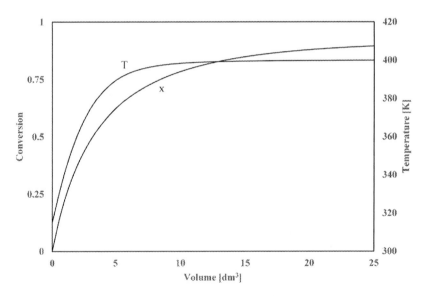

FIGURE 2.62 Profiles for reactor temperature and conversion.

Figure 2.62 also presents the temperature profile within the reactor. In this case, since the reaction is endothermic, energy must be added to the system for the reaction to take place. In this case, the amount of heat given into the system is greater than the amount required by the reaction and therefore, the temperature increases along the reactor.

Reactor Design

2.11.3 CO-CURRENT HEATING/COOLING FLUID SYSTEM

In several scenarios, it is not possible to have the process working adiabatically, even with a constant removal of heat. This can be due to the high temperature which can deteriorate the raw materials, the catalyst, produce deactivation, or eventually produce damage to the reactor material itself. In these cases, it is imperative to have a heating/cooling agent flowing outside of the reactor.

The system that we have now is similar to the one presented in Figure 2.60; however, the schematic representation is shown in Figure 2.63 where the constant supply/removal of heat is now dependent on the energy and flow of the external fluid "f".

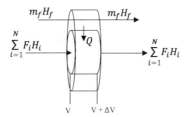

FIGURE 2.63 Differential plug flow representation for energy calculations.

The solution for the mole balance and the energy balance, inside the reactor is identical to the one carried out before; the only difference is that we now have a second energy balance in the external fluid. Therefore, looking at the external fluid, for the volume selected, the energy balance is:

$$\text{Energy in} - \text{Energy out} = \text{Accumulation}$$

In this case, for a steady state scenario:

$$m_f H_f \big|_V - m_f H_f \big|_{V+\Delta V} - Q = 0 \tag{2.252}$$

where $Q = U\varpi\,(T_f - T)\Delta V$ and reorganizing, we have:

$$\lim_{\Delta V \to 0} \frac{m_f H_f \big|_{V+\Delta V} - m_f H_f \big|_V}{\Delta V} = U\varpi\left(T - T_f\right) \tag{2.253}$$

This further gives:

$$\frac{d\left(m_f H_f\right)}{dV} = U\varpi\left(T - T_f\right) \tag{2.254}$$

which on expanding gives:

$$m_f \frac{d\left(H_f\right)}{dV} + H_f \frac{d\left(m_f\right)}{dV} = U\varpi\left(T - T_f\right) \tag{2.255}$$

While the flow, m_f, is constant, its derivative is then zero, and as done before, we know that:

$$\frac{dH_i}{dV} = C_{Pi}\frac{dT}{dV} \qquad (2.256)$$

Substituting 2.256 into 2.255, simplifying and rearranging:

$$\frac{dT_f}{dV} = \frac{U\varpi(T-T_f)}{m_f C_{Pf}} \qquad (2.257)$$

Equation 2.257 then needs to be solved simultaneously with the energy balance for the rector and the mole balance in order to obtain the conversion and reactor temperature profiles as well as the temperature profile for the external heating/cooling fluid.

2.11.4 Counter Current Heating/Cooling Fluid System

The derivation of the equation for this problem is similar to the one for the previous case; the expression will be exactly alike for heating as well as for cooling, but Equation 2.257 changes into:

$$\frac{dT_f}{dV} = -\frac{U\varpi(T-T_f)}{m_f C_{Pf}} \qquad (2.258)$$

The major difference is the procedure to solve it. Most software, including Polymath, require the user to provide an initial value to solve the differential equations; in the case of counter-current flow, the data for initial T_f is known at the end of the process (final value for the volume). Therefore, to solve it, the shooting approach needs to be implemented. We need to propose a final value for the temperature for the heating/cooling and solve the problem. When this is done, we can then compare the final temperature value of the simulation with the initial temperature value that we have, and when these two numbers are close enough, the process is finished. This approach is used in the problem presented in Chapter 4.

Example 2.14
We will now solve the same problem as in the previous example but using a co-current and a counter-current flow, with the same flow rate of m_f = 30 [mol/min], Cp_f = 180 [J/(mol* K)], all the remaining information is the same as in the previous example. Therefore, the script is as follows:

Reactor Design

Script:

```
#Mole balance
    d(Fa)/d(V) = ra
    Fa(0) = 15
    d(Fb)/d(V) = rb
    Fb(0) = 15
    d(Fc)/d(V) = rc
    Fc(0) = 0
#Energy balance
    d(T)/d(V) = (Uw*(Tf−T)+ra*DH)/((Fa*Cpa)+(Fb*Cpb)+(Fc*Cpc))
    T(0) = 315
    d(Tf)/d(V) = (Uw*(T−Tf))/(mf*Cpf)
    Tf(0) = 400
#Auxiliary equations
    ra = −kf*(Ca^2*Cb−(Cc/K))
    rb = ra/2
    rc = ra/−2
    kf = koo*exp(−Ea/(R*T))
    K = K1*EXP((−DH/R)*((1/T)-(1/T1)))
    Ca = Fa/v
    Cb = Fb/v
    Cc = Fc/v
    v = v0*(Ft/Ft0)*(T/T0)
    Ft = Fa+Fb+Fc
    Ft0 = Fa0+Fb0+Fc0
    DH = Hc−(2*Ha+Hb)
    x = (Fa0−Fa)/Fa0
#Constants and initial values
    Fa0 = 15
    Fb0 = 15
    Fc0 = 0
    T0 = 315
    T1 = 298
    Uw = 1,300
    R = 8.3143
    v0 = 1.1
    Ea = 3,000
    Ha = 4,500
```

Hb = −3,500
Hc = 9,500
koo = 0.0052
K1 = 0.2
Cpa = 120
Cpb = 90
Cpc = 150
Cpf = 180
mf = 30
#Independent variable
V(0) = 0
V(f) = 25

Solving this script, the plot shown in Figure 2.64 is obtained, presenting the variation of the flow for each component as a function of the volume. The tendency observed is similar to the previous case with a constant heat.

Figure 2.65 presents the variation in the reaction temperature as well as the variations on the external fluid's temperature and conversion. It can be seen that the conversion is very similar to the values obtained for a constant heat supply. The temperature of the external fluids decreases until it reaches the value of the reaction temperature and it stays there for the rest of the process, as expected.

When comparing Figures 2.62 and 2.65, it cannot be seen any significant difference when evaluating the conversion profiles. However, the reaction temperature shows a decrease from 400 [K] when having a constant fluids temperature, to 370 when having a co-current flow.

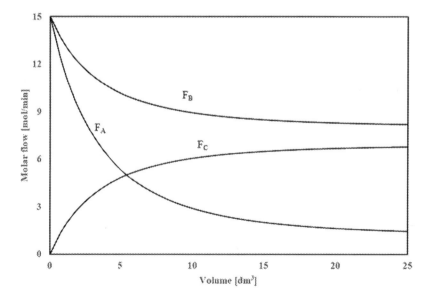

FIGURE 2.64 Molar flow as function of the reactor volume.

Reactor Design

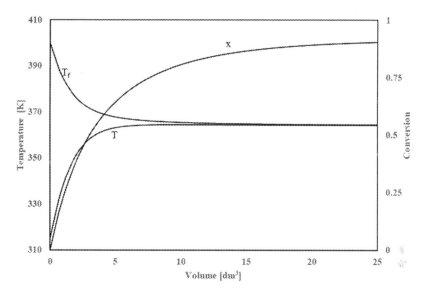

FIGURE 2.65 Variations on the conversion as well as the temperature for the reactor and the external warming fluid.

In order to see a clearer effect of the initial temperature of the external fluids, four different temperatures were tested, *i.e.*, 300, 400, 600, and 900 [K], respectively. The effect on the reaction temperature can be seen in Figure 2.66. It is clear that a temperature that is lower than the reaction temperature will produce a decrease in the reaction temperature due to the endothermicity of the process, making the final temperature not as high as expected. As T_f increases, the reaction temperature profile increases, reaching a higher final value.

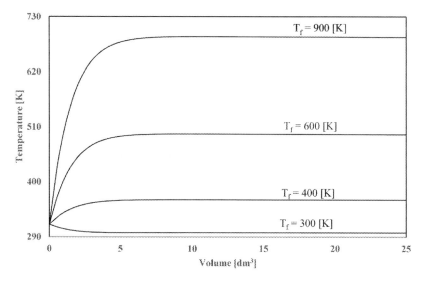

FIGURE 2.66 The effect of the initial temperature of the external fluid on the reaction temperature along the reactor volume.

Counter-current flow in heat exchangers have the benefit that the final temperature of the heating/cooling fluids can be higher than the final temperature of the reaction mixture, which was not the case for a heat exchanger with co-current flow.

Similarly, we will solve the same problem as in the previous example but with a counter-current flow, which requires trial and error approach and will be shown in Chapter 4.

Solving the system, we obtained Figures 2.67 and 2.68, showing the molar flows and the temperature profiles and the conversion are presented as functions of the volume, respectively.

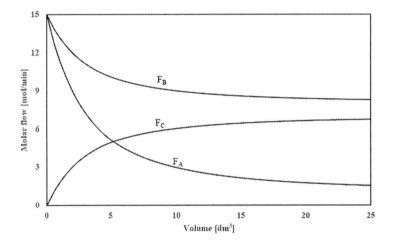

FIGURE 2.67 Profiles of the molar flow as a function of the volume.

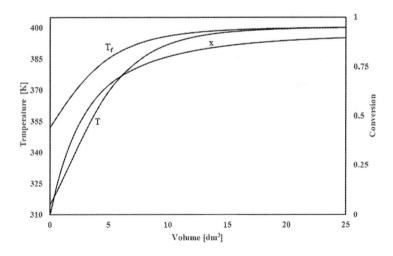

FIGURE 2.68 Conversion and temperatures profile as a function of the volume.

Figure 2.67 does not show big changes from the previous molar flow profiles; however, the temperature profile presented in Figure 2.68 is quite different due to the

Reactor Design

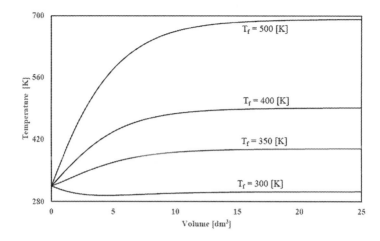

FIGURE 2.69 The initial temperature effect of the external fluids on the reaction temperature along the reactor volume.

fact that now the external fluid is flowing in the opposite direction. This is very important since now the ΔT between the two streams will be significant and therefore, there will be a strong driving force (Q) between them.

As for the previous case, it is important to see the effect of the initial temperature of the external fluid. Figure 2.69 shows the effect on the reaction temperature by varying the initial temperature of the external fluid for four different values. The larger the initial value, the higher the final temperature that can be reached within the reactor.

2.11.5 CSTR with Heat Transfer

So far, we have been discussing the plug flow systems; however, there are other systems that are in flow mode and the second ideal scenario for that is the CSTR reactor. We will present here the energy balance for a CSTR reactor (Figure 2.70) and solve this system for different case scenarios.

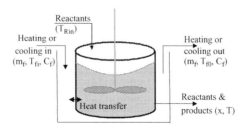

FIGURE 2.70 Schematic of CSTR with a heat exchanger system.

The energy equation (with neglected work of the shaft) to be solved simultaneously with the mole balance is:

$$0 = \dot{Q} - \sum_{i=1}^{P} F_i^0 C_{P_i}(T - T_{i0}) - F_A^0 x \left[\left(\Delta H_{RX}(T_R)\right) + \sum_{i=1}^{P} \frac{\upsilon_i}{|\upsilon_A|} C_{P_i}(T - T_R) \right] \quad (2.259)$$

In order to solve this equation for temperature (T), we need to know the value or equation for $\dot{Q}$. In the case of a CSTR, it is important to see that there is a transfer of heat among the fluids through the wall; therefore, we can analyze the energy that is transferred from the external fluids. This amount of energy must be equal to the energy that the reaction medium gets or gives, depending on heating or cooling, respectively.

If we look at the change in energy from the cooling/heating fluids surrounding the reactor, we know that there is energy that goes in and out, and energy that is transferred to the reaction medium. This can be summarized as:

$$\text{Energy in} - \text{Energy out} = 0 \quad (2.260)$$

where the *energy out* has two terms, the one due to the fluids moving out of the reactor and one due to the heat transfer to the reaction medium through the wall. Thus, Equation 2.260 can be expressed as:

$$m_f C_{p_f}(T_{fi} - T_{f0}) = \dot{Q} \quad (2.261)$$

We can then express $\dot{Q}$ as a function of temperature difference, similar to that done for a heat exchanger, using $\dot{Q} = UA\Delta T_{LMTD}$, where U is the overall heat transfer coefficient, A is the area for heat transfer, and ΔT_{LMTD} is the logarithmic mean temperature difference. In a heat exchanger, the temperature difference between hot and cold is used based on whether there is co-current or counter-current flow. This situation does not apply to the given problem, because the equation we will obtain is independent of the direction of flow. This is due to the fact that the inlet and outlet temperature of the reaction medium is the same, T. It is very common to mistake this with the temperature that the reactants have before entering the reactor. Since one of the ideal conditions in the CSTR model is uniform temperature and concentration, which are equal to those at the exit, therefore, T is the same in all parts of the reactor. With this in mind, we can rewrite the equation for the heat using:

$$\dot{Q} = UA \frac{(T_{fi} - T_{f0})}{\ln\left(\dfrac{T_{fi} - T}{T_{f0} - T}\right)} \quad (2.262)$$

Substituting Equation 2.262 into 2.261:

$$m_f C_{p_f}(T_{fi} - T_{f0}) = UA \frac{(T_{fi} - T_{f0})}{\ln\left(\dfrac{T_{fi} - T}{T_{f0} - T}\right)} \quad (2.263)$$

Reactor Design

Solving for T_{f0}:

$$T_{f0} = T + (T_{fi} - T) e^{\left(-\frac{UA}{m_f C_{pf}}\right)} \qquad (2.264)$$

Equation 2.264 can then be substituted into Equation 2.261 to obtain:

$$m_f C_{pf} \left[T_{fi} - \left(T + (T_{fi} - T) e^{\left(-\frac{UA}{m_f C_{pf}}\right)} \right) \right] = \dot{Q} \qquad (2.265)$$

Arranging it, we have:

$$m_f C_{pf} (T_{fi} - T) \left(1 - e^{\left(-\frac{UA}{m_f C_{pf}}\right)} \right) = \dot{Q} \qquad (2.266)$$

Equation 2.266 will then be used for the determination of the heat as a function of temperature that can be then substituted into Equation 2.259 (energy balance).

Note: A simplification of this process can be done if we have a high flow (m_f) and therefore, the inlet and outlet temperatures of the external fluids are the same. In that case, we can rewrite as:

$$Q = UA(T_f - T) \qquad (2.267)$$

which will lead to a simpler equation for the temperature effects.

Equation 2.259 can then be represented in a generic form as:

$$0 = m_f C_{pf} (T_{fi} - T) \left(1 - e^{\left(-\frac{UA}{m_f C_{pf}}\right)} \right) - \sum_{i=1}^{P} F_i^0 C_{P_i} (T - T_{i0})$$
$$- F_A^0 x \left[\left(\Sigma \Delta H_{RX} (T_R) \right) + \sum_{i=1}^{P} \frac{\upsilon_i}{|\upsilon_A|} C_{P_i} (T - T_R) \right] \qquad (2.268)$$

When using Equation 2.267 for the heat transfer, Equation 2.259 takes the form:

$$0 = UA(T_f - T) - \sum_{i=1}^{P} F_i^0 C_{P_i} (T - T_{i0})$$
$$- F_A^0 x \left[\left(\Delta H_{RX} (T_R) \right) + \sum_{i=1}^{P} \frac{\upsilon_i}{|\upsilon_A|} C_{P_i} (T - T_R) \right] \qquad (2.269)$$

Example 2.15

The reaction in liquid phase of 3 moles of chemical A and 2 moles of chemical B to produce 2 moles of C and 1 mole of D takes place non-isothermally in a CSTR reactor. The reaction is an elementary reaction with stoichiometric feed. The initial concentration of chemical A is 60 [moles/l] and the volumetric flow is of 2 [l/min]. In order to heat up the process, an external fluid was used with an inlet temperature of 565 [K]. Plot the variation of conversion as a function of temperature, both, from the mole and the energy balance, and calculate the conversion achieved and the temperature reached when the reactor has a volume of 3 [m³].

Additional information:

$Cp_A = 150$ [J/(mol*K)]; $Cp_B = 190$ [J/(mol*K)]; $Cp_C = 140$ [J/(mol*K)];
$Cp_D = [140$ J/(mol*K)]; $U = 1,000$ [W/(m²*K)]; $A = 20$ [m²];
$H_A = 4,500$ [J/mol]; $H_B = 6,000$ [J/mol]; $H_C = -9,500$ [J/mol]; $H_D = -5,500$ [J/mol];
$T_r = 298$ [K]; $T_0 = 298$ [K]; and $k = 0.52 \exp\left(\dfrac{-3,000}{8.3143 T}\right)$ [l⁴/(min*mol⁴)].

Solution

To solve this problem, we need to solve the mole balance and the energy balance simultaneously. An easy approach to obtain this solution is to plot the different balances and to see when they cross each other.

We will start with a simplified version of the problem, where then the mole balance (the same for both cases) will be:

$$V = \frac{F_A^0 - F_A}{-r_A} = \frac{F_A^0 - \left(F_A^0 - F_A^0 x\right)}{k_1 C_A^3 C_B^2} = \frac{C_A^0 v x}{k_{00} e^{\left(\frac{-E_A}{RT}\right)} \left(C_A^0 - C_A^0 x\right)^3 \left(C_B^0 - \frac{2}{3} C_A^0 x\right)^2} \quad (2.270)$$

This can be rearranged as:

$$T = \frac{-E_A}{R} \frac{1}{\ln\left(\dfrac{C_A^0 v x}{V k_{00} \left(C_A^0 - C_A^0 x\right)^3 \left(C_B^0 - \dfrac{2}{3} C_A^0 x\right)^2}\right)} \quad (2.271)$$

From Equation 2.271, we can then plot T versus x.

The simplified energy balance, Equation 2.269, can be reorganized as follows:

$$T = \frac{UAT_f + F_A^0 C_{P_A}(T_0) + F_B^0 C_{P_B}(T_0) - F_A^0 x \Delta H_{RX}(T_R) + T_R \left(\dfrac{v_C}{|v_A|} C_{P_C} + \dfrac{v_D}{|v_A|} C_{P_D} + \dfrac{v_B}{|v_A|} C_{P_B} + \dfrac{v_A}{|v_A|} C_{P_A}\right)}{UA + F_A^0 C_{P_A} + F_B^0 C_{P_B} + \dfrac{v_C}{|v_A|} C_{P_C} + \dfrac{v_D}{|v_A|} C_{P_D} + \dfrac{v_B}{|v_A|} C_{P_B} + \dfrac{v_A}{|v_A|} C_{P_A}} \quad (2.272)$$

To solve this problem, Equations 2.271 and 2.272 must be satisfied at the same time. But it is not so simple to substitute one in the other; therefore, this might require either the use of a mathematical software that can do this analysis or the

Reactor Design

use of a table in, for example, Excel, for a trial and error methodology. The trial and error methodology (Table 2.9) as well as the graphical interpretation are provided here. We will try for several values of conversion and calculate the temperature based on the mole balance and energy balance. When both T obtained are close in value, that will be the solution.

TABLE 2.9
Trial and Error Solution

x	T [K] (Eq. 2.262)	T [K] (Eq. 2.263)
0.1	222.74	409.94
0.2	235.15	421.24
0.3	246.65	432.55
0.4	259.18	443.86
0.5	274.04	455.17
0.6	293.31	466.47
0.7	321.02	477.78
0.8	368.21	489.09
0.9	486.92	500.40
0.91	511.68	501.53
0.906	501.17	501.09

Table 2.9 shows the variation of the temperature for both equations as a function of conversion (at least numerically); this allows us to see that the answer is closer to $x = 0.906$ than to $x = 0.91$. When conversion = 0.906, the difference among the two temperatures is considered small enough and the result is accepted as correct. Based on the information in Table 2.9, we can then represent this information in Figure 2.71. It is clear that the solution is the intersection of both plots, given the final conversion and the final temperature.

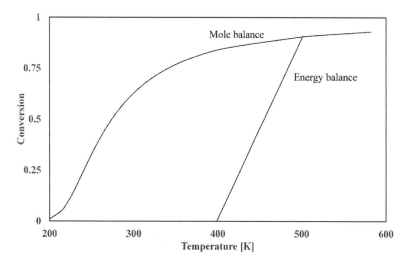

FIGURE 2.71 Graphical solution using plot of the variation of conversion with temperature.

As mentioned before, this solution was carried out using a simplified approach, which is quite accurate and reliable. But in order to know how different this is from the complete solution, we will now solve the same problem, Equation 2.268, for the energy balance. The mole balance remains the same, following Equation 2.270 and therefore, Equation 2.271.

From Equation 2.268, we can explicitly obtain T as a function of conversion.

$$T = \frac{m_f C_{pf} T_{fi}\left(1 - e^{\left(-\frac{UA}{m_f C_{pf}}\right)}\right) + F_A^0 C_{P_A} T_0 + F_B^0 C_{P_B} T_0 - F_A^0 x \Delta H_{RX}(T_R) + T_R\left(\frac{\upsilon_C}{|\upsilon_A|} C_{P_C} + \frac{\upsilon_D}{|\upsilon_A|} C_{P_D} + \frac{\upsilon_B}{|\upsilon_A|} C_{P_B} + \frac{\upsilon_A}{|\upsilon_A|} C_{P_A}\right)}{m_f C_{pf}\left(1 - e^{\left(-\frac{UA}{m_f C_{pf}}\right)}\right) + F_A^0 C_{P_A} + F_B^0 C_{P_B} + \left(\frac{\upsilon_C}{|\upsilon_A|} C_{P_C} + \frac{\upsilon_D}{|\upsilon_A|} C_{P_D} + \frac{\upsilon_B}{|\upsilon_A|} C_{P_B} + \frac{\upsilon_A}{|\upsilon_A|} C_{P_A}\right)}$$

(2.273)

The major difference from the previous scenario is that in this case, we have the flow of the external fluid as a part of the equation, as well as its C_p values. This means that in order to obtain similar results with the previous case, the flow needs to be extremely large. Therefore, we will solve this problem for six different flows. We will use a similar approach as in the previous part of the example, but we will only show the final values. Table 2.10 shows these results, *i.e.*, as the value of m_f increases, the result obtained is closer to that obtained by the simplified method.

TABLE 2.10
Solution for Different Flows

m_f [kg/min]	T [K] (Eq. 2.262)	x
20	274	0.5
100	397	0.837
1,000	486	0.9
10,000	499	0.9055
100,000	500	0.90555
1,000,000	501	0.906

2.12 NON-ISOTHERMAL, NON-STEADY STATE REACTORS

In the previous section, we worked with steady state, non-isothermal ideal flow reactors; we will now look at the non-steady state non-isothermal system; this can be applied for flow in the start-up or shut-down part of the operation but it is more common to be used with the batch and semi-batch system since those are by definition, a time-dependent process.

Reactor Design

2.12.1 BATCH REACTOR

For the batch reactor system, we have obtained the mole balance:

$$\frac{dN_i}{dt} = -\upsilon_i r_A V + F_i^0 - F_i \tag{2.274}$$

where $r_A = f(T, x)$ and $F_i^0 = F_i = 0$. Therefore, if the process is non-isothermal, we need an expression to quantify the variation of temperature with time. For that, we need to take a look at the energy balance (Equation 2.209), and since it is considered that there is no flow in or out, we have:

$$\frac{dE_{sys}}{dt} = \dot{Q} - W_s \tag{2.275}$$

E_{sys} is the total energy of the system; that energy can be divided as the contribution of the different components in the system as $E_{sys} = \sum_{i=1}^{n} N_i E_i$. Taking into consideration that there is no significant change in the kinetics neither the potential energy, E_i is mainly a contribution of the internal energy, U_i, which in terms of enthalpy gives:

$$\frac{dE_{sys}}{dt} = \frac{d\sum_{i=1}^{n} N_i(H_i - PV_i)}{dt} = \dot{Q} - W_s \tag{2.276}$$

Reorganizing:

$$\frac{d\sum_{i=1}^{n} N_i H_i}{dt} - \frac{Pd\sum_{i=1}^{n} N_i V_i}{dt} = \dot{Q} - W_s \tag{2.277}$$

Taking into consideration that $\sum_{i=1}^{n} N_i V_i = V = constant$, and the derivative of a constant is zero, Equation 2.277 can be rearranged as:

$$\frac{\sum_{i=1}^{n} N_i dH_i}{dt} + \frac{\sum_{i=1}^{n} H_i dN_i}{dt} = \dot{Q} - W_s \tag{2.278}$$

Substituting Equation 2.274:

$$\sum_{i=1}^{n} N_i \frac{dH_i}{dt} + \sum_{i=1}^{n} H_i V r_i = \dot{Q} - W_s \tag{2.279}$$

We also know that $\frac{r_i}{\nu_i} = \frac{r_A}{\nu_A}$, if the reaction is then normalized to chemical A, this can be expressed as $r_i = -\nu_i r_A$ which can then be substituted into 2.279. In addition, $\frac{dH_i}{dt} = C_{P_i}\frac{dT}{dt}$ as presented before. This will allow us to write expression 2.279 as:

$$\sum_{i=1}^{n} N_i C_{P_i} \frac{dT}{dt} + \sum_{i=1}^{n} H_i V - \nu_i r_A = \dot{Q} - W_s \tag{2.280}$$

We can reorganize them, as done for previous cases, and assume that the work done by the shaft can be neglected:

$$\frac{dT}{dt}\left(\sum_{i=1}^{n} N_i C_{P_i}\right) + V(-r_A)\sum_{i=1}^{n} \nu_i H_i = \dot{Q} \tag{2.281}$$

Equation 2.281 can be rearranged to give a more explicit equation for the variation of temperature with time, using the definition that $\sum_{i=1}^{n} \nu_i H_i = \Delta H_{RX}$:

$$\frac{dT}{dt} = \frac{\dot{Q} - V(-r_A)\Delta H_{RX}}{\sum_{i=1}^{n} N_i C_{P_i}} \tag{2.282}$$

We can now solve Equations 2.282 and 2.274 to solve the time dependency of the conversion and temperature for a batch reactor process.

2.12.2 Semi-Batch Reactor

We can now expand this analysis for a semi-batch system; as previously mentioned in a semi-batch system, we have a flow in but no flow out. Therefore, Equation 2.275 needs to be expanded:

$$\frac{dE_{sys}}{dt} = \dot{Q} - W_s + \sum_{i=1}^{n} \dot{F}_i^0 H_i. \tag{2.283}$$

We applied the same assumptions for the energy term as for the batch reactor, and then we can reexpress the energy of the system in terms of the internal energy (neglecting the potential and kinetic energy):

Reactor Design

$$\frac{dE_{sys}}{dt} = \frac{d\sum_{i=1}^{n} N_i(H_i - PV_i)}{dt} = \dot{Q} - W_s + \sum_{i=1}^{P} F_i H_i \Big|_{in} \tag{2.284}$$

Rearranging Equation 2.284, neglecting the work of the shaft as well as knowing that $\dfrac{r_i}{\nu_i} = \dfrac{r_A}{\nu_A}$, if the reaction is then normalized to chemical A, this can be expressed as $r_i = -\nu_i r_A$ and $\dfrac{dH_i}{dt} = C_{P_i} \dfrac{dT}{dt}$. Considering Equation 2.274 and that $F_i = 0$, Equation 2.284 becomes:

$$\frac{dT}{dt} = \frac{\dot{Q} - V(-r_A)\Delta H_{RX} - \sum_{i=1}^{n} F_i^0 C_{pi}(T - T_i^0)}{\sum_{i=1}^{n} N_i C_{P_i}} \tag{2.285}$$

It is important to point out that this expression, obtained for a semi-batch with flow into the system, is the same that will be obtained and used if we do the general energy balance for a CSTR.

Example 2.16
The reversible elementary reaction of chemicals A and B into C takes place in a semi-batch adiabatic reactor, in liquid phase. A second-order reaction takes place for chemical A and a first-order reaction takes place for chemicals B and C.

A is the only liquid in the reactor at the beginning of the process and only B is added to the reaction system with a flow of $v_0 = 0.1$ [mol/min]. The concentration of A in the reactor is 1.5 [mol/l] while the concentration of B at the inlet flow is 0.8 [mol/l]. The initial volume of the reactor is 5 [l].

The process is adiabatic, $Q = 0$, and the enthalpy of reaction can be estimated as the reference enthalpy of reaction (the decrease/increase of the enthalpy at a different temperature can be neglected).

Additional information:
$Cp_A = 140$ [J/(mol*K)]; $Cp_B = 120$ [J/(mol*K)]; $Cp_C = 180$ [J/(mol*K)];
$H_A = 4,500$ [J/mol]; $H_B = 3,500$ [J/mol]; $H_C = -5,500$ [J/mol]; $T_B^0 = 300$ [K];

$T_0 = 300$ [K]; $k_1 = 1\exp\left(\dfrac{-3,300}{8.3143T}\right)$ [l²/(min*mol²)]; $k_2 = 5\exp\left(\dfrac{-4,500}{8.3143T}\right)$ [1/min].

Plot the variation of concentration as a function of time as well as the conversion and temperature profile.

Solution
In order to solve this problem, we need to take a look at the semi-batch reactor as we presented it before, but adding the energy balance. Let's start first with the

reaction rate. It is clearly a reversible reaction, and we have the reaction order for each component and some of the initial concentrations. Based on this, we can write:

$$r_A = -k_1 C_A^2 C_B + k_2 C_C \quad (2.286)$$

with $k_1 = 0.01 \exp\left(\dfrac{-3,300}{8.3143T}\right)$ [l²/(min*mol²)] and

$k_2 = 0.005 \exp\left(\dfrac{-2,500}{8.3143T}\right)$ [1/min].

For each component, we will need to solve the mole balance for a batch reactor, except for component B that is also being fed into the system. Doing an analogous analysis as with the semi-batch in Section 2.12.2, we obtain:

$$\frac{dC_A}{dt} = r_A - \frac{v^0}{V} C_A \quad (2.287)$$

$$\frac{dC_B}{dt} = r_B + \frac{v^0}{V}\left(C_B^0 - C_B\right) \quad (2.288)$$

$$\frac{dC_C}{dt} = r_C - \frac{v^0}{V} C_C \quad (2.289)$$

In order to obtain the reaction rate, we use the relation:

$$\frac{r_A}{-2} = \frac{r_B}{-1} = \frac{r_C}{1} \quad (2.290)$$

Since the reaction is running adiabatically, we need the energy balance accordingly. From Equation 2.285, we can set $Q = 0$ and obtain:

$$\frac{dT}{dt} = \frac{V(-r_A)(-\Delta H_{RX}) - \sum_{i=1}^{n} F_i^0 C_{pi}\left(T - T_i^0\right)}{\sum_{i=1}^{n} N_i C_{P_i}} \quad (2.291)$$

which can be expanded for our case into:

$$\frac{dT}{dt} = \frac{V(-r_A)(-\Delta H_{RX}) - F_B^0 C_{pB}\left(T - T_B^0\right)}{N_A C_{P_A} + N_B C_{P_B} + N_C C_{P_C}} \quad (2.292)$$

We can then put all these expressions, plus the additional information, into Polymath to obtain the following script:

Script:

```
#Mole balance
d(Ca)/d(t) = ra-(v0*Ca/V)
Ca(0) = 1.5
```

Reactor Design

$d(Cb)/d(t) = rb+((v0/V)*(Cb0-Cb))$
$Cb(0) = 0$
$d(Cc)/d(t) = rc-(v0*Cc/V)$
$Cc(0) = 0$
#Energy balance
$d(T)/d(t) = (Q-(Fb0*Cpb*(T-Tb0))-((DHrx)*(-ra*V)))/((Na*Cpa)+(Nb*Cpb)+(Nc*Cpc))$
$T(0) = 300$
#Volume balance
$d(V)/d(t) = v0$
$V(0) = 5$
#Auxiliary equations
$rb = (1/2)*ra$
$rc = (-1/2)*ra$
$ra = -k1*Ca\wedge2*Cb+k2*Cc$
$k1 = k100*EXP(-Ea1/(R*T))$
$k2 = k200*EXP(-Ea2/(R*T))$
$Fb0 = Cb0*v0$
$DHrx = Hc0-(2*Ha0+Hb0)$
$x = (Ca0-Ca)/Ca0$
#Constants and initial values
$Cb0 = 0.8$
$Ca0 = 1.5$
$v0 = 0.1$
$k100 = 1$
$k200 = 5$
$Ea1 = 3,300$
$Ea2 = 4,500$
$R = 8.3143$
$Tb0 = 300$
$Cpa = 140$
$Cpb = 120$
$Cpc = 180$
$Na = Ca*V$
$Nb = Cb*V$
$Nc = Cc*V$
$Hc0 = -5,500$
$Ha0 = 4,500$
$Hb0 = 3,500$
$Q = 0$

#Independent variable
t(0) = 0
t(f) = 500

By solving the script, we can plot the variation of the concentrations, temperature, and conversion as a function of time.

Figure 2.72 shows the variation of the concentrations of A, B, and C and shows that while chemical A is constantly decreasing, chemical B has a maximum. This

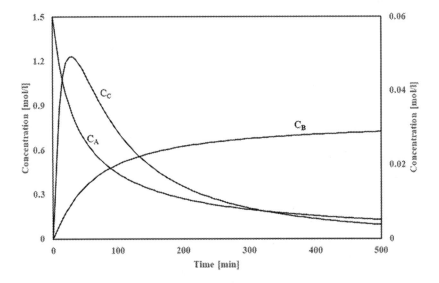

FIGURE 2.72 Concentration profile for an adiabatic semi-batch reactor.

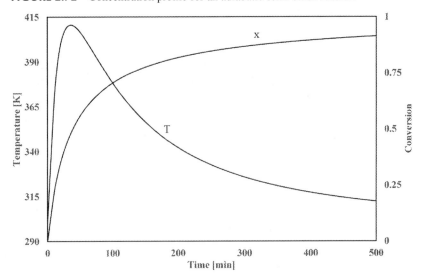

FIGURE 2.73 Temperature and conversion profiles for an adiabatic semi-batch reactor.

Reactor Design

is due to the fact that the amount of B increases due to the semi-batch mode and it is consumed due to the reaction itself leading to an equilibrium situation. The opposite behavior is seen for chemical C, *i.e.*, its amount increases until reaching the equilibrium state. This can be corroborated with the plot of conversion in Figure 2.73, that after some time it reaches the equilibrium conversion. The temperature profile increases as expected due to this reaction being exothermic.

The variations in the volume are not plotted, but as expected, they are linear as a function of time. This can be very useful if we need to know when the reactor reaches 90% of the total volume and therefore, the input of chemicals must be stopped so as to not cause an overflow (or increase the pressure) too much as that could destroy the equipment.

2.13 MULTIPLE REACTION SYSTEM IN A PLUG FLOW REACTOR

The multiple reactions in Sections 2.9 have been presented only from a mole balance perspective. We will present here how the energy balance is modified due to the presence of more than one reaction, and eventually how the non-steady-state balance equation is also different when more than one reaction takes place simultaneously.

The energy balance for multiple reactions can be developed in an analogous way to that one for one reaction, leading to the following:

$$\frac{dT}{dV} = \frac{\sum_{i=1}^{m}(-r_{ij})(-\Delta H_{RXij}(T)) - U\varpi(T - T_f)}{\sum_{j=1}^{n} F_j C_{P_j}} \quad (2.293)$$

where we have *m* reactions taking place simultaneous and *n* species involved.

This expression, together with the general mole balance, can be solved to obtain the steady-state solution for a multireaction system. This system can have reactions in parallel, series, or a combination of them.

Example 2.17
The following equilibrium gas reactions take place in a PFR:

$$2A + B \leftrightarrows C + D \quad (2.294)$$

$$B \leftrightarrows 2E + F \quad (2.295)$$

$$A + F \leftrightarrows 2G \quad (2.296)$$

As shown in Equations 2.294 to 2.296, we have a complex system of reactions in parallel and in series. Therefore, we need to solve this problem all at once. Chemicals A and B are fed into the reactor and the reaction is carried out in a tube of 25 l of volume. The initial flows are $F_A^0 = 15$ [mol/min], $F_B^0 = 8$ [mol/min], and the total initial concentration is $C_{T0} = 15$ [mol/l].

Plot the variations of all the flows as a function of the volume as well as the temperature profile, along the volume of the reactor and discuss the results.

Additional information:

Cp_A = 80 [J/(mol*K)]; Cp_B = 95 [J/(mol*K)]; Cp_C = 85 [J/(mol*K)];
Cp_D = 92 [J/(mol*K)]; Cp_E = 120 [J/(mol*K)]; Cp_F = 132 [J/(mol*K)];
Cp_G = 99 [J/(mol*K)]; ΔH_{1A} = −2,500 [J/mol]; ΔH_{2B} = 1,500 [J/mol];
ΔH_{3A} = −2,000 [J/mol]; $U\varpi$ = 15 [J/(min*K)]; T_0 = 450 [K]; T_f = 400 [K];

$$k_1 = 0.12 \exp\left(\frac{-2{,}000}{8.3143T}\right) \; [l^2/(min*mol^2)];$$

$$k_2 = 0.14 \exp\left(\frac{-1{,}500}{8.3143T}\right) \; [l/(mol*min)];$$

$$k_3 = 0.15 \exp\left(\frac{-1{,}300}{8.3143T}\right) \; [1/min];$$

$$k_4 = 0.014 \exp\left(\frac{-1{,}100}{8.3143T}\right) \; [l^2/(min*mol^2)];$$

$$k_5 = 0.12 \exp\left(\frac{-1{,}400}{8.3143T}\right) \; [l/(min*mol)];$$

$$k_6 = 0.09 \exp\left(\frac{-1{,}100}{8.3143T}\right) \; [l/(min*mol)].$$

Solution

First, the reaction rates for each of the three above-mentioned reactions will be deduced and then they will be used to obtain the reaction rate for each of the components involved. For this, we will set reactant A as the main chemical for Reactions 2.294 and 2.296 and B for Reaction 2.295.

Therefore, we can express the following:

$$r_{1A} = -k_1 C_A^2 C_B + k_2 C_C C_D \tag{2.297}$$

$$r_{2B} = -k_3 C_B + k_4 C_E^2 C_F \tag{2.298}$$

$$r_{3A} = -k_5 C_A C_F + k_6 C_G^2 \tag{2.299}$$

Now, based on these expressions, we have:

$$\frac{r_{1A}}{-2} = \frac{r_{1B}}{-1} = \frac{r_{1C}}{1} = \frac{r_{1D}}{1} \tag{2.300}$$

$$\frac{r_{2B}}{-1} = \frac{r_{2E}}{2} = \frac{r_{2F}}{1} \tag{2.301}$$

$$\frac{r_{3A}}{-1} = \frac{r_{3F}}{-1} = \frac{r_{3G}}{2} \tag{2.302}$$

$$\frac{dF_i}{dV} = r_i \tag{2.303}$$

Reactor Design

We can then write the following balances:

$$\frac{dF_A}{dV} = r_A = r_{1A} + r_{3A} \tag{2.304}$$

$$\frac{dF_B}{dV} = r_B = \frac{1}{2}r_{1A} + r_{2B} \tag{2.305}$$

$$\frac{dF_C}{dV} = r_C = -\frac{1}{2}r_{1A} \tag{2.306}$$

$$\frac{dF_D}{dV} = r_D = -\frac{1}{2}r_{1A} \tag{2.307}$$

$$\frac{dF_E}{dV} = r_E = -2r_{2B} \tag{2.308}$$

$$\frac{dF_F}{dV} = r_F = -r_{2B} + r_{3A} \tag{2.309}$$

$$\frac{dF_G}{dV} = r_G = -2r_{3A} \tag{2.310}$$

We can then substitute Equations 2.297, 2.298, and 2.299 into Equations 2.304 to 2.310, respectively, and accordingly obtain the mole balances. However, these equations (see script for all of them, we will present only one here) are functions of concentration and flow of the different components:

$$\frac{dF_A}{dV} = r_A = r_{1A} + r_{3A} = -k_1 C_A^2 C_B + k_2 C_C C_D - k_5 C_A C_F + k_6 C_G^2 \tag{2.311}$$

Therefore, we need to express the concentration as a function of the flows. As it was presented in Section 2.7 for a gas phase reaction, if pressure is constant, we can then write:

$$C_i = C_{T0}\left(\frac{F_i}{F_T}\right)\left(\frac{T_0}{T}\right) \tag{2.312}$$

This can be substituted for all the concentrations, and using Arrhenius expression, we can obtain the mole balance for all components (see script). As an example, only chemical A is shown here:

$$\frac{dF_A}{dV} = -k_{100}e^{\left(\frac{-E_{a1}}{RT}\right)}\left(C_{T0}\left(\frac{F_A}{F_T}\right)\left(\frac{T_0}{T}\right)\right)^2\left(C_{T0}\left(\frac{F_B}{F_T}\right)\left(\frac{T_0}{T}\right)\right)$$
$$+k_{200}e^{\left(\frac{-E_{a2}}{RT}\right)}\left(C_{T0}\left(\frac{F_C}{F_T}\right)\left(\frac{T_0}{T}\right)\right)\left(C_{T0}\left(\frac{F_D}{F_T}\right)\left(\frac{T_0}{T}\right)\right)$$
$$-k_{500}e^{\left(\frac{-E_{a5}}{RT}\right)}\left(C_{T0}\left(\frac{F_A}{F_T}\right)\left(\frac{T_0}{T}\right)\right)\left(C_{T0}\left(\frac{F_F}{F_T}\right)\left(\frac{T_0}{T}\right)\right) \quad (2.313)$$
$$+k_{600}e^{\left(\frac{-E_{a6}}{RT}\right)}\left(C_{T0}\left(\frac{F_F}{F_T}\right)\left(\frac{T_0}{T}\right)\right)^2$$

So far, we have managed to present the mole balance of the problem; we can solve this problem in different ways, isothermally ($T = T_0 =$ constant), adiabatically ($Q = 0$), with an external fluid at constant temperature ($T_f =$ constant), or with a fluid with variable temperature profile.

We will solve in this case for the scenario when $T_f =$ constant. Therefore, Equation 2.293 needs to be expressed for this particular problem; in this case, the energy balance is:

$$\frac{dT}{dV} = \frac{\left[(-r_{1A})(-\Delta H_{RX1A}(T)) + (-r_{2B})(-\Delta H_{RX2B}(T)) + (-r_{3A})(-\Delta H_{RX3A}(T))\right] - U\varpi(T - T_f)}{(F_A C_{PA} + F_B C_{PB} + F_C C_{PC} + F_D C_{PD} + F_E C_{PE} + F_F C_{PF} + F_G C_{PG})} \quad (2.314)$$

With this expression, the problem can be solved and in order to do this, we need the additional data.

Script:

#Mole balance
```
    d(Fa)/d(V) = ra
    Fa(0) = 15
    d(Fb)/d(V) = rb
    Fb(0) = 8
    d(Fc)/d(V) = rc
    Fc(0) = 0
    d(Fd)/d(V) = rd
    Fd(0) = 0
    d(Fe)/d(V) = re
    Fe(0) = 0
    d(Ff)/d(V) = rf
    Ff(0) = 0
    d(Fg)/d(V) = rg
```

Reactor Design

Fg(0) = 0

#Energy balance

d(T)/d(V) = (((−r1a)*(25,000))+((−r2b)*(15,000))+((−r3a)*(−20,000))
−(15*(T−400)))/((Fa*Cpa)+(Fb*Cpb)+(Fc*Cpc)+(Fd*Cpd)+(Fe*Cpe)
+(Ff*Cpf)+(Fg*Cpg))

T(0) = 450

#Auxiliary equations

ra = r1a+r3a

rb = 1/2*r1a+r2b

rc = −1/2*r1a

rd = −1/2*r1a

re = −2*r2b

rf = −r2b+r3a

rg = −2*r3a

r1a = −k1*Ca^2*Cb+k2*Cc*Cd

r2b = −k3*Cb+k4*Ce^2*Cf

r3a = −k5*Ca*Cf+k6*Cg^2

k1 = 0.12*exp(−2000/(R*T))

k2 = 0.14*exp(−1500/(R*T))

k3 = 0.15*exp(−1300/(R*T))

k4 = 0.14*exp(−1100/(R*T))

k5 = 0.12*exp(−1400/(R*T))

k6 = 0.09*exp(−1100/(R*T))

Ca = Cto*(Fa/FT)*(450/T)

Cb = Cto*(Fb/FT)*(450/T)

Cc = Cto*(Fc/FT)*(450/T)

Cd = Cto*(Fd/FT)*(450/T)

Ce = Cto*(Fe/FT)*(450/T)

Cf = Cto*(Ff/FT)*(450/T)

Cg = Cto*(Fg/FT)*(450/T)

FT = Fa+Fb+Fc+Fd+Fe+Ff+Fg

#Constants and initial values

R = 8.3143

Cto = 15

Cpa = 80

Cpb = 95

Cpc = 85

Cpd = 92

Cpe = 120
Cpf = 132
Cpg = 99
#Independent variable
V(0) = 0
V(f) = 25

Solving the script, we can obtain the desired results. Figure 2.74 shows the variation of all the flows as a function of the volume of the reactor. If the reactor is a cylinder, then it can also be presented as a function of the length of the tube by knowing the diameter of the equipment.

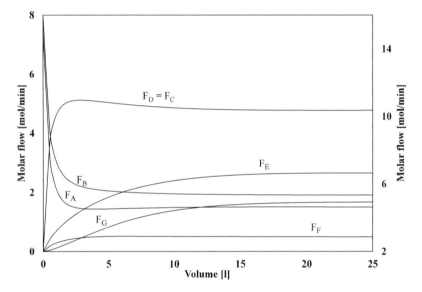

FIGURE 2.74 Variations of the molar flow as a function of volume. Molar flow of chemical A must be read on the right axis, and for all other chemicals, refer to the axis on the left.

Special attention should be given to the flows of C and D. Within Figure 2.74, no difference among the tendencies of these two flows is observed. This is because both of them behave exactly the same way, and therefore, their graphical presentation is overlapping. Additionally, since chemical B is also being consumed in reaction 2, its profile decreases faster than that one of A. However, C and D present a maximum in their curve, *i.e.*, they convert back into raw materials due to the high demand of chemical B from the second reaction, in order to reach a global equilibrium of all the processes. The profile for F could also have a maximum since it is being produced in reaction 2 and consumed in reaction 3. It does not show this tendency based on the physicochemical parameters of the problem, but it can potentially have this tendency. Chemical G shows a tendency that is increasing with a very small slope, like a delay time. This is because G will only be produced after chemical F is generated; therefore, if reactions 1 and 2 are slow, then G will need more time before it can be produced.

Figure 2.75 presents the variation of temperature as a function of the volume. It can be seen from the data that reactions 1 and 2 are exothermic and reaction 3 is endothermic. Based on the order of magnitude of these values, it can be predicted that the system will have an exothermic behavior. Figure 2.75 confirms this observation; a significant increase of the temperature at the beginning of the reaction, is observed. Due to the presence of an external fluid at a constant and lower temperature (50 K below the entrance temperature of the raw materials), heat is constantly removed from the system; the higher the difference between the reaction temperature and the fluid, the more energy removed. As the reactions move forward in the reactor, less and less reactants are available, and the reaction approaches closer to the equilibrium. When the equilibrium is reached, no reaction takes place and therefore, no heat is being produced. The removal of heat, that is constant, will continue and eventually (if the reactor is long enough), it will make the unreacted raw materials and products reach a final temperature of 400 [K]. The variation of the heat being removed is also presented in Figure 2.75.

In Figure 2.75, the *y*-axis has been adapted in such a way that the tendency of the removed heat can be seen, and it is clear that it follows the same tendency as the temperature. When looking at the units of the removed heat, it is important to mention that we are plotting the term $U\varpi(T - T_f)$ alone.

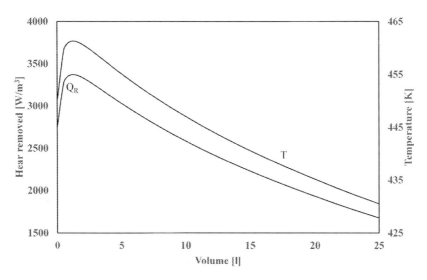

FIGURE 2.75 The trend of the heat being removed from the reactor and reaction temperature as a function of volume.

2.14 MULTIPLE REACTIONS IN A BATCH OR SEMI-BATCH REACTOR

So far, we looked at the impact of temperature in a plug flow, in steady state, for multiple reactions. We will now look at the temperature effects in a batch or semi-batch system for multiple reactions.

Similarly, as we did for the previous section, the general energy equation can be derived for multiple reactions in a similar manner as for one reaction. To refresh the memory, for one reaction, we have:

$$\frac{dT}{dt} = \frac{\dot{Q} - V(-r_A)\Delta H_{RX} - \sum_{i=1}^{n} F_i^0 C_{pi}(T - T_i^0)}{\sum_{i=1}^{n} N_i C_{P_i}} \qquad (2.315)$$

We can then expand Equation 2.315 to consider n species involved in m reactions:

$$\frac{dT}{dt} = \frac{\dot{Q} - \sum_{j=1}^{m}\left(V(-r_{ij})\Delta H_{RXij}(T)\right) - \sum_{i=1}^{n} F_i^0 C_{pi}(T - T_i^0)}{\sum_{i=1}^{n} N_i C_{P_i}} \qquad (2.316)$$

This is a generic expression for a batch or semi-batch process in which the term for heat has not yet been defined.

For a batch reactor, the term $\sum_{i=1}^{n} F_i^0 C_{pi}(T - T_i^0)$ will disappear since all the flows are zero and since nothing goes in or out.

For the heat term, $\dot{Q}$, we need to determinate what type of system we have, adiabatic ($Q = 0$), or constant heat (Q = constant value) or with a fluid around (like in a CSTR) and its flow. If the flow in the external fluid is very high, we can assume that the temperature of that fluid is constant (T_f = constant value), among others.

As we did for the CSTR with variations of temperature at steady state, we will use the same expression for the values of heat, as follows:

$$m_f C_{pf}(T_{fi} - T)\left(1 - e^{\left(-\frac{UA}{m_f C_{pf}}\right)}\right) = \dot{Q} \qquad (2.317)$$

Then substituting Equation 2.317 into 2.316, we can obtain a generic expression for a batch/semi-batch that can be reduced to all the difference scenarios accordingly.

$$\frac{dT}{dt} = \frac{\left(m_f C_{pf}(T_{fi} - T)\left(1 - e^{\left(-\frac{UA}{m_f C_{pf}}\right)}\right)\right) - \sum_{j=1}^{m}\left(V(-r_{ij})\Delta H_{RXij}(T)\right) - \sum_{i=1}^{n} F_i^0 C_{pi}(T - T_i^0)}{\sum_{i=1}^{n} N_i C_{P_i}} \qquad (2.318)$$

Reactor Design

Now, we can solve the problem for multiple reactions taking place in a batch/semi-batch reactor.

Example 2.18
In a batch reactor, we have two reactions taking place simultaneously. The system is in liquid phase and the initial reactants are only chemicals A and B with concentrations of 15 and 30 [mol/l], respectively. These are elementary reactions at equilibrium:

$$A + 2B \leftrightarrows C + 2D \tag{2.319}$$

$$B + C \leftrightarrows E \tag{2.320}$$

The system presented in Equations 2.319 and 2.320 is a complex system where chemical B is running in parallel and chemical C is in series. Therefore, to solve this problem, we need to be able to solve for all the chemicals simultaneously. The system works adiabatically ($Q = 0$). Plot the variation of the concentration of all chemicals and the variation of temperature, as functions of time and discuss the results.

Additional information:

$Cp_A = 180$ [J/(mol*K)]; $Cp_B = 185$ [J/(mol*K)]; $Cp_C = 175$ [J/(mol*K)]; $Cp_D = 180$ [J/(mol*K)]; $Cp_E = 190$ [J/(mol*K)]; $\Delta H_{1A} = -3{,}520$ [J/mol]; $\Delta H_{2B} = -2{,}530$ [J/mol]; $T_0 = 300$ [K]; $V = 30$ [l];

$$k_1 = 0.00022 \exp\left(\frac{-2{,}050}{8.3143 T}\right) \ [l^2/(min*mol^2)];$$

$$k_2 = 0.00044 \exp\left(\frac{-1{,}150}{8.3143 T}\right) \ [l^2/(min*mol^2)];$$

$$k_3 = 0.00038 \exp\left(\frac{-1{,}950}{8.3143 T}\right) \ [l/(min*mol)];$$

$$k_4 = 0.00009 \exp\left(\frac{-1{,}845}{8.3143 T}\right) \ [1/min].$$

Solution
For both of the above-mentioned reactions, we need to develop the reaction rates; for this, we will set reactant A as the main chemical for the first reaction and B for the second reaction.

Therefore, we can express the following:

$$r_{1A} = -k_1 C_A C_B^2 + k_2 C_C C_D^2 \tag{2.321}$$

$$r_{2B} = -k_3 C_B C_C + k_4 C_E \tag{2.322}$$

Now, based on these expressions, we have:

$$\frac{r_A}{-1} = \frac{r_B}{-2} = \frac{r_C}{1} = \frac{r_D}{2} \tag{2.323}$$

$$\frac{r_{2B}}{-1} = \frac{r_{2C}}{-1} = \frac{r_{2E}}{1} \tag{2.324}$$

$$\frac{dC_i}{dt} = r_i \tag{2.325}$$

We can then write the following balances:

$$\frac{dC_A}{dt} = r_A = r_{1A} \tag{2.326}$$

$$\frac{dC_B}{dt} = r_B = 2 * r_{1A} + r_{2B} \tag{2.327}$$

$$\frac{dC_C}{dt} = r_C = -r_{1A} \tag{2.328}$$

$$\frac{dC_D}{dt} = r_D = -2 * r_{1A} \tag{2.329}$$

$$\frac{dC_E}{dt} = r_E = -r_{2B} \tag{2.330}$$

We can then substitute Equations 2.321 and 2.322 into Equations 2.326 to 2.330, respectively, and accordingly obtain the mole balances.

$$\frac{dC_A}{dt} = -k_1 C_A C_B^2 + k_2 C_C C_D^2 \tag{2.331}$$

$$\frac{dC_B}{dt} = \frac{1}{2}\left(-k_1 C_A C_B^2 + k_2 C_C C_D^2\right) + \left(-k_3 C_B C_C + k_4 C_E\right) \tag{2.332}$$

$$\frac{dC_C}{dt} = \left(k_1 C_A C_B^2 - k_2 C_C C_D^2\right) + \left(-k_3 C_B C_C + k_4 C_E\right) \tag{2.333}$$

$$\frac{dC_D}{dt} = \left(\frac{1}{2}\left(k_1 C_A C_B^2 - k_2 C_C C_D^2\right)\right) \tag{2.334}$$

$$\frac{dC_E}{dt} = \left(k_3 C_B C_C - k_4 C_E\right) \tag{2.335}$$

Reactor Design

In order to solve this problem adiabatically, we need to add the energy balance, simultaneously with the mole balance presented above. The general energy balance can be seen in Equation 2.316 which can be adapted to our problem as follows:

$$\frac{dT}{dt} = \frac{-V*\left[\left((-r_{1A})\Delta H_{RX1A}(T)\right)+\left((-r_{2B})\Delta H_{RX2B}(T)\right)\right]}{N_A C_{P_A} + N_B C_{P_B} + N_C C_{P_C} + N_D C_{P_D} + N_E C_{P_E}} \quad (2.336)$$

With this expression, we can then solve the problem, and in order to do this, we need the additional data. We can then put Equations 2.331 to 2.336 into Polymath and solve them. The script is:

Script:

```
#Mole balance
    d(Ca)/d(t) = r1a
    Ca(0) = 15
    d(Cb)/d(t) = 2*r1a+r2b
    Cb(0) = 30
    d(Cc)/d(t) = -r1a+r2b
    Cc(0) = 0
    d(Cd)/d(t) = -2*r1a
    Cd(0) = 0
    d(Ce)/d(t) = -r2b
    Ce(0) = 0
#Energy balance
    d(T)/d(t) = (-V*(((-r1a)*(-3520))+((-r2b)*(-2530)))+(Q))/((Na*Cpa)+(Nb
        *Cpb)+(Nc*Cpc)+(Nd*Cpd)+(Ne*Cpe))
    T(0) = 300
#Auxiliary equations
    r1a = -k1*Ca*(Cb^2)+k2*Cc*(Cd^2)
    r2b = -k3*Cb*Cc+k4*Ce
    k1 = 0.00022*exp(-2050/(R*T))
    k2 = 0.00044*exp(-1150/(R*T))
    k3 = 0.00038*exp(-1950/(R*T))
    k4 = 0.00009*exp(-1845/(R*T))
    Na = Ca/V
    Nb = Cb/V
    Nc = Cc/V
    Nd = Cd/V
    Ne = Ce/V
    Q = 0
```

```
#Constants
    R = 8.3143
    Cpa = 180
    Cpb = 185
    Cpc = 175
    Cpd = 180
    Cpe = 190
    V = 30
#Independent variable
    t(0) = 0
    t(f) = 300
```

We can now solve the problem and obtain the desired results. Figure 2.76 presents the variations of all chemicals concentrations as a function of the reaction time.

Figure 2.76 shows that the concentration of raw materials decreases in a similar way at the beginning. After some time, chemical A is consumed at a slower pace than chemical B as the latter is also being consumed by the second reaction. Even though chemical B is being consumed, the equilibrium in reaction 1 does not strongly shift toward the product, which is why the product C is also consumed in the second reaction. Therefore, they are balancing to some extent, keeping reaction 1 close to the equilibrium.

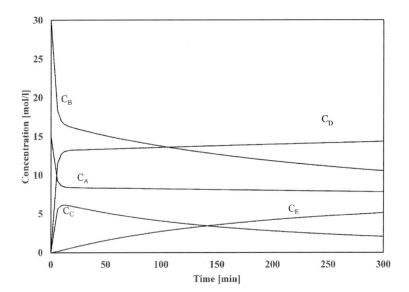

FIGURE 2.76 Variations of the concentration.

Since chemical C is produced and consumed, it should have a maximum concentration, and this can easily be seen in Figure 2.76. After a certain reaction time period, its consumption via the second reaction will overtake its production. For

Reactor Design

the last product (E), a small delay on its concentration profile can be seen. This is also associated with the appearance of this product; chemical C must be produced and afterward, consumed; therefore, it will take some time for E to start being produced.

Figure 2.77 shows the temperature profile for this case, where the temperature increases to unrealistic numbers (4,600 [K]).

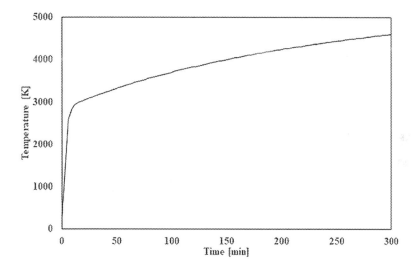

FIGURE 2.77 Variations of temperature with reaction time.

Mathematically the model can be solved and will give these results. There is no physical relevance as it is impossible to have a system where we can have these high temperatures without putting at risk the integrity of the chemicals (they might suffer cracking or decomposition) as well as putting at risk the safety of the equipment where the reaction takes place. Therefore, in order to allow this reaction to take place in a safe system, we might not be able to run it adiabatically, but it is possible either isothermally or with some heat exchanger to remove the heat being produced. These possibilities are left as problems at the end of the chapter for the student to work with and explore.

NOTES

1. Where α, β, δ, and Ω are the stoichiometric coefficients.
2. The dependence of r_A on concentration, pressure, and temperature is presented in this chapter and its experimental determination will be presented in Chapter 3.
3. The effect of temperature on k are not presented for simplicity and an arbitrary value has been chosen to show the tendency of the concentration profile.
4. This C_A^0 correspond to the concentration in the flow, not in the reactor, that is zero and is the initial value for the differential equation's solution.
5. We recommend reading reference [2] for details of this approach.
6. Based on the definition of conversion, x_2 is the conversion achieved in the second reactor based on the input flow.
7. We recommend the reader to read [1–4] for complimentary information on this subject.

8 Typically, the initial concentration of the products is zero; however, in order to present a generic case, we keep them in the analysis.
9 We have added an inert component in this evaluation since such components are regularly involved but do not participate in the reaction. However, their presence does affect the volume in a batch and the flow in a flow reactor.
10 Fogler [2] has carried out this derivative and its reading is encouraged.

REFERENCES

1. Westerterp, K.R., van Swain, W.P.M., Beenackers, A.A.C.M. *Chemical Reactor Design and Operation*. (2001). John Wiley & Sons. ISBN: 0471917303.
2. Fogler, H.S. *Elements of Chemical Reaction Engineering*. 5th edition. (2016). Pearson Education Inc. ISBN: 9780133887518.
3. Levenspiel, O. *Chemical Reaction Engineering*. 3rd edition. (1999). John Wiley & Sons. ISBN: 9780471254249..
4. Froment, G.F., Bischoff, K.B., De Wilde, J. *Chemical Reactor Analysis and Design*. 3rd edition. (2011). John Wiley & Sons. ISBN: 9780470565414.
5. Gallucci, F., Fernandez, E., Corengia, P., van Sint Annaland, M. "Recent advances on membranes and membrane reactors". *Chemical Engineering Science*. 92. (2013). 40–66.
6. Jia, H., Xu, H., Sheng, X., Yang, X., Shen, W., Goldbach, A. "High-temperature ethanol steam reforming in PdCu membrane reactor". *Journal of Membrane Science*. 605. (2020). 118083.
7. Lee, J., Kim, E.Y., Chang, B.J., Han, M., Lee, P.S., Moon, S.Y. "Mixed-matrix membrane reactors for the destruction of toxic chemicals". *Journal of Membrane Science*. 605. (2020). 118112.
8. Su, Z., Luo, J., Li, X., Pinelo, M. "Enzyme membrane reactors for production of oligosaccharides: a review on the interdependence between enzyme reaction and membrane separation". *Separation and Purification Technology*. 243. (2020). 116840.
9. Liu, L., Qiao, J., Zhang, H., Qi, L. "Fabrication of a porous polymer membrane enzyme reactor and its enzymatic kinetics study in an artificial kidney model". *Talanta*. 216. (2020). 120963.
10. Ergun, S. "Fluid flow through packed columns". *Journal of Chemical Engineering Progress*. 48. (1952). 89–94.
11. Marchetti, J.M., Errazu, A.F. "Biodiesel production from acid oils and ethanol using a solid basic resin as catalyst". *Biomass & Bioenergy*. 34(3). (2010). 272–277.
12. Marchetti, J.M., Pedernera, M.N., Schbib, N.S. "Production of biodiesel from acid oil using sulfuric acid as catalyst: kinetics study". *International Journal of Low Carbon Technologies*. 6(1). (2011). 38–43.
13. Borgnakke, C., Sonntag, R.E. *Fundamentals of Thermodynamics*. 8th edition. John Wiley & Sons. ISBN: 9781119382843. (2017).
14. Engel, T., Reid, P. *Thermodynamics, Statistical Thermodynamics & Kinetics*. 3rd edition. (2013). Pearson. ISBN: 9780321766182.

PROBLEMS

PROBLEM 1

Solve the following reaction rate expression in order to obtain the variation of the concentration, C_A, as a function of time for a generic reaction of order n. In this

Reactor Design

equation, substitute n with 0, 1, and 2, and present the profile of concentration of A as a function of time.

Generic reaction rate:

$$r_A = \frac{dC_A}{dt} = -kC_A^N$$

PROBLEM 2
An irreversible first-order reaction of chemical A to give B, takes place in a batch reactor, where the reaction rate is defined as $r_A = -k_1 C_A$. If the initial concentration of A is 1 [mol/l], plot C_A versus time when k_1 values are 1, 10, 15, and 20 [1/min]. What is the effect of decreasing the reaction rate constant?

PROBLEM 3
The production of a desired product P takes place in a semi-batch reactor. Chemical A is present inside the reactor with an initial concentration of 1.2 [mol/l] and is added to the reactor at a flow of 1 [mol/min] with a concentration in the flow in of 0.5 [mol/l]. The second reactant (B) is only added with a volumetric flow of 2.5 [l/min] in a concentration of 1 [mol/l]. Experimentally, it has been found out that this reaction is a second-order from A and first-order from B.

   a. How long will it take to fill up 90% of the reactors volume if initially there is 10 [l] and the total volume is of 120 [l].
   b. Plot C_A, C_B, and C_P as a function of time and discuss the results.
      $k = 10$ [l²/(min*mol²)] and the reaction time is 50 [min].

PROBLEM 4

   a. Solve Example 2.2 using the generic reaction rate: $r_A = \frac{dC_A}{dt} = -kC_A^N$, obtain an expression for the space time (τ).
   b. Calculate τ for $N = 0$, $N = 1$, and $N = 3$. Compare these results together with the result from Example 2.2 for $N = 2$ and then explain/discuss the results.

PROBLEM 5
The reaction A → B, is irreversible and elementary; therefore, it is a first-order reaction with respect to A and is being carried out in a plug flow system. The initial concentration of reactant A is 2 [mol/l] while B is zero. What is the space time required to achieve 65% conversion? If the volumetric flow is 3 [l/min], and the kinetics parameter is 0.5 [mol/(l*min)]. Calculate the value of space time (τ) for these conditions; compare this result with similar results (for $N = 1$) from the previous problem and discuss.

PROBLEM 6
The transformation of A into B (similar to the one from the previous problem) is being carried out in a plug flow system with a space time value of 5.3 [min]. If the

initial concentration of reactant A is 2 [mol/l] while B is zero and A is being fed into the system with a volumetric flow is 3 [l/min], the reaction takes place in liquid phase and the kinetics parameter is 0.5 [mol/(l*min)]. What will be the concentration of A at the end of the reactor, and what conversion can be achieved?

PROBLEM 7
The chemical transformation of A to B (A → B) is being carried out in a PFR. The generic reaction rate is: $r_A = -kC_A^N$.

   a. For the generic case, solve the mole balance for a PFR and obtain the variation of C_A as a function of the space time (τ).
   b. For the generic solution from a), substitute $N = 0, 2$, and 3 and compare the results.
   c. Plot all cases in (b).

PROBLEM 8
The production of the desired and valuable material, C, is being done when A and B reacts following an elementary irreversible reaction in liquid phase. The reaction is being carried out in a PBR with 30 kg of catalysts. The reaction under consideration, and the reaction rate are:

$$A + 2B \rightarrow C$$

$$r_A^w = -k_w C_A C_B^2$$

The reaction constant is 31 [l³/(kg*min*mol²)], and the volumetric flow is 34 [l/min]. Initially, the reactor is being fed A and B at rates of 10 and 25 [mol/min], respectively; while C is not present at the beginning of the reaction.

   a. What are the molar flows for each chemical (A, B, and C) when leaving the reactor? (use any software you prefer to solve the differential equations)
   b. Plot the variation of all the flows as a function of the amount of catalysts.
   c. The same reactor as in the previous case is being used for the same reaction, with the modification that C can decomposed into some undesired products (U) following a first-order reaction based on C, ($k_2 = 6$ [l/(kg*min)]). For this new scenario, calculate all the flows at the end of the reactor, plot all the flows as a function of the catalyst amount. Are we operating at the optimal conditions to produce the most of C? if not, how can we find the optimal amount of catalysts to produce the highest amount of C?

PROBLEM 9
In order to improve the selectivity and production of the chemical D, a membrane reactor is being used. The reactor allows the reaction of A and B into D and U, in liquid phase, following these reactions:

Reactor Design

$$A + B \rightarrow U \quad r_U = k_U C_A C_B$$

$$A + B \rightarrow D \quad r_D = k_D C_A^2 C_B^{1/2}$$

a. Based on the definition of selectivity (r_D/r_U), will you add any chemical through the membrane? If yes, which one? Will you remove any chemical through the membrane? If yes, which one?
b. If the reactor's volume is 50 dm³, plot the selectivity as a function of the volume. Are we operating in the optimal condition? If no, is there an optimal condition? If yes, where is the optimal? B is present in the sweeping gas with a concentration of $C_{BSW} = 1$ [mol/l] and $k_{mem} = 0.5$ [l/min], $v = 15$ [l/min], $F_A^0 = 10$ [mol/min], and $F_B^0 = 1$ [mol/min].
c. Plot all flows (F_i) as a function of volume and discuss the results.
d. Calculate the variation of the conversion of A versus volume. Is this a good representation of the amount of D being produced? Or what is this conversion actually showing?

PROBLEM 10
For the generic reaction rate equation, from Problem 1 ($r_A = \dfrac{dC_A}{dt} = -kC_A^N$), using the definition of conversion, obtain a general expression (based on x), for this irreversible N-order reaction.

PROBLEM 11
The elementary liquid phase reaction, A + B → C, will be carried out in different reactor configurations. The process will achieve a conversion of 60% in the first reactor and a 90% conversion on the second reactor. If the initial concentration of A and B are identical, i.e., 1.8 [mol/l], the volumetric flow is 1.5 [l/min], and the reaction is carried out isothermally with a reaction constant of 0.8 [l/(mol*min)], which of the following reactor configurations will give the smallest reactor volume?

a. PFR followed by a CSTR
b. CSTR followed by a PFR.
c. Only one large PFR.
d. Only one large CSTR.

PROBLEM 12
A gas phase, elementary, irreversible reaction is carried out at constant temperature (500 [K]) and constant pressure in a PFR with a volume of 10 [l]. This reaction is:

$$2A + 3B \rightarrow 3C + 1D$$

$$r_A = -k_1 C_A^2 C_B^3$$

The initial flow of components A and B are 1.6 and 2.4 [mol/min] while C and D have no initial flow, the initial volumetric flow is of 2.5 [l/min], and the reaction constant is of 12 [l⁴/(mol⁴*min)].

a. Plot the variation of the conversion as a function of the volume.
b. Plot all the flows as a function of the volume.
c. If the reaction is a reversible, plot the new variation of conversion and flows as function of the volume for the new system. k_{rev} = 3 [l³/(mol³*min)].

PROBLEM 13
The transformation of A and B into C is given by an elementary gas phase reversible reaction, A + B → 3C. The process takes place in a batch reactor where the pressure is kept at 10 [bar] and the temperature is kept at 350 [K]. Initially, in the reactor, we have 0.76, 1.15, and 0.1 [mol/l] as concentrations of A, B, and C, respectively. The reaction is carried out for 25 min.

a. Plot the variation of C_i as a function of time.
b. Plot the time variations of the conversion.
c. Recalculate the conversion when C_C^0 takes the values 0, 0.1, 0.5, and 1 [mol/l].

Additional data:

$$k_1 = 0.12 \left(\frac{l}{mol * min} \right)$$

$$k_2 = 0.01 \left(\frac{l^2}{mol^2 * min} \right)$$

PROBLEM 14
A PBR is being used for a gas phase, elementary, reversible reaction, with an initial volumetric flow of 15 [l/min]. The reaction under study is A + B ⇆ 2C + 2D. The reaction kinetic equation is $r_A = -k_1 C_A C_B + k_2 C_C^2 C_D^2$. The total amount of catalysts employed is 10 [kg].

Additional data:
The catalyst diameter is 0.001 [m], and the reactor pipe diameter is 0.025 [m], gas viscosity is 0.000031 [kg/(m*s)]], Superficial mass velocity is 0.002 [kg/(m²*s)], initial flows are 3, 2, 0, and 0 [mol/min] for chemicals A, B, C, and D, respectively. The pressure at the entrance of the reactor is 4,500 [kPa] and the catalysts porosity is 0.35. The densities of the catalyst and the gas are 2,500 and 1.8 [kg/m³], respectively and the reaction rate constants are k_1 = 12 [l²/(kg*mol*min)] and k_2 = 3 [l²/(kg*mol*min)].

Reactor Design

a. Plot all the flows as a function of the amount of catalysts.
b. Plot the conversion as a function of the amount of catalysts.
c. Recalculate F_A and x_A when there is pressure drop in the system, compare these results with those from (a) and (b).
d. If the initial pressure is changed, i.e., P_0 = 4,500, 5,000, 5,500, and 6,000 [kPa]. Plot the variation of conversion as a function of the amount of catalysts for all cases and compare.
e. If the superficial mass velocity is changed with the values of 0.005, 0.004, 0.003, and 0.002 [kg/(m²*s)], plot the variations of conversion.

PROBLEM 15

The reaction of α moles of A and β moles of B takes place to produce C and D, following the reaction, 2A + B → 1C + 3D.

a. If 20 moles of A and 10 moles of B react to give a final conversion of 65%, how much of A, B, C, and D will be at the end of the reaction?
b. If 20 moles of A and 20 moles of B react to give a final conversion of 65%, how much of A, B, C, and D will be at the end of the reaction?
c. To the case described in (b), we add 10 moles of an inert component. Calculate the moles at the end of the reaction when the conversion is still 65%.

PROBLEM 16

A liquid phase irreversible elementary reactions, in series, take place in a batch reactor. The reaction under consideration is $A \xrightarrow{k_1} 2B \xrightarrow{k_2} C$. This reaction is being carried out in order to optimize the production of chemical B, also, to improve the selectivity toward this high value chemical.

Initially, we have only A in the reactor with a concentration of 5 [mol/l], while the reaction rates are: k_1 = 0.1 [1/min] and k_2 = 0.03 [l/(mol*min)].

a. Plot the variation of the concentration of all chemical versus time. What is the optimal time to produce the highest amount of B?
b. Plot the conversion of A as a function of time. When does the conversion have the highest value? Is this optimal equal to the on previously obtained? Explain.
c. Consider that the reaction from A to be reversible but not the one from B to C, the new reaction system will be:

$$A \underset{k_3}{\overset{k_1}{\rightleftarrows}} 2B \xrightarrow{k_2} C$$

With k_3 = 0.01 [l/(mol*min)], plot all concentrations versus time for this new case and compare the results with part (a); is the optimal for this new scenario at the same time as previous case? Is the optimal concentration also the same as for the previous case? Discuss the results.

d. Identically to part (c), but now both reactions are reversible, so the new scenario is:

$$A \underset{k_3}{\overset{k_1}{\rightleftarrows}} 2B \underset{k_4}{\overset{k_2}{\rightleftarrows}} C$$

With $k_4 = 0.03$ [1/min].

PROBLEM 17

The gas phase reaction for the production of chemical D is carried out via the chemical transformation of A and B in a PFR, with a total volume of 30 dm³ following this reaction:

$$A + 2B \xrightarrow{k_1} D \quad r_A^1 = -k_1 C_A C_B^2$$

However, these reactants also produce a low-priced chemical U via the following reaction:

$$A + B \xrightarrow{k_2} U \quad r_A^2 = -k_1 C_A C_B$$

Chemicals D and U are not present at the beginning of the reaction. Due to the selling price of D in comparison to U, it is desired to improve the production of D and/or its selectivity.

a. Plot all F_i flows as a function of the reactor volume.
b. Plot the selectivity as a function of the reactor volume.
c. At what volume is the F_D optimal and maximum? What is the selectivity value at that optimal value?
d. If the reaction toward the undesired product is reversible, what new value of F_D will be optimal, what new value of selectivity will be optimal, and at what reactor volume will this take place?
e. If the reaction toward the undesired and the desired product are both reversible, what new value of F_D will be optimal, what new value of selectivity will be optimal, and at what reactor volume will this take place?
f. If the reactions used for case (c) are in liquid phase instead of gas phase, when will F_D be optimal? What will be the selectivity for that maximum and where in the reactor volume will it take place?

Additional data:

$F_A^0 = 48 \left[\dfrac{mol}{min}\right]$; $F_B^0 = 35 \left[\dfrac{mol}{min}\right]$; $v^0 = 3 \left[\dfrac{l}{min}\right]$; $k_1 = 0.005 \left[\dfrac{l^2}{mol^2 * min}\right]$;

$k_2 = 0.01 \left[\dfrac{l}{mol * min}\right]$; $k_3 = 0.03 \left[\dfrac{1}{min}\right]$; $k_4 = 0.01 \left[\dfrac{1}{min}\right]$.

Reactor Design

PROBLEM 18
You are starting in a new job and your new boss gives you the following elementary reactions that take place in liquid phase, in a PFR of the processing plant:

$$\text{Reaction 1: } A + B \xrightarrow{k_1} D \quad r_A^1 = -k_1 C_A C_B$$
$$\text{Reaction 2: } A + 2B \xrightarrow{k_2} U \quad r_A^2 = -k_2 C_A C_B^2$$
$$\text{Reaction 3: } A + D \xrightarrow{k_3} F \quad r_A^3 = -k_3 C_A C_D$$

Your duty is to find out if the maximum amount of D, under the given conditions, can be achieved within the total volume, 30 [l], of the reactor. Also, at what volume will the highest selectivity toward D be obtained? Chemicals A and B are fed into the reactor with equal molar flow of 48 [mol/min] each, while D, U, and F have zero flow. The volumetric flow is set to 2 [l/min], the process is isothermal and isobaric. The reactions constants are given as follows:

$$k_1 = 0.004 \left[\frac{l}{mol*min} \right]; \quad k_2 = 0.001 \left[\frac{l^2}{mol^2*min} \right]; \quad k_3 = 0.1 \left[\frac{l}{mol*min} \right]$$

PROBLEM 19
Within the theory of this book, we developed the relationship for the equilibrium constant (K) as a function of the conversion (x) for a first-order reversible reaction. We encourage the reader to look for it if required. Obtain the relationship for the equilibrium constant (K) as a function of the conversion (x) for:

a. $2A \rightleftarrows B$, with $C_B^0 = 0$, and $C_A^0 = 1$ [mol/l]
b. $2A \rightleftarrows 2B$, with $C_B^0 = 0$, and $C_A^0 = 1$ [mol/l]
c. For (a), re-solve the problem when $C_A^0 = 1$ [mol/l] and $C_B^0 = \frac{1}{2} C_A^0$.
d. For (a), re-solve the problem when $C_A^0 = 1$ [mol/l] and $C_B^0 = 2 C_A^0$.

PROBLEM 20
At equilibrium, describe how equilibrium constants will depend on the conversion for the reaction $2A \rightleftarrows B$, where B is not present at the beginning of the reaction and if this reaction takes place in:

a. Liquid phase.
b. Gas phase.

If needed, plot the results by giving arbitrary numbers and discuss with different scenarios (no solutions are given for this, ask your teaching assistant for advice).

PROBLEM 21
A PFR is being used to carry out an adiabatic, liquid phase elementary reaction, $2A \rightarrow B$. This reaction takes place in a tubular cylinder of 200 [l] of volume. The initial conditions are a flow of A of 12 [mol/l] and zero for B. How much B will be produced at the end of the reactor? What is the temperature profile along the reactors

volume and how is the conversion profile along the reactors volume? How is the conversion changing with temperature and how are the molar flows profile along the reactors volume?

Additional data:

$$r_A = -k_1 C_A^2; \quad \frac{r_A}{-2} = \frac{r_B}{1}; \quad v = 3.25\left[\frac{l}{min}\right]; \quad k_1 = 0.001 e^{\left(\frac{-1,100}{8.3143*T}\right)}\left[\frac{l}{mol*min}\right];$$

$$C_{PA} = 160\left[\frac{J}{mol*K}\right]; C_{PB} = 150\left[\frac{J}{mol*K}\right]; T_r = T_0 = 300\ [K]; \Delta H = -13,000\left[\frac{J}{mol}\right].$$

PROBLEM 22

How many reactors are required for the equilibrium reaction of $2A \underset{k_2}{\overset{k_1}{\rightleftarrows}} B + C$ to reach a final conversion of 89% when all the reactors start at the same temperature $T_R = T_0 = 500\ [K]$.

You know that the equilibrium reaction at $T = 298\ [K]$ is 0.35 and that

$\Delta H = -35,000\left[\frac{J}{mol}\right]$. It is also known that reactant A is only presented at the beginning of the reaction with a molar flow of 2.4 [mol/min] and that $C_{PA} = 120\left[\frac{J}{mol*K}\right]$,

$C_{PB} = 120\left[\frac{J}{mol*K}\right]$, and $C_{Pc} = 140\left[\frac{J}{mol*K}\right]$.

PROBLEM 23

In a PFR, a reversible elementary reaction takes place for the production of two valuable chemicals.

$$2A + B \underset{k_2}{\overset{k_1}{\rightleftarrows}} C + 2D \quad r_A = -k_1 C_A^2 C_B + k_2 C_C C_D^2$$

The total volume of the plug reactor is 45 [dm³].

a. If heat is being added/removed with a $U\varpi = 800\ [J/(min*K*dm^3)]$ at a constant T_f. Plot all the flows involved (F_i), the conversion and reaction temperature, as a function of reactor volume.
b. If a co-current flow is being used with $\dot{m}_f = 180$ [mol/min] and $Cp_F = 80\ [J/(mol*K)]$. Plot all the flows involved (F_i), the conversion and reaction temperature as a function of reactor volume.
c. If a counter-current flow is being used with $\dot{m}_f = 180$ [mol/min] and $Cp_F = 80\ [J/(mol*K)]$. Plot all the flows involved (F_i), the conversion and reaction temperature as a function of reactor volume.
d. Plot all the conversions together and discuss the results.
e. Plot all the reactor temperature trends together, compare and discuss the results.

Reactor Design

Additional data:

$$v^0 = 1.75 \left[\frac{l}{\min}\right]; k_1 = 0.01 e^{\left(\frac{-4,200}{8.3143*T}\right)} \left[\frac{l^2}{mol^2 * \min}\right]; k_2 = 0.002 e^{\left(\frac{-2,100}{8.3143*T}\right)} \left[\frac{l^2}{mol^2 * \min}\right];$$

$$C_{PA} = 120 \left[\frac{J}{mol*K}\right]; C_{PB} = 130 \left[\frac{J}{mol*K}\right]; C_{PC} = 120 \left[\frac{J}{mol*K}\right]; C_{PD} = 150 \left[\frac{J}{mol*K}\right];$$

$$T_0 = 350\ [K];\ \Delta H = -45,000 \left[\frac{J}{mol}\right];\ T_f^0 = 400\ [K];\ F_A^0 = 30 \left[\frac{mol}{\min}\right];\ F_B^0 = 15 \left[\frac{mol}{\min}\right];$$

$$F_C^0 = F_D^0 = 0.$$

PROBLEM 24

A CSTR is being used to carry out the production of two main chemicals, C and D. A liquid phase elementary reaction occurs, where 2A and 2B combine to produce 1C and 1D. Only A and B are fed to the CSTR, with a molar flow of 30 [mol/min] and both have a volumetric flow of 1.5 [l/min]. Knowing that the total volume of the system is 15 [l], calculate the temperature and conversion that can be achieved within this volume when the system has an external flow cooling/heating of the CSTR with a molar flow of 0.8 [mol/min]. If you need to iterate, show the iteration procedure and plot the temperature–conversion and/or temperature–volume and/or conversion–volume used to narrow down the domain where this problem might have a solution.

Additional data:

$$k_1 = 0.001 e^{\left(\frac{-4,500}{8.3143*T}\right)} \left[\frac{l^3}{mol^3 * \min}\right];\ k_2 = 0.001 e^{\left(\frac{-3,500}{8.3143*T}\right)} \left[\frac{l}{mol * \min}\right];\ T_R = 298\ [K];$$

$$C_{PA} = 120 \left[\frac{J}{mol*K}\right];\ C_{PB} = 140 \left[\frac{J}{mol*K}\right];\ C_{PC} = 110 \left[\frac{J}{mol*K}\right];\ C_{PD} = 150 \left[\frac{J}{mol*K}\right];$$

$$T_0 = 400\ [K];\ \Delta H = -30,000 \left[\frac{J}{mol}\right];\ T_f = 350\ [K];\ F_A^0 = 30 \left[\frac{mol}{\min}\right];\ F_B^0 = 30 \left[\frac{mol}{\min}\right];$$

$$F_C^0 = F_D^0 = 0;\ C_{Pf} = 120 \left[\frac{J}{mol*K}\right];\ U = 1000 \left[\frac{W}{m^2*K}\right];\ A = 2.5\ [m^2].$$

PROBLEM 25

A batch reactor is being used to carry out the elementary, reversible reaction of 2A reacting to give 1 B. The reaction rate is $r_A = -k_1 C_A^2 + k_2 C_B$. Initially, in the batch system, we have only C_A with a concentration of 24 [mol/l]. The total volume of the reactor is 25 [l] while the initial temperature is $T_0 = 298$ [K]. The reaction rate constant follows the Arrhenius equations as $k_1 = 0.01 e^{\left(\frac{-4,200}{8.3143*T}\right)} \left[\frac{l}{mol*\min}\right];$

$$k_2 = 0.002 e^{\left(\frac{-2,100}{8.3143*T}\right)} \left[\frac{1}{\min}\right];\quad \Delta H = -45,000 \left[\frac{J}{mol}\right];\quad C_{PA} = 120 \left[\frac{J}{mol*K}\right];$$

$$C_{PB} = 130 \left[\frac{J}{mol*K}\right].$$

How is the variation of conversion, temperature, C_A, and C_B over the reaction time (0–120 min) when:

a. $Q = 0$ (adiabatic process)
b. $Q = -12{,}000$ [W]
c. $Q = UA(T_F - T)$, with $U = 120\ [\dfrac{W}{m^2 * K}]$ and $A = 100$ [m²] and $T_F =$ constant but has values 200, 300, 400, 500, 600, and 800 [K]?

PROBLEM 26

The conversion of chemical A into P takes place via a series of elementary reactions in series and in parallel, in a PFR, in liquid phase. The process only has chemical A as a reactant and therefore, A is the only one flowing into the system (total volume of 75 [dm³]) with a molar flow of 25 [mol/min].

The reactions involved are as follows:

Reaction 1: $A \underset{k_2}{\overset{k_1}{\rightleftarrows}} B \quad r_A^1 = -k_1 C_A + k_2 C_B, \quad \Delta H_A^1 = -25{,}000 \left[\dfrac{J}{mol}\right]$

Reaction 2: $A + B \underset{k_4}{\overset{k_3}{\rightleftarrows}} U \quad r_A^2 = -k_3 C_A C_B + k_4 C_U, \quad \Delta H_A^2 = -15{,}000 \left[\dfrac{J}{mol}\right]$

Reaction 3: $2A + B \underset{k_6}{\overset{k_5}{\rightleftarrows}} P \quad r_A^3 = -k_5 C_A^2 C_B + k_6 C_P, \quad \Delta H_A^3 = 12{,}000 \left[\dfrac{J}{mol}\right]$

Reaction 4: $A + 2B \underset{k_8}{\overset{k_7}{\rightleftarrows}} W \quad r_A^4 = -k_7 C_A C_B^2 + k_8 C_W, \quad \Delta H_A^4 = 17{,}000 \left[\dfrac{J}{mol}\right]$

Plot the variations of the flow for all components, the conversion, and the reaction temperature, as functions of the volume of the reactor. Do this for the following cases:

a. If the process is working isothermally at $T = 500$ [K].
b. Solve the problem when the reaction temperature varies along the reactor, but the cooling/heating fluids are at a constant temperature of $T_F^0 = 500$ [K].
c. Solve the problem when the reaction temperature varies along the reactor, but the cooling/heating fluids flow co-current to the reaction system. The initial temperature is $T_F^0 = 500$ [K]. For this case, plot a comparison of the reaction temperature and the external fluids temperature as well.
d. Solve the problem when the reaction temperature varies along the reactor, but the cooling/heating fluids flow counter-current to the reaction system. Therefore, the temperature of the input for the external fluids is $T_F^0 = 500$ [K]. For this case, plot a comparison of the reaction temperature and the external fluids temperature as well.

Additional data:

$$k_1 = 1e^{\left(\frac{-3,800}{8.3143*T}\right)}\left[\frac{1}{min}\right]; \quad k_2 = 0.1e^{\left(\frac{-2,500}{8.3143*T}\right)}\left[\frac{1}{min}\right]; \quad k_3 = 3e^{\left(\frac{-2,400}{8.3143*T}\right)}\left[\frac{l}{mol*min}\right];$$

$$k_4 = 0.8e^{\left(\frac{-3,400}{8.3143*T}\right)}\left[\frac{1}{min}\right]; \quad k_5 = 7.1e^{\left(\frac{-3,800}{8.3143*T}\right)}\left[\frac{l^2}{mol^2*min}\right]; \quad k_6 = 1e^{\left(\frac{-4,200}{8.3143*T}\right)}\left[\frac{1}{min}\right];$$

$$k_7 = 3e^{\left(\frac{-6,800}{8.3143*T}\right)}\left[\frac{l^2}{mol^2*min}\right]; \quad k_8 = 0.5e^{\left(\frac{-2,000}{8.3143*T}\right)}\left[\frac{1}{min}\right];$$

$$C_{PA} = 120\left[\frac{J}{mol*K}\right]; \quad C_{PB} = 140\left[\frac{J}{mol*K}\right]; \quad C_{PU} = 110\left[\frac{J}{mol*K}\right];$$

$$C_{PP} = 150\left[\frac{J}{mol*K}\right]; \quad C_{PW} = 90\left[\frac{J}{mol*K}\right]; \quad C_{Pf} = 160\left[\frac{J}{mol*K}\right];$$

$$T^0 = 500\ [K]; \quad m_f = 60\left[\frac{mol}{min}\right]; \quad v = 1.50\left[\frac{l}{min}\right]; \quad U\varpi = 1500\left[\frac{J}{dm^3*min*K}\right].$$

PROBLEM 27

In a batch reactor, the transformation of A and B into two undesired products and one desired product takes place in liquid phase. At the beginning of the reaction, both reactants are present and there is no product. The initial concentrations of A and B are 20 [mol/l], each. The reactions involved are shown as follows:

Reaction 1: $A + B \underset{k_2}{\overset{k_1}{\rightleftarrows}} W \quad r_A^1 = -k_1 C_A C_B + k_2 C_W, \quad \Delta H_A^1 = 15,000\left[\frac{J}{mol}\right]$

Reaction 2: $2A + B \underset{k_4}{\overset{k_3}{\rightleftarrows}} P \quad r_A^2 = -k_3 C_A^2 C_B + k_4 C_P, \quad \Delta H_A^2 = -25,000\left[\frac{J}{mol}\right]$

Reaction 3: $A + 2B \underset{k_6}{\overset{k_5}{\rightleftarrows}} 2U \quad r_A^3 = -k_5 C_A C_B^2 + k_6 C_U^2, \quad \Delta H_A^3 = 35,000\left[\frac{J}{mol}\right]$

Plot the variations of the concentration for all chemicals involved as well as the reaction temperature and conversion as a function of time when:

a. The reactor is isothermal $T = 400$ [K].
b. The reaction is being carried out adiabatically, $T^0 = 400$ [K].
c. The reaction temperature varies due to the presence of external fluids. The initial outside temperatures are $T_F^0 = 300$ [K], $T_F^0 = 400$ [K], $T_F^0 = 500$ [K], and $T_F^0 = 600$ [K].
d. Plot the comparison of the reactor temperature for all initial cases from (c).
e. Compare all the reaction temperatures (for case (c)) using 400 [K], compare all the values of conversion, and discuss the results.

Additional data:

$$k_1 = 0.01 e^{\left(\frac{-3,200}{8.3143*T}\right)} \left[\frac{l}{mol*min}\right]; \quad k_2 = 0.008 e^{\left(\frac{-3,700}{8.3143*T}\right)} \left[\frac{1}{min}\right];$$

$$k_3 = 0.006 e^{\left(\frac{-3,200}{8.3143*T}\right)} \left[\frac{l^2}{mol^2*min}\right]; \quad k_4 = 0.0008 e^{\left(\frac{-2,100}{8.3143*T}\right)} \left[\frac{1}{min}\right];$$

$$k_5 = 0.0004 e^{\left(\frac{-3,500}{8.3143*T}\right)} \left[\frac{l^2}{mol^2*min}\right]; \quad k_6 = 0.000026 e^{\left(\frac{-2,300}{8.3143*T}\right)} \left[\frac{l}{mol*min}\right];$$

$$C_{PA} = 140 \left[\frac{J}{mol*K}\right]; \quad C_{PB} = 150 \left[\frac{J}{mol*K}\right]; \quad C_{PU} = 110 \left[\frac{J}{mol*K}\right];$$

$$C_{PP} = 140 \left[\frac{J}{mol*K}\right]; \quad C_{PW} = 120 \left[\frac{J}{mol*K}\right]; \quad C_{Pf} = 140 \left[\frac{J}{mol*K}\right];$$

$$m_f = 10 \left[\frac{mol}{min}\right]; \quad UA = 1200 \left[\frac{J}{min*K}\right].$$

3 Reaction Kinetics

3.1 INTRODUCTION

In the previous sections in Chapter 2, we have been able to design a reactor in order to know the volume, amount of catalyst, or time required to obtain a certain quantity of the desired product; we also learnt about selectivity when having multiple reactions. However, these concepts were only introduced by necessity and not fully developed. Even more, throughout the book so far, we have used the concepts of reaction rate and reaction kinetics without knowing how they are obtained or developed. This chapter will allow us to introduce some concepts and definitions that were assumed to be known and develop them a little further. We will also present different approaches on how to obtain kinetics expressions based on experimental data; this part is also very relevant for a good understanding of Chapter 4. Finally, when having a solid catalyst, deactivation as well as mass transfer limitations will be presented.

3.2 ELEMENTARY REACTIONS

Elementary reactions, as shown below, are the simplest types of reactions to deal with:

$$\alpha A + \beta B \leftrightarrows \delta C + \gamma D \tag{3.1}$$

where this reaction takes place as it is presented in Equation 3.1. It cannot be subdivided into any other reaction, and the stoichiometric coefficients indicate the reaction order for the forward and backward reaction. Therefore, for this case, the reaction rate for chemical A will be:

$$r_A = -k_1 C_A^\alpha C_B^\beta + k_2 C_C^\delta C_D^\gamma \tag{3.2}$$

where the forward reaction has an order of α + β and the reverse reaction of γ + δ.

Several of our most known reactions are elementary reactions, the combustion of oxygen by our bodies and the production of CO_2, is a very simple elementary reaction that is very relevant to our daily life.

Elementary reactions can also be classified based on the number of molecules that are participating in it, for instance, for irreversible reactions, we can have unimolecular, bimolecular, or termolecular reactions. This gives us information about the reaction under consideration; however, it is not specific enough to know if a bimolecular reaction has two molecules of the same or two different types. Reactions presented in Equations 3.3 and 3.4 are both bimolecular:

$$A + A \rightarrow Products \tag{3.3}$$

$$A + B \rightarrow \text{Products} \tag{3.4}$$

As mentioned, any elementary reaction provides information of the reaction rate for both the forward and the reverse reaction. We will see later on that there are other reactions, such as non-elementary reactions, that could have an elementary reaction rate order. This means that even if it is not an elementary reaction, after the full analysis of the process, we arrive at a reaction rate expression that satisfies the reaction as if it was considered elementary.

3.3 NON-ELEMENTARY REACTIONS

Most of the reactions that we know are not elementary reactions. This means that we know and generally analyze the overall reactions but when looking into details, especially from a kinetics point of view, the reaction takes place in two or more steps. A typical case that most chemical engineers can relate to is a polymerization reaction, in which a monomer is chemically attached to several hundreds, thousands, or even hundred thousand monomers, before the reaction is terminated. Another common reaction that is carried out in several steps is the Fischer–Tropsch synthesis to produce liquid fuels from methane.

Other common examples of non-elementary reactions are those that take place over heterogeneous catalyst. In these cases, it is common that one or more of the reactants (that could be in liquid or gas phase) interact with the solid catalyst to be adsorbed (physically or chemically). Then, the reaction takes place and a desorption of the products might take place. A simplified scheme can be seen in Equations 3.5 to 3.7:

$$A + s \rightarrow As \tag{3.5}$$

$$As + B \rightarrow Ps \tag{3.6}$$

$$Ps \rightarrow P + s \tag{3.7}$$

where s is a site over the catalyst and As and Ps are the raw material and the product that are adsorbed over the catalytic material. Even though this scheme is simplified, most of the heterogeneous reactions (liquid–solid, gas–solid) will go through a pathway that will include at least two of the steps hereby presented, if not more.

Non-elementary reactions can always be subdivided into elementary reactions, and these steps are the ones that we must analyze in order to understand the mechanism and kinetics involved. It is important to know that the there are cases where the overall reaction and the final kinetics obtained might mislead us to believe that the reaction is pseudo-elementary. However, the reality is that there are several chemical steps (non-elementary process), or chemical interactions of the molecules that cancel with each other out, giving a final equation that looks like if the process was elementary.

For an overall reaction $A + B \rightarrow P$, that we assumed consists of three elementary steps, *i.e.*, an adsorption, a reaction, and a desorption step; the activation energy variation along the reaction pathways changes in comparison with the same reaction taking place in an elementary step. In order to obtain the reaction rate expression, we need to have more information regarding the steps involved in this process. Mainly

Reaction Kinetics

identify which one has the highest activation energy since that will be the step that controls the process.

Figure 3.1 shows the energy variation along the reaction path for an elementary reaction while Figure 3.2 shows the same for a non-elementary reaction having three steps (as in the previous example).

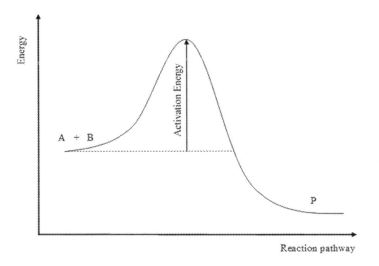

FIGURE 3.1 Energy variation for an elementary reaction.

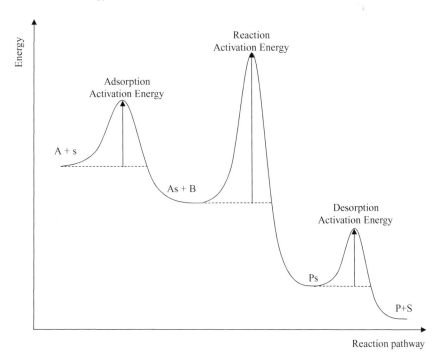

FIGURE 3.2 Energy variation for a non-elementary reaction.

It can be seen from both Figures 3.1 and 3.2 that the energy for the products is lower than the energy for the raw materials; this means that the reaction will occur and release some energy. If the reactions under considerations were reversible, we can see that each of the activation energies involved when going from reactants to products are smaller than when moving in the opposite direction. Thus, less energy is needed to move from left to right than it is to move from right to left.

3.4 MULTIPLE REACTIONS

So far, we have presented elementary and non-elementary reactions, and in Chapter 2, we have presented multiple reactions and studied them in different reaction systems, and we will discuss them briefly here. Multiple reactions can include: (i) a combination of elementary reactions, (ii) a combination of non-elementary reactions, and (iii) a mix of elementary and non-elementary reactions. This categorization is independent if the reactions are in parallel, in series, or in a complex mechanism. For example, imagine the reaction A → B → C; this is a series reaction. This series could be considered as two elementary steps, from A to B and from B to C, or it could be an elementary reaction from A to B, with two or more reaction steps to transform B to C.

In order to solve any of these problems, as we did in Chapter 2 and will do in Chapter 4, we need to obtain the kinetics expression. So far, we have given the reaction rate as data; however, this can be obtained from experimental information and by proposing different reaction mechanisms and checking it with the experimental data. A full problem is presented in Chapter 4 and we strongly recommend the reader to go through it carefully.

3.5 EVALUATION OF EXPERIMENTAL DATA

In order to obtain the reaction kinetics for any given chemical reaction, we need to have experimental information of such process. This can be the variation of conversion, concentration (if we have a liquid), pressure (if we have a gas), or number of moles, as a function of time. In order to do that, the desired reaction is carried out in a batch process since this will allow us to get time-based data. Once, the data, as shown in Table 3.1, is obtained, then, we need to develop different methodologies to process such information and to find information like the reaction rate constant, activation energy, and reaction order. Table 3.1 presents the variation of the concentration of reactant A for an elementary irreversible reaction where reactant A is being consumed.

Once the data have been obtained, we can then try to get the reaction rate from it. There are several ways to obtain a reaction rate and the methodologies and approaches used, and their efficiencies, vary based on the type of reaction that needs to be evaluated.

In Table 3.1, we have data for a simple reaction where only chemical A is being consumed. Then, we can use two very well-known and simple methods that can help us obtain the reaction order and the value of the reaction constant. These two

Reaction Kinetics

TABLE 3.1
Experimental Data

Time [min]	Concentration of A [mol/l]
0.00	20.00
0.25	11.00
0.50	6.00
0.75	3.00
1.00	1.50
1.25	2.00
1.50	1.60
1.75	1.00
2.00	1.20
2.25	0.60
2.50	0.90
2.75	1.20
3.00	0.90

methods are the integration method and the differential method. We will solve this example using both methods so that we can show the benefits of each of them and can compare the results.

3.5.1 Integration Method

As the name tells us, this methodology takes into account the differential equation for a batch reactor and integrates it in order to obtain a reaction rate. The mole balance to be solved is:

$$\frac{dC_A}{dt} = r_A \tag{3.8}$$

with

$$r_A = -kC_A^\alpha \tag{3.9}$$

We can then substitute Equation 3.9 into 3.8, solve the integration, and obtain a generic equation that can be used to compare with the experimental data:

$$\frac{C_A^{(1-\alpha)} - C_A^{0(1-\alpha)}}{(1-\alpha)} = -kt \tag{3.10}$$

This is a generic expression that is valid for any value of α except for 1. In the case of $\alpha = 1$, Equation 3.10 cannot be applied and Equations 3.8 and 3.9 must be solved again for that special case. The result in such situation is:

$$C_A = C_A^0 \exp^{(-kt)} \tag{3.11}$$

In order to use Equation 3.10 or to know that we need to use Equation 3.11, we have to assume the reaction order, as α is an unknown. With today's technology, we can easily make a solver problem, in Excel, for example, and allow the software to predict α; however, this means some extra work. Otherwise, one can solve this problem for different values of α (normally for integer values so the mathematics is simpler) and compare each result with the experimental information to see which one fits the best. We could use this methodology for all possible vales of α; however, if this approach does not give any result for orders 0, 1, and 2, is highly recommended to change to another methodology.

Figure 3.3 shows the experimental data from Table 3.1 and predicted reaction order for $\alpha = 0$, 1, and 2. It can be easily seen that orders 0 and 2 do not fully predict the data while order 1 is extremely accurate, indicating that this is a first-order reaction.

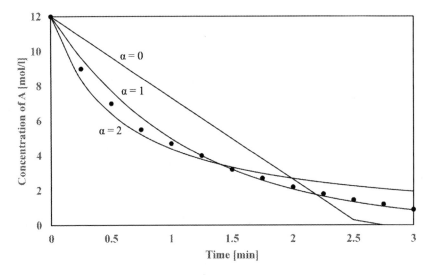

FIGURE 3.3 Comparison of reaction order and data using integration methodology.

This methodology has the benefit of being quite straightforward and easy to implement. Here, one equation is solved, and the result is compared with the data. The limitations of this method are that the order of reaction is an input and therefore, an assumption. Furthermore, it is not so easy to use it when we have a more complex reaction with two or more reactants or if we have an equilibrium reaction. We will present an alternative method for two reactants later on.

3.5.2 Differential Method

This method tries to solve the reactor balance without carrying out the integration; it tries to predict the values of the derivative of the concentration as a function of time and then solve the problem.

Taking into account Equation 3.9 and substituting in Equation 3.8, we get:

$$\frac{dC_A}{dt} = -kC_A^\alpha \tag{3.12}$$

In order to solve this problem, we need to rearrange Equation 3.12; this can easily be done by taking the logarithm on each side of the equality, leading to:

$$\ln\left(-\frac{dC_A}{dt}\right) = \ln(k) + \alpha \ln(C_A) \tag{3.13}$$

Here, we can see the first differences with the integration method. If we plot $\ln\left(\frac{-dC_A}{dt}\right)$ versus $\ln(C_A)$, we obtain a straight line, from which the slope gives the reaction order and from the intercept, we can indirectly obtain the value of k. For this method, the reaction order is a result and not an input; thus, simplifying the process in that matter. However, the complication of this methodology is on how to calculate the derivative term. Even more, as mentioned for the integration, this approach cannot be used if we have more complex scenarios or reversible reactions.

To calculate the differential term, there are several approaches; it can be done by: (i) the finite difference method, which is a very common, established, and reliable technique; (ii) interpolating the data, obtaining a mathematical expression, and then treating it as such; (iii) graphical integration using the Simpsons methodology, and (iv) commercial mathematical software that could implement any of the previously mentioned possibilities, among others. Manual approaches are getting less relevant since easy calculations can be done faster with mathematically oriented software. Nevertheless, it is important for an engineer to have a better understanding of the fundamentals behind what the software is doing. Therefore, the use of an interpolation-based method will be presented here and then treated mathematically as an equation. Note that interpolation can be easily done using software like Excel, Origin, Mathematica, and MATLAB®.

When choosing the type of interpolation equation, it is important to understand what the data are showing us. For the given experimental data, we can have a perfect regression if we use a polynomial equation of a degree that is the number of data minus 1. However, this will not follow the tendency of the experimental data, and in many cases, it will predict the value of the experimental data perfectly but will be completely wrong for all the rest of the predicted information. Therefore, the selection of the interpolation needs to be careful and accurate, since the calculations that follow will depend on this. For instance, for the data in Table 3.1, a polynomial of degree 3 can be selected which will give a $R_2 = 0.996$; however, as it can be seen from Figure 3.4, the slope in the end does not fit well.

A better tendency for these data is to use an exponential approach; Figure 3.5 shows that this is a more accurate representation of the data and shows a very similar R^2 coefficient.

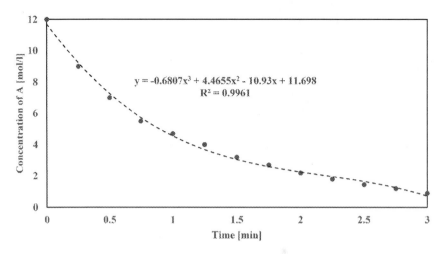

FIGURE 3.4 Interpolation using a third-degree polynomial equation.

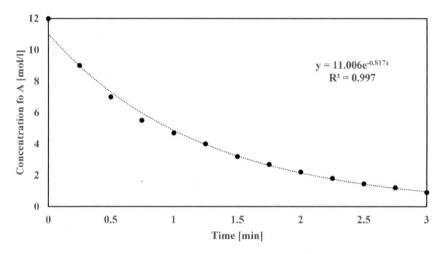

FIGURE 3.5 Interpolation using an exponential equation.

It is important to mention that for this particular case, we already know that a first-order, and therefore, an exponential expression fits the data very accurately. However, this information is not generally available beforehand and this will not be the case if we have a second- or third-order reaction.

Once the mathematical expression for the interpolation of the data is obtained (Equation 3.14), we can then carry out the derivation of this new expression as a function of C_A:

$$C_A = 11.006 \exp^{(-0.817t)} \tag{3.14}$$

Reaction Kinetics

The derivation gives:

$$\frac{dC_A}{dt} = -8.9919 \exp^{(-0.817t)} \quad (3.15)$$

With Equation 3.15, we can produce Table 3.2:

We can now plot the $\ln(-dC_A/dt)$ versus $\ln(C_A)$ to produce Figure 3.6, from where it can be seen that it follows a straight line as expected and the slope (0.9969) gives the value of the reaction order. This value is extremely close to 1, as predicted before.

TABLE 3.2
Results for Calculations Data

Time [min]	Concentration [mol/l]	$-dC_A/dt$	$\ln(-dC_A/dt)$	$\ln(C_A)$
0.00	12	8.99	2.19	2.48
0.25	9	7.33	1.99	2.19
0.50	7	5.97	1.78	1.94
0.75	5.5	4.87	1.58	1.70
1.00	4.7	3.97	1.38	1.54
1.25	4	3.23	1.17	1.38
1.50	3.2	2.64	0.97	1.16
1.75	2.7	2.15	0.76	0.99
2.00	2.2	1.75	0.56	0.78
2.25	1.8	1.43	0.35	0.58
2.50	1.45	1.16	0.15	0.37
2.75	1.2	0.95	−0.05	0.18
3.00	0.9	0.77	−0.25	−0.105

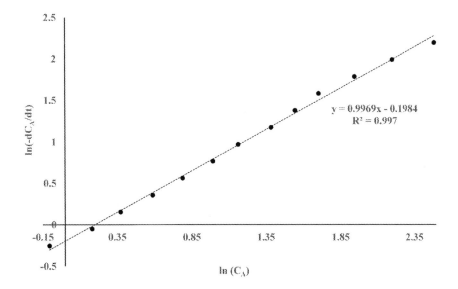

FIGURE 3.6 Interpolation of data to obtain reaction order and reaction constant.

From the intercept, we can obtain the value of k = 0.82 [1/min], which is in good agreement with the value that can be obtained from the integration methodology.

As it has been shown, this methodology takes a little more of calculations and time. However, with today's technology and computer power, this time is considerably reduced and there is not much difference in these methods based on calculation time. Nevertheless, as mentioned, these approaches are suitable for very simple reactions and irreversible cases mainly. If a two-reactant reaction takes place (between two different molecules), the problem requires some strategies to obtain the reaction order for each reactant and the reaction constant. We will not get into the details of this methodology, as this is well presented in other books; however, we will present a methodology for a generalized case of a generic reaction. This general methodology can be applied for a two-reactant irreversible case.

3.5.3 GENERIC METHOD

When dealing with a more complex reaction, irreversible but non-elementary, reversible and elementary, or multiple, reversible, and non-elementary reaction (see Chapter 4 for a full problem on this matter), we need to be able to have a procedure that can help us find the reaction kinetics in all possible cases.

The methodology that is being currently used is based on experience and experimental data, to propose a reaction mechanism, to assign the reaction steps and the controlling step, solve using those assumptions and compare the results with the experimental data using a regression algorithm like the Marquardt. This will allow us to obtain the reaction parameters that best fits our model and the data. This method will be developed in more detail in the upcoming sections where it will be introduced for a simple reaction system as well as for a complex reaction system. In addition, we strongly recommend going through details in Chapter 4, where this approach is being used quite extensively and in detail for a complex reaction, following this example will help understand how this works.

3.6 KINETICS MODELING FOR SIMPLE REACTIONS

First of all, it is important to point out that a reaction is defined as simple or complex, to some extent, arbitrarily. We will deal here with elementary and non-elementary reactions, where one or two reactants are being consumed and one or two products are being produced. These reactions could, at the same time, be reversible. More complex scenarios will be addressed in the following section, when the basic theory has already been understood.

Let's consider a simple elementary irreversible reaction where A → B. For this type of scenario, we can easily recommend using the integration or the derivation method previously explained, since this reaction is very simple. However, in order to follow a generic approach, we will need to solve the mole balance for reactant A:

$$\frac{dC_A}{dt} = r_A = -kC_A \tag{3.16}$$

giving

Reaction Kinetics

$$C_A = C_A^0 e^{(-kt)} \qquad (3.17)$$

Furthermore, compared with the experimental data from the lab, we can estimate the values for the reaction constant. This estimation can easily be done in Excel, for example, with the solver.

While this approach is too simple and has been presented before, it is presented again to establish the sequence of steps that should be followed when dealing with more complex reactions.

We will now make a separation among 1-phase systems and multiphase systems.

3.6.1   1-Phase Irreversible Reaction

The simplest system that can be studied is a reaction taking place in 1-phase, for instance, in liquid phase. Assuming it is an irreversible reaction that takes place in 3 reaction steps, consider the following reaction mechanism:

$$\text{Step 1}: A \rightarrow A^* \qquad (3.18)$$

$$\text{Step 2}: A^* \rightarrow B^* \qquad (3.19)$$

$$\text{Step 3}: B^* \rightarrow B \qquad (3.20)$$

This simple mechanism has three elementary reactions, where chemical A is electronically modified into A*, which could be a reorganization of the electrical charge; then this intermediate is then transformed into B*, which finally is modified into product B. Since the reaction is irreversible, we can apply the Pseudo-Steady-State Hypothesis method (PSSH). This approach considers that the concentration of all intermediate components, which in our case will be A* and B*, over time, are zero. This means that the intermediates are being produced and consumed at the same rate.

Let's first establish the reaction rate for the desired product:

$$r_3 = k_3 C_{B^*} \qquad (3.21)$$

In order to obtain the desired reaction kinetic expression, we need to obtain the variation of C_{B^*}; for that, we will use the PSSH methodology for both intermediates:

$$\frac{dC_{A^*}}{dt} = 0 = k_1 C_A - k_2 C_{A^*} \qquad (3.22)$$

$$\frac{dC_{B^*}}{dt} = 0 = k_2 C_{A^*} - k_3 C_{B^*} \qquad (3.23)$$

From 3.32 and 3.23, we can deduce:

$$C_{A^*} = \frac{k_1}{k_2} C_A \qquad (3.24)$$

$$C_{B^*} = \frac{k_2}{k_3} C_{A^*} \qquad (3.25)$$

Therefore:

$$C_{B^*} = \frac{k_1}{k_3} C_A \qquad (3.26)$$

Substituting 3.26 into Equation 3.21 to obtain the expression for the desired reaction:

$$r_3 = k_1 C_A \qquad (3.27)$$

Equation 3.27 is then the reaction rate expression that needs to be compared with the experimental data to see if the fitting is good. If the data are fitted by the model, then the mechanism proposed could be considered as a possible representation of the process. If the data are not fitted, then a new mechanism, with new intermediate components needs to be proposed and a similar analysis needs to be carried out.

3.6.2 1-Phase Reversible Reaction

In the previous part, we analyzed a 1-phase irreversible reaction that was the simplest case for a mechanism that we wanted to study. Now, we will move into a similar process but with a reversible reaction; this will introduce some level of complexity to the analysis.

Let's assume that reactant A will be isomerized into product B, following:

$$\text{Step 1:} \quad A \rightleftarrows A^* \qquad (3.28)$$
$$\text{Step 2:} \quad A^* \rightleftarrows B^* \qquad (3.29)$$
$$\text{Step 3:} \quad B^* \rightleftarrows B \qquad (3.30)$$

This problem needs to be treated similarly to the previous one; the analysis for this case is more complex than for the previous scenario; however, it is simpler, and it will lead to simpler reaction mechanism due to the absence of a 2-phase system, as we will see in the upcoming cases of heterogeneous reactions. To solve this problem, we will take a look at two different approaches, *i.e.*, the equilibrium method and the PSSH. We will start with the equilibrium approach which considers that one of the steps requires the most energy and therefore, it is the controlling step of the process, while the other steps involved happen so fast that their reactions rate can be considered to be zero, or that the reaction is at an equilibrium state.

3.6.2.1 Equilibrium Method

For each step we can obtain a reaction rate:

$$r_1 = -k_1 C_A + k_2 C_{A^*} \qquad (3.31)$$

Reaction Kinetics

$$r_2 = -k_3 C_{A^*} + k_4 C_{B^*} \tag{3.32}$$

$$r_3 = -k_5 C_{B^*} + k_6 C_B \tag{3.33}$$

We need to develop the reaction rate when each of the possible steps is controlling and once the final equation has been developed, we can compare with the experimental data.

3.6.2.1.1 Scenario 1

Step 1 is the controlling one, consequently, we will consider steps 2 and 3 to be faster and therefore, be considered in equilibrium:

$$r_A = r_1 = -k_1 C_A + k_2 C_{A^*} \tag{3.34}$$

while

$$0 = r_2 = -k_3 C_{A^*} + k_4 C_{B^*} \tag{3.35}$$

$$0 = r_3 = -k_5 C_{B^*} + k_6 C_B \tag{3.36}$$

From Equations 3.35 and 3.36, we obtain:

$$C_{A^*} = \frac{k_4}{k_3} C_{B^*} \tag{3.37}$$

$$C_{B^*} = \frac{k_6}{k_5} C_B \tag{3.38}$$

Substituting 3.38 into 3.37, we get:

$$C_{A^*} = \frac{k_4}{k_3} \frac{k_6}{k_5} C_B \tag{3.39}$$

We can now substitute 3.39 into 3.34:

$$r_A = -k_1 C_A + k_2 \frac{k_4}{k_3} \frac{k_6}{k_5} C_B \tag{3.40}$$

Equation 3.40 is the reaction rate when step 1 is the controlling step. This expression needs to then be compared with experimental information to see if the data can be fitted with this model. Otherwise, a new controlling step, and eventually new mechanism, must be proposed and developed.

3.6.2.1.2 Scenario 2

Step 2 is the controlling one; thus, steps 1 and 3 to be faster and therefore, be considered in equilibrium:

$$r_A = r_2 = -k_3 C_{A^*} + k_4 C_{B^*} \tag{3.41}$$

while

$$0 = r_1 = -k_1 C_A + k_2 C_{A^*} \tag{3.42}$$

$$0 = r_3 = -k_5 C_{B^*} + k_6 C_B \tag{3.43}$$

From Equations 3.42 and 3.43, we obtain:

$$C_{A^*} = \frac{k_1}{k_2} C_A \tag{3.44}$$

$$C_{B^*} = \frac{k_6}{k_5} C_B \tag{3.45}$$

We can now substitute 3.44 and 3.45 into 3.41:

$$r_A = -k_3 \frac{k_1}{k_2} C_A + k_4 \frac{k_6}{k_5} C_B \tag{3.46}$$

Equation 3.46 represents the reaction rate when step 2 is controlling.

3.6.2.1.3 Scenario 3

Finally, if none of the previous steps was good enough when fitting the data, we should then try analyzing the process using the last step as the one with the highest energy and therefore, the limiting one.

By doing that, we are taking into consideration that steps 1 and 2 are in equilibrium, or they are much faster in comparison with step 3:

$$r_A = r_3 = -k_5 C_{B^*} + k_6 C_B \tag{3.47}$$

while

$$0 = r_1 = -k_1 C_A + k_2 C_{A^*} \tag{3.48}$$

$$0 = r_2 = -k_3 C_{A^*} + k_4 C_{B^*} \tag{3.49}$$

From Equations 3.48 and 3.49, we obtain:

$$C_{A^*} = \frac{k_1}{k_2} C_A \tag{3.50}$$

$$C_{B^*} = \frac{k_3}{k_4} C_{A^*} \tag{3.51}$$

We can then substitute Equation 3.50 into 3.51:

$$C_{B^*} = \frac{k_3}{k_4} \frac{k_1}{k_2} C_A \tag{3.52}$$

Reaction Kinetics

We can now substitute Equation 3.52 into 3.47:

$$r_A = -k_5 \frac{k_3}{k_4} \frac{k_1}{k_2} C_A + k_6 C_B \qquad (3.53)$$

Equation 3.53 represents the reaction rate when the controlling step is step 3.

As previously mentioned, now that all the possible reaction rates have been developed, if none of them satisfactorily fits the data, a new mechanism should be proposed, new kinetics expression should be developed, and a new comparison among the model and the data should be carried out.

It is crucial to not overlook the point that we have obtained the reaction expression based on the reactant A. Therefore, if the experimental data are based on the desired product, we need to convert the reaction rate from reactant to produce; to do that, we need to remember:

$$\frac{r_A}{\nu_A} = \frac{r_P}{\nu_P} \qquad (3.54)$$

3.6.2.2 PSSH Method

As done for the 1-phase irreversible reaction, we will apply the PSSH methodology here as well. Using the same principles, the concentration of intermediate components over time are zero, therefore, they are produced and consumed at the same pace.

We first need to write the reaction rate equation to produce the desired product:

$$r_B = k_5 C_{B^*} - k_6 C_B \qquad (3.55)$$

In order to solve this problem using the PSSH methodology, we need to solve:

$$\frac{dC_{A^*}}{dt} = 0 = k_1 C_A - k_2 C_{A^*} - k_3 C_{A^*} + k_4 C_{B^*} \qquad (3.56)$$

$$\frac{dC_{B^*}}{dt} = 0 = k_3 C_{A^*} - k_4 C_{B^*} - k_5 C_{B^*} + k_6 C_B \qquad (3.57)$$

This is a system of two equations with two unknowns; therefore, it can be solved and will give you one and only one solution. We encourage the student to do the math in order to check the steps involved when reaching the answer. Solving these two equations, we obtain:

$$C_{B^*} = \left(\frac{k_1 k_3 C_A + k_2 k_6 C_B + k_3 k_6 C_B}{k_2 k_4 + k_2 k_5 + k_3 k_5} \right) \qquad (3.58)$$

Substituting this into Equation 3.55:

$$r_B = k_5 \left(\frac{k_1 k_3 C_A + k_2 k_6 C_B + k_3 k_6 C_B}{k_2 k_4 + k_2 k_5 + k_3 k_5} \right) - k_6 C_B \qquad (3.59)$$

Equation 3.59 is the reaction rate for the production of the desired chemical B.

This expression is different from the one obtained from the equilibrium method; this is because the assumptions behind each method are different.

So far, we have solved 1-phase systems, let's assumed that we have a liquid–liquid reaction but it takes place on the surface of a solid catalyst; this new system is a 2-phase system. A similar and common case is a gas–solid system.

3.6.3 2-Phase Irreversible Reaction

We will start with what we think is a simple case. In this reaction, chemical A is transform into chemical C in a liquid–solid reaction irreversible in a 3-step process. The steps involved are adsorption of the reactant, reaction over the catalyst, as well as desorption of the desired product.

Adsorption of the reactant:

$$A + S \rightarrow AS \tag{3.60}$$

Reaction:

$$AS \rightarrow CS \tag{3.61}$$

Desorption of the product:

$$CS \rightarrow C + S \tag{3.62}$$

In order to solve this problem, as we have done for the previous scenarios, we will use the PSSH method.

3.6.3.1 PSSH Method

For the previously mentioned reaction scheme, we need to then write the reaction rate for the desired product C:

$$r_C = k_3 C_{CS} \tag{3.63}$$

In order to solve this expression, we need to find out the expression for the C_{CS}. Using the PSSH method, we then, need to solve:

$$\frac{dC_{CS}}{dt} = 0 = k_2 C_{AS} - k_3 C_{CS} \tag{3.64}$$

$$\frac{dC_{AS}}{dt} = 0 = k_1 C_A C_S - k_2 C_{AS} \tag{3.65}$$

Now we can solve for both of the adsorbed chemicals, or intermediates, to obtain:

$$C_{AS} = \frac{k_1}{k_2} C_A C_S \tag{3.66}$$

Reaction Kinetics

$$C_{CS} = \frac{k_2}{k_3} C_{AS} \tag{3.67}$$

We can substitute Equation 3.67 into 3.66 in order to get:

$$C_{CS} = \frac{k_1}{k_3} C_A C_S \tag{3.68}$$

Now that we have an expression that can be substituted into Equation 3.63 to produce the reaction rate for the production of chemical C:

$$r_C = k_1 C_A C_S \tag{3.69}$$

Equation 3.69 represents the production of chemical C as a function of the concentration of chemical A as well as the concentration of the active sites of the catalyst.

The concentration of the active sites, C_S, is related to the total amount of catalyst that is added to the reaction itself. Therefore, we need to make balance of the catalytic sites. This can be generically represented by the following balance:

$$C_S^0 = C_S + \sum_{i=1}^{n} C_{iS} \tag{3.70}$$

where C_S is the available amount of catalyst, C_{iS} is the amount of catalyst that is currently being used and C_S^0 is the total initial amount of catalyst.

For our previous case, this is:

$$C_S^0 = C_S + C_{AS} + C_{CS} \tag{3.71}$$

Substituting Equations 3.66 and 3.68 into 3.71:

$$C_S^0 = C_S + \frac{k_1}{k_2} C_A C_S + \frac{k_1}{k_3} C_A C_S \tag{3.72}$$

This expression can then be reorganized to obtain:

$$C_S = \frac{C_S^0}{\left(1 + C_A \left(\frac{k_1}{k_2} + \frac{k_1}{k_3}\right)\right)} \tag{3.73}$$

Equation 3.73 can be substituted into Equation 3.69 to give the final reaction rate expression:

$$r_C = \frac{k_1 C_A C_S^0}{\left(1 + C_A \left(\frac{k_1}{k_2} + \frac{k_1}{k_3}\right)\right)} \tag{3.74}$$

Equation 3.74 is the final expression that should be compared toward the experimental data in order to see if it fits. If the mathematical model does not fit the data, a new mechanism should be proposed, and a similar analysis should be carried out.

The alternative method, the equilibrium approach, cannot be used for this type of reaction since it is irreversible.

3.6.4 2-Phase Reversible Reaction

We will now study a heterogeneous reaction (2-phase system) for the production of chemical C. In this case, we will have a reversible non-elementary reaction taking place over a solid catalyst. The global reaction is as follows:

$$A \leftrightarrows C \tag{3.75}$$

However, as it is a non-elementary reaction, this global reaction is the sum up of all elementary steps. The steps involved could be unknown, and therefore, a reaction mechanism must be presented, solved, and compared with the experimental data. For this purpose, we will introduce a mechanism here, assuming the following reaction steps:

Adsorption of the reactant:

$$A + S \leftrightarrows AS \tag{3.76}$$

Reaction:

$$AS \leftrightarrows CS \tag{3.77}$$

Desorption of the product:

$$CS \leftrightarrows C + S \tag{3.78}$$

Since now we are looking at a system where we have three steps, we need to establish which one of these steps control the reaction. Recalling Figure 3.2, the step which requires the most energy for the full reaction to occur, is the controlling one. This means that we have three options and therefore, three final answers that need to be compared with the experimental data, where one of them might fit the data.

In order to obtain the reaction rates, we need to not only decide which step is controlling, but also what methodology we are willing to use to solve it. There are typically two approaches, the equilibrium method and the PSSH method. We will try to solve the problem using both methodologies.

3.6.4.1 Equilibrium Method

This approach is based on the premise that all the steps that are *not* controlling, are so fast that they are in equilibrium.[1] Therefore, for our case, we will need to solve the problem by considering each of the different steps as controlling and the rest in equilibrium.

Reaction Kinetics

Before starting, the reaction rates for each step are given by:

For adsorption:

$$r_{ads} = -k_1 C_A C_S + k_2 C_{AS} \tag{3.79}$$

For reaction:

$$r_{rx} = -k_3 C_{AS} + k_4 C_{CS} \tag{3.80}$$

For desorption:

$$r_{des} = -k_5 C_{CS} + k_6 C_C C_S \tag{3.81}$$

Now that all the steps have been explicitly presented and their reaction rate established, we can then start solving the problem for each possible scenario.

3.6.4.1.1 Scenario 1

For this case, we will assume that the controlling step is the adsorption of the reactant A, and therefore, the reaction and desorption steps are so fast that they can be considered to be in equilibrium; this leads to:

$$r_A = r_{ads} = -k_1 C_A C_S + k_2 C_{AS} \tag{3.82}$$

while

$$0 = r_{rx} = -k_3 C_{AS} + k_4 C_{CS} \tag{3.83}$$

$$0 = r_{des} = -k_5 C_{CS} + k_6 C_C C_S \tag{3.84}$$

From Equations 3.83 and 3.84, the following can be obtained:

$$C_{AS} = \frac{k_4}{k_3} C_{CS} \tag{3.85}$$

$$C_{CS} = \frac{k_6}{k_5} C_C C_S \tag{3.86}$$

Then, substituting 3.86 into 3.85:

$$C_{AS} = \frac{k_4}{k_3} \frac{k_6}{k_5} C_C C_S \tag{3.87}$$

Then, Equation 3.87 can be substituted into 3.82:

$$r_A = -k_1 C_A C_S + k_2 \frac{k_4}{k_3} \frac{k_6}{k_5} C_C C_S \tag{3.88}$$

Equation 3.88 is the reaction rate when the adsorption is controlling. It is important to notice that there is a concentration of the catalysts, C_S, which is information that we do not have but that we can obtain based on the total initial amount of catalyst, C_S^0.

Doing the site balance as done before, we need to solve:

$$C_S^0 = C_S + \sum_{i=1}^{n} C_{iS} \tag{3.89}$$

where C_S is the available amount of catalyst, C_{iS} is the amount of catalyst that is currently being used, and C_S^0 is the total initial amount of catalyst.

For our previous case, this is:

$$C_S^0 = C_S + C_{AS} + C_{CS} \tag{3.90}$$

Substituting Equations 3.86 and 3.87:

$$C_S^0 = C_S + \frac{k_4 \, k_6}{k_3 \, k_5} C_C C_S + \frac{k_6}{k_5} C_C C_S \tag{3.91}$$

Equation 3.91 can be rearranged in terms of C_S; so, it can be substituted in Equation 3.88:

$$C_S = \frac{C_S^0}{1 + \left(\dfrac{k_4 \, k_6}{k_3 \, k_5} + \dfrac{k_6}{k_5} \right) C_C} \tag{3.92}$$

Substituting it into Equation 3.88:

$$r_A = \left(-k_1 C_A + k_2 \frac{k_4 \, k_6}{k_3 \, k_5} C_C \right) \frac{C_S^0}{1 + \left(\dfrac{k_4 \, k_6}{k_3 \, k_5} + \dfrac{k_6}{k_5} \right) C_C} \tag{3.93}$$

Equation 3.93 represents the reaction rate for the reactant A when the adsorption is the controlling step. This expression needs to be compared with experimental data and if that is the case, this might be the current controlling step we need. If the data cannot be fitted, then a new controlling step needs to be considered.

3.6.4.1.2 Scenario 2

For this case, we will assume that the controlling step is the reaction step. Therefore, both adsorption and desorption are fast and considered to be in equilibrium. Typically, the reaction step involves the breakdown and production of new chemical bonds.

Reaction Kinetics

These steps are very energy intense and for most of the cases, they are the controlling steps. However, this is not true for all cases; so it is important to not lose sight of the larger picture.

Therefore:

$$r_A = r_{rx} = -k_3 C_{AS} + k_4 C_{CS} \tag{3.94}$$

while

$$0 = r_{ads} = -k_1 C_A C_S + k_2 C_{AS} \tag{3.95}$$

$$0 = r_{des} = -k_5 C_{CS} + k_6 C_C C_S \tag{3.96}$$

From Equations 3.95 and 3.96, the following can be obtained:

$$C_{AS} = \frac{k_1}{k_2} C_A C_S \tag{3.97}$$

$$C_{CS} = \frac{k_6}{k_5} C_C C_S \tag{3.98}$$

Then, we can obtain:

$$r_A = -k_3 \frac{k_1}{k_2} C_A C_S + k_4 \frac{k_6}{k_5} C_C C_S \tag{3.99}$$

Equation 3.99 is the reaction rate when the reaction step is controlling. Similarly, to the previous case, we need to obtain the functionality of C_S with initial concentration as well as with measurable data. Using Equation 3.90 and substituting into Equations 3.97 and 3.98, we obtain:

$$C_S^0 = C_S + \frac{k_1}{k_2} C_A C_S + \frac{k_6}{k_5} C_C C_S \tag{3.100}$$

Equation 3.100 can be rearranged in terms of C_S:

$$C_S = \frac{C_S^0}{1 + \frac{k_1}{k_2} C_A + \frac{k_6}{k_5} C_C} \tag{3.101}$$

Substituting it into Equation 3.99:

$$r_A = \left(-k_3 \frac{k_1}{k_2} C_A + k_4 \frac{k_6}{k_5} C_C \right) \frac{C_S^0}{1 + \frac{k_1}{k_2} C_A + \frac{k_6}{k_5} C_C} \tag{3.102}$$

Equation 3.102 represents the reaction rate for the reactant A when the reaction step is controlling. As before, this model should be compared with the experimental data and see if the model fits. If it does not, we should then consider the third alternative controlling step, i.e., the desorption step.

3.6.4.1.3 Scenario 3

For this case, we will assume that the controlling step is the desorption of the product generated. As before, we will therefore, assume that the other steps, adsorption, and reaction are much faster and therefore, can be considered in equilibrium.

Therefore:

$$r_A = r_{des} = -k_5 C_{CS} + k_6 C_C C_S \tag{3.103}$$

while

$$0 = r_{ads} = -k_1 C_A C_S + k_2 C_{AS} \tag{3.104}$$

$$0 = r_{rx} = -k_3 C_{AS} + k_4 C_{CS} \tag{3.105}$$

From Equations 3.104 and 3.105, the following can be obtained:

$$C_{AS} = \frac{k_1}{k_2} C_A C_S \tag{3.106}$$

$$C_{CS} = \frac{k_3}{k_4} C_{AS} \tag{3.107}$$

Then, we can substitute Equation 3.104 into 3.107:

$$C_{CS} = \frac{k_3}{k_4} \frac{k_1}{k_2} C_A C_S \tag{3.108}$$

Then, we can obtain:

$$r_A = -k_5 \frac{k_3}{k_4} \frac{k_1}{k_2} C_A C_S + k_6 C_C C_S \tag{3.109}$$

Equation 3.109 is the reaction rate when the desorption is the controlling step. Similarly, to the previous case, we need to obtain the functionality of C_S with initial concentration as well as with measurable data. Using Equation 3.90 and substituting in it Equations 3.106 and 3.108, we obtain:

$$C_S^0 = C_S + \frac{k_1}{k_2} C_A C_S + \frac{k_3}{k_4} \frac{k_1}{k_2} C_A C_S \tag{3.110}$$

Equation 3.110 can be rearranged in terms of C_S:

$$C_S = \frac{C_S^0}{1 + \frac{k_1}{k_2}C_A + \frac{k_3}{k_4}\frac{k_1}{k_2}C_\ell} \tag{3.111}$$

Substituting it into Equation 3.109:

$$r_A = \left(-k_5 \frac{k_3}{k_4}\frac{k_1}{k_2} C_A + k_6 C_C\right) \left(\frac{C_S^0}{1 + \left(\frac{k_1}{k_2} + \frac{k_3}{k_4}\frac{k_1}{k_2}\right)C_A}\right) \tag{3.112}$$

Equation 3.112 represents the reaction rate for the reactant A when the desorption step is controlling. As before, this model should be compared with the experimental data to see if the model fits.

After all the possible controlling steps for this mechanism have been developed and compared with the data, we might have a possible mechanism that could represent our experimental values. If none of them fits the data, then it is important to propose a full new mechanism and start all over again with the analysis. In Chapter 4, we have done this procedure as a full-extended example and recommend a careful reading of it.

A second methodology introduced previously is the PSSH method. We will try to solve the problem again using this approach.

3.6.4.2 PSSH Method

We will now solve the same problem, as we just did, but using the PSSH approach instead.

Based on the same mechanism proposed in Equations 3.72 to 3.75 for adsorption, reaction, and desorption, respectively, we can the write the reaction rate for the desired product C:

$$r_C = k_5 C_{CS} - k_6 C_C C_S \tag{3.113}$$

Applying the PSSH method means the concentration of the intermediates components over time is zero, they are produced and consumed very fast.

In our case, this means that the variation of the concentration of C_{CS} and C_{AS} with time are zero:

$$\frac{dC_{CS}}{dt} = 0 = k_3 C_{AS} - k_4 C_{CS} - k_5 C_{CS} + k_6 C_C C_S \tag{3.114}$$

$$\frac{dC_{AS}}{dt} = 0 = k_1 C_A C_S - k_2 C_{AS} - k_3 C_{AS} + k_4 C_{CS} \tag{3.115}$$

Reorganizing Equation 3.115, we can obtain an equation for C_{AS}:

$$C_{AS} = \frac{k_1 C_A C_S + k_4 C_{CS}}{(k_2 + k_3)} \quad (3.116)$$

We substitute 3.116 into 3.114 and solve it to obtain an expression for C_{CS}:

$$C_{CS} = \frac{(k_1 k_3 C_A C_S + k_2 k_6 C_C C_S + k_3 k_6 C_C C_S)}{(k_2 k_4 + k_2 k_5 + k_3 k_6)} \quad (3.117)$$

Equation 3.117 can be substituted into Equation 3.116 in order to have an equation for C_{AS}:

$$C_{AS} = \frac{k_1 C_A C_S}{(k_2 + k_3)} + \left(\frac{k_4}{(k_2 + k_3)} \left(\frac{(k_1 k_3 C_A C_S + k_2 k_6 C_C C_S + k_3 k_6 C_C C_S)}{(k_2 k_4 + k_2 k_5 + k_3 k_6)} \right) \right) \quad (3.118)$$

In order to have a final expression for the reaction rate, we need to substitute Equation 3.117 into Equation 3.113:

$$r_C = k_5 \left(\frac{(k_1 k_3 C_A C_S + k_2 k_6 C_C C_S + k_3 k_6 C_C C_S)}{(k_2 k_4 + k_2 k_5 + k_3 k_6)} \right) - k_6 C_C C_S \quad (3.119)$$

Equation 3.119 can be reorganized into Equation 3.120 in order to take as common factor the concentration of C_s:

$$r_C = C_S \left(k_5 \left(\frac{(k_1 k_3 C_A + k_2 k_6 C_C + k_3 k_6 C_C)}{(k_2 k_4 + k_2 k_5 + k_3 k_6)} \right) - k_6 C_C \right) \quad (3.120)$$

In order to find the final expression for the reaction rate in terms of information that can be measured or that is known, we need to solve the balance of the catalytic sites so that we can substitute C_S for an expression based on measurable and known data.

For this, we will use Equation 3.90, which is:

$$C_S^0 = C_S + C_{AS} + C_{CS} \quad (3.121)$$

We will substitute Equations 3.117 and 3.118 into 3.121 to get:

$$C_S^0 = C_S + \frac{k_1 C_A C_S}{(k_2 + k_3)} + \left(\frac{k_4 C_{CS}}{(k_2 + k_3)} \left(\frac{(k_1 k_3 C_A C_S + k_2 k_6 C_C C_S + k_3 k_6 C_C C_S)}{(k_2 k_4 + k_2 k_5 + k_3 k_6)} \right) \right)$$
$$+ \left(\frac{(k_1 k_3 C_A C_S + k_2 k_6 C_C C_S + k_3 k_6 C_C C_S)}{(k_2 k_4 + k_2 k_5 + k_3 k_6)} \right) \quad (3.122)$$

Reaction Kinetics

Rearranging:

$$C_S = \frac{C_S^0}{1+\frac{k_1 C_A C_S}{(k_2+k_3)}+\left(\frac{k_4 C_{CS}}{(k_2+k_3)}\left(\frac{(k_1 k_3 C_A C_S + k_2 k_6 C_C C_S + k_3 k_6 C_C C_S)}{(k_2 k_4 + k_2 k_5 + k_3 k_6)}\right)\right)+\left(\frac{(k_1 k_3 C_A C_S + k_2 k_6 C_C C_S + k_3 k_6 C_C C_S)}{(k_2 k_4 + k_2 k_5 + k_3 k_6)}\right)} \quad (3.123)$$

We can then substitute Equation 3.123 into 3.120:

$$r_C = \left(k_5 \left(\frac{(k_1 k_3 C_A + k_2 k_6 C_C + k_3 k_6 C_C)}{(k_2 k_4 + k_2 k_5 + k_3 k_6)}\right) - k_6 C_C\right)$$

$$* \left(\frac{C_S^0}{1+\frac{k_1 C_A C_S}{(k_2+k_3)}+\left(\frac{k_4 C_{CS}}{(k_2+k_3)}\left(\frac{(k_1 k_3 C_A C_S + k_2 k_6 C_C C_S + k_3 k_6 C_C C_S)}{(k_2 k_4 + k_2 k_5 + k_3 k_6)}\right)\right)+\left(\frac{(k_1 k_3 C_A C_S + k_2 k_6 C_C C_S + k_3 k_6 C_C C_S)}{(k_2 k_4 + k_2 k_5 + k_3 k_6)}\right)}\right) \quad (3.124)$$

Equation 3.124 represents the reaction rate for the production of chemical C when using the PSSH methodology instead of the equilibrium approach. As mentioned before, the two different methodologies produce two different results since the assumptions considered are different.

As for all previous cases, it is important to point out that this expression must be compared with the experimental information and evaluate the goodness of the fitting. If that is not the case, a new mechanism must be proposed, a similar analysis conducted, and a new comparison carried out. This must be repeated as many times as it is required in order to have a mechanism, and therefore, a reaction rate, that can satisfactorily model the data.

3.7 KINETIC MODELING OF COMPLEX SYSTEMS

So far, we have looked at what we consider simple reactions, *i.e.*, those reactions that include very little and few steps, if any. However, there are reactions, most of them, in series or in parallel, that take place simultaneously. As presented in Chapter 2, this can be a system like:

1. A + B → C + D
2. A + C → F + G
3. D + G → E
4. B + E → H

where different chemicals can be either only consumed, only produced, or consumed and produced; giving a system with a mix of reactions in parallel and in series.

In order to solve this problem, we need to know if the process is reversible or not, if the reactions are elementary reactions or not, and so on. The more complex the system, the more complex the model is as well.

3.7.1 Homogeneous Systems

Consider the following set of reactions:

$$A + B \rightleftharpoons C + D \quad (3.125)$$
$$D + B \rightleftharpoons C + E \quad (3.126)$$
$$E + B \rightleftharpoons C + F \quad (3.127)$$

This set of parallel and series reactions is a typical case to produce, for instance, biodiesel from a pure triglyceride in the presence of an alcohol.

We will assume that each reaction involve is an elementary step; therefore, we can present a reaction rate for each reaction step involve:

$$r_1 = (k_1 C_A C_B - k_2 C_C C_D) \quad (3.128)$$
$$r_2 = (k_3 C_D C_B - k_4 C_C C_E) \quad (3.129)$$
$$r_3 = (k_5 C_E C_B - k_6 C_C C_F) \quad (3.130)$$

Therefore, to study the variation of any chemical we need to solve the generic expression:

$$\frac{dC_i}{dt} = \sum_{j=1}^{n} \left(\frac{r_{ij}}{\upsilon_{ij}}\right) = \frac{r_{i1}}{\upsilon_{i1}} + \frac{r_{i2}}{\upsilon_{i2}} + \ldots + \frac{r_{ij}}{\upsilon_{ij}} + \ldots + \frac{r_{in}}{\upsilon_{in}} \quad (3.131)$$

For our specific case, it will lead to:

$$\frac{dC_A}{dt} = \sum_{j=1}^{3} \left(\frac{r_{Aj}}{\upsilon_{Aj}}\right) = -r_{A1} \quad (3.132)$$

$$\frac{dC_B}{dt} = \sum_{j=1}^{3} \left(\frac{r_{Bj}}{\upsilon_{Bj}}\right) = -r_{B1} - r_{B2} - r_{B3} \quad (3.133)$$

$$\frac{dC_C}{dt} = \sum_{j=1}^{3} \left(\frac{r_{Cj}}{\upsilon_{Cj}}\right) = r_{C1} + r_{C2} + r_{C3} \quad (3.134)$$

$$\frac{dC_D}{dt} = \sum_{j=1}^{3} \left(\frac{r_{Dj}}{\upsilon_{Dj}}\right) = r_{D1} - r_{D2} \quad (3.135)$$

$$\frac{dC_E}{dt} = \sum_{j=1}^{3} \left(\frac{r_{Ej}}{\upsilon_{Ej}}\right) = r_{E2} - r_{E3} \quad (3.136)$$

$$\frac{dC_F}{dt} = \sum_{j=1}^{3} \left(\frac{r_{Fj}}{\upsilon_{Fj}}\right) = r_{F3} \quad (3.137)$$

Reaction Kinetics

Substituting 3.128, 3.129, and 3.130 into 3.132–3.137, we get the following set of differential equations to be solved simultaneously:

$$\frac{dC_A}{dt} = -k_1 C_A C_B + k_2 C_C C_D \tag{3.138}$$

$$\frac{dC_B}{dt} = -k_1 C_A C_B + k_2 C_C C_D - k_3 C_D C_B + k_4 C_C C_E - k_5 C_E C_B + k_6 C_C C_F \tag{3.139}$$

$$\frac{dC_C}{dt} = k_1 C_A C_B - k_2 C_C C_D + k_3 C_D C_B - k_4 C_C C_E + k_5 C_E C_B - k_6 C_C C_F \tag{3.140}$$

$$\frac{dC_D}{dt} = k_1 C_A C_B - k_2 C_C C_D - k_3 C_D C_B + k_4 C_C C_E \tag{3.141}$$

$$\frac{dC_E}{dt} = k_3 C_D C_B - k_4 C_C C_E - k_5 C_E C_B + k_6 C_C C_F \tag{3.142}$$

$$\frac{dC_F}{dt} = k_5 C_E C_B - k_6 C_C C_F \tag{3.143}$$

Considering that $k_i = k_{i\infty} e^{\left(-E_a/RT\right)}$ with $i = 1-6$, we can then solve the desired set of equations. In this case, we have arbitrarily given values to the parameters in order to show the tendency of the concentration profiles; this can be seen in Figure 3.7.

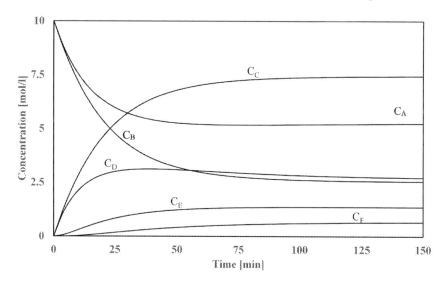

FIGURE 3.7 Concentration profiles.

In order to see if this mechanism with the corresponding assumption is correct, we need to compare the result from the model with the experimental data. If there is a good fit, the mechanism could be representing the reality of the reaction. However, in this case, since we have so many parameters, it is not as simple as it can be done for the simpler cases. Therefore, to solve this type of problem, we can either use a commercial software where a multidifferential equation solver is already in, or we can write our own code to solve simultaneous differential equations using, for example, a Marquardt algorithm.

We have presented the previous case that is quite similar to another one in Section 2.9.3. This is done since now we can introduce a modification from the previous case. We can now change the assumptions that these are elementary steps and we can then introduce that each of the steps (1, 2, and 3) are actually non-elementary and each one of them might follow a set of elementary steps. For simplification we will consider that only step 1 is a set of two elementary steps (1A and 1B) while steps 2 and 3 will remain as elementary. The new reaction system is:

$$1-\alpha) A \leftrightarrows A^* \tag{3.144}$$

$$1-\beta) A^* + B \leftrightarrows C + D \tag{3.145}$$

$$2) D + B \leftrightarrows C + E \tag{3.146}$$

$$3) E + B \leftrightarrows C + F \tag{3.147}$$

Equations 3.144 and 3.145 together represent what it used to be reaction 3.125. So, for reaction 1, we have two possible steps and therefore, two options of a controlling steps for reaction 1, while reactions 2 and 3 remains the same as before. To get the expression for each of the reaction rates, we will use the equilibrium methodology.

The reaction rates for this new process are:

$$r_{1\alpha} = \left(k_1 C_A - k_2 C_{A^*}\right) \tag{3.148}$$

$$r_{1\beta} = \left(k_3 C_{A^*} C_B - k_4 C_C C_D\right) \tag{3.149}$$

$$r_2 = \left(k_5 C_D C_B - k_6 C_C C_E\right) \tag{3.150}$$

$$r_3 = \left(k_7 C_E C_B - k_8 C_C C_F\right) \tag{3.151}$$

If reaction 1-α controls, then the reaction 3.145 (1-β) is in equilibrium and can be used to obtain the expression for C_{A^*}:

$$0 = r_{1\beta} = \left(k_3 C_{A^*} C_B - k_4 C_C C_D\right) \tag{3.152}$$

Then:

$$C_{A^*} = \frac{k_4 C_C C_D}{k_3 C_B} \tag{3.153}$$

We can substitute Equation 3.153 into Equation 3.148 to get:

$$r_1 = r_{1\alpha} = \left(k_1 C_A - k_2 \frac{k_4 C_C C_D}{k_3 C_B}\right) \tag{3.154}$$

Reaction Kinetics

Equation 3.154 together with Equations 3.150 and 3.151 provide the reaction rate expression for the new system when equation 1-α is controlling. In order to solve the problem, we can now substitute Equations 3.150, 3.151, and 3.154 into Equations 3.132–3.137 and solve the system of differential equations. We will show the result for this case only since the important part of this chapter is to obtain the reaction rate expression. For the implementation, we strongly recommend the reader to take a close look at Chapter 4. Figure 3.8 shows the new concentration profiles for the new mechanism with reaction 1-α as the controlling steps. The numerical values have been kept as similar as possible to those used in the previous case.

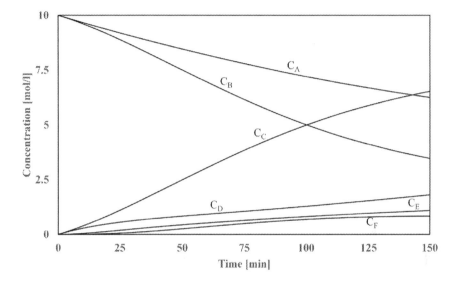

FIGURE 3.8 Concentration profiles.

Similarly, we need to solve the problem when expression 1-β is the controlling step, while the rest of the equation remain the same. Therefore, we have:

$$r_{1\alpha} = \left(k_1 C_A - k_2 C_{A^*}\right) \tag{3.155}$$

$$r_{1\beta} = \left(k_3 C_{A^*} C_B - k_4 C_C C_D\right) \tag{3.156}$$

$$r_2 = \left(k_5 C_D C_B - k_6 C_C C_E\right) \tag{3.157}$$

$$r_3 = \left(k_7 C_E C_B - k_8 C_C C_F\right) \tag{3.158}$$

We need to have an expression for C_{A^*} to be substitute into Equation 3.156. Using the equilibrium method, then Equation 3.155 is considered to be in equilibrium and its reaction rate then is zero. From there, we can get:

$$C_{A^*} = \frac{k_1 C_A}{k_2} \tag{3.159}$$

We can substitute Equation 3.159 into 3.156 and obtain:

$$r_1 = r_{1\beta} = \left(\frac{k_1 k_3 C_A C_B}{k_2} - k_4 C_C C_D\right) \tag{3.160}$$

Equation 3.160, together with expressions 3.157 and 3.158 can be substituted into Equations 3.132–3.137 and the differential equations system can be solved. Figure 3.9 shows the variation of the concentration as a function of time when expression 1-β is the limiting step.

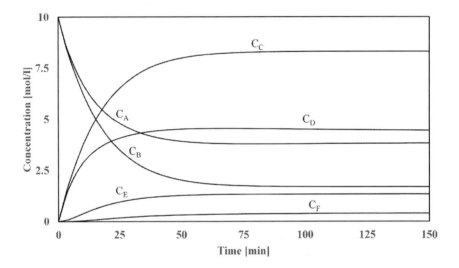

FIGURE 3.9 Concentration profiles.

Figures 3.7, 3.8, and 3.9, show the concentration variation for the different chemicals for the same reaction when different steps are limiting or controlling. It is not easier to see the difference among them since they are a lot of chemicals involved. Therefore, in Figure 3.10, we have selected one of the chemicals to show the different profiles when the different mechanisms are employed. It is easier to see that there are different assumptions behind each model and therefore, the mathematical solution is different. To know which one of them is the one that is representing the experimental data, a regression among the different models and data must be done to adjust the parameters and to see the fitting of the models. The one that fits the best, not only mathematically, but also physically (meaning that the result has physical meaning for the problem under consideration) is the one that is most likely to represent what is going on in our reactor.

Reaction Kinetics

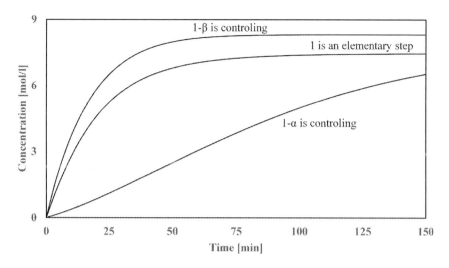

FIGURE 3.10 Comparison of the concentration profile for chemical C.

A more complex mechanism can be presented for the same three reaction steps, in this case, we will present that each of the steps is built of two elementary steps. Therefore, our previous reaction system:

$$A + B \leftrightarrows C + D \tag{3.161}$$

$$D + B \leftrightarrows C + E \tag{3.162}$$

$$E + B \leftrightarrows C + F \tag{3.163}$$

While now be as follows:

$$1\text{-}\alpha) \; A \leftrightarrows A^* \tag{3.164}$$

$$1\text{-}\beta) \; A^* + B \leftrightarrows C + D \tag{3.165}$$

$$2\text{-}\alpha) \; D \leftrightarrows D^* \tag{3.166}$$

$$2\text{-}\beta) \; D^* + B \leftrightarrows C + E \tag{3.167}$$

$$3\text{-}\alpha) \; E \leftrightarrows E^* \tag{3.168}$$

$$3\text{-}\beta) \; E^* + B \leftrightarrows C + F \tag{3.169}$$

For this new reaction system, we have two possible controlling steps (α and β) for each reaction step (1, 2, and 3). Therefore, we will have several reaction rates to evaluate against the experimental data that is a combinatory of all these possibilities.

Let's first write the reaction rate for all the steps involved:

$$r_{1_\alpha} = \left(k_1 C_A - k_2 C_{A^*} \right) \tag{3.170}$$

$$r_{1\beta} = \left(k_3 C_{A^*} C_B - k_4 C_C C_D\right) \quad (3.171)$$

$$r_{2\alpha} = \left(k_5 C_D - k_6 C_{D^*}\right) \quad (3.172)$$

$$r_{2\beta} = \left(k_7 C_{D^*} C_B - k_8 C_C C_E\right) \quad (3.173)$$

$$r_{3\alpha} = \left(k_9 C_E - k_{10} C_{E^*}\right) \quad (3.174)$$

$$r_{3\beta} = \left(k_{11} C_{E^*} C_B - k_{12} C_C C_F\right) \quad (3.175)$$

Based on Equations 3.170 to 3.175, we can obtain the reaction rate for steps 1, 2, and 3, for all possible combinations of controlling steps. We can separate the problem into three different steps and obtain all the reaction rates for each step and then combine them. This is the approach we will use here.

Step 1:
For step one, we have already done the analysis; the result that we have gotten in the previous section are valid for this section as well, therefore:

If step 1-α controls:

$$r_1 = r_{1\alpha} = \left(k_1 C_A - k_2 \frac{k_4 C_C C_D}{k_3 C_B}\right) \quad (3.176)$$

If step 1-β controls:

$$r_1 = r_{1\beta} = \left(\frac{k_1 k_3 C_A C_B}{k_2} - k_4 C_C C_D\right) \quad (3.177)$$

Step 2:
We need to find similar equations for the cases when 2-α and 2-β are the controlling ones. If step 2-α is the controlling step, it means that Equation 3.173 is in equilibrium and the reaction rate is equal to zero. Then, in order to solve the reaction rate as presented below:

$$r_2 = r_{2\alpha} = \left(k_5 C_D - k_6 C_{D^*}\right) \quad (3.178)$$

we need to find the concentration of C_{D^*}, which is obtained by using the equilibrium method into Equation 3.173:

$$0 = r_{2\beta} = \left(k_7 C_{D^*} C_B - k_8 C_C C_E\right) \quad (3.179)$$

Reorganizing, we can get:

Reaction Kinetics

$$C_{D^*} = \frac{k_8 C_C C_E}{k_7 C_B} \tag{3.180}$$

Equation 3.180 can be substituted into Equation 3.178 to obtain the reaction rate when step 2-α is the controlling step:

$$r_2 = \left(k_5 C_D - \frac{k_6 k_8 C_C C_E}{k_7 C_B} \right) \tag{3.181}$$

Similarly, we need to solve the problem when step 2-β is the controlling step; then, we have the reaction rate to be:

$$r_2 = r_{2_\beta} = \left(k_7 C_{D^*} C_B - k_8 C_C C_E \right) \tag{3.182}$$

And using the equilibrium method, we can obtain the expression for C_{D^*} from Equation 3.172 equal to zero:

$$0 = r_{2_\alpha} = \left(k_5 C_D - k_6 C_{D^*} \right) \tag{3.183}$$

Reorganizing it:

$$C_{D^*} = \frac{k_5 C_D}{k_6} \tag{3.184}$$

We can then substitute Equation 3.184 into Equation 3.181 to get:

$$r_2 = \left(\frac{k_5 k_7 C_D C_B}{k_6} - k_8 C_C C_E \right) \tag{3.185}$$

Equation 3.185 is the reaction rate for step 2 when the substep 2-β is the controlling step.

Step 3:
As the last part of this problem, we will get the reaction rate for step 3 when both steps 3-α and 3-β are controlling. The analysis is analog to those presented for steps 1 and 2.

We will start with step 3-α as the controlling one, so:

$$r_3 = r_{3_\alpha} = \left(k_9 C_E - k_{10} C_{E^*} \right) \tag{3.186}$$

In order to obtain an expression for C_{E^*} we will use the equilibrium method for the 3-β step. This means that Equation 3.175 will be in equilibrium and therefore, the reaction rate $r_{3_\beta} = 0$.

$$0 = \left(k_{11} C_{E^*} C_B - k_{12} C_C C_F \right) \tag{3.187}$$

Reorganizing it:

$$C_{E^*} = \frac{k_{12}C_C C_F}{k_{11}C_B} \tag{3.188}$$

Equation 3.188 can be substituted into Equation 3.186:

$$r_3 = \left(k_9 C_E - \frac{k_{10}k_{12}C_C C_F}{k_{11}C_B} \right) \tag{3.189}$$

Equation 3.189 is the reaction rate for step 3 when the step 3-α is the controlling step.

Finally, we will obtain the reaction rate when step 3-β is the controlling step.

$$r_3 = r_{3_\beta} = \left(k_{11} C_{E^*} C_B - k_{12} C_C C_F \right) \tag{3.190}$$

In order to solve Equation 3.190, we need to find an equation for C_{E^*}, this will be done as before, using the equilibrium method and we will be applied to step 3-α. Then, Equation 3.174 will be equal to zero:

$$0 = r_{3_\alpha} = \left(k_9 C_E - k_{10} C_{E^*} \right) \tag{3.191}$$

Reorganizing it:

$$C_{E^*} = \frac{k_9 C_E}{k_{10}} \tag{3.192}$$

Then, we can substitute Equation 3.192 into 3.190:

$$r_3 = \left(\frac{k_{11} k_9 C_E C_B}{k_{10}} - k_{12} C_C C_F \right) \tag{3.193}$$

Equation 3.193 is the reaction rate for step 3 when step 3-β is the controlling step.

We have now obtained all the reactions rates for all the possible controlling steps.

With this result, we can obtain the reaction rate for steps 1, 2, and 3, that needs to be used into Equations 3.132 to 3.137 to solve the differential equation system. These six differential equations can be solved simultaneously using a regression algorithm in order to estimate the kinetics parameters and make the model best fit the experimental data. Table 3.3 presents a summary of all the combinations that needs to be evaluated.

In Table 3.4, we have substituted the reaction rate into Table 3.3 so that the user can easily see the combinations to be used.

Reaction Kinetics

TABLE 3.3
All Possible Combinations for the Reaction Rates

Alternative	Controlling Step for r_1	Controlling Step for r_2	Controlling Step for r_3
1	$r_{1-\alpha}$	$r_{2-\alpha}$	$r_{3-\alpha}$
2	$r_{1-\alpha}$	$r_{2-\alpha}$	$r_{3-\beta}$
3	$r_{1-\alpha}$	$r_{2-\beta}$	$r_{3-\alpha}$
4	$r_{1-\alpha}$	$r_{2-\beta}$	$r_{3-\beta}$
5	$r_{1-\beta}$	$r_{2-\alpha}$	$r_{3-\alpha}$
6	$r_{1-\beta}$	$r_{2-\alpha}$	$r_{3-\beta}$
7	$r_{1-\beta}$	$r_{2-\beta}$	$r_{3-\alpha}$
8	$r_{1-\beta}$	$r_{2-\beta}$	$r_{3-\beta}$

TABLE 3.4
Reaction Rates

Option	Reaction Rate for r_1	Reaction Rate for r_2	Reaction Rate for r_3
1	$r_1 = \left(k_1 C_A - k_2 \dfrac{k_4 C_C C_D}{k_3 C_B} \right)$	$r_2 = \left(k_5 C_D - \dfrac{k_6 k_8 C_C C_E}{k_7 C_B} \right)$	$r_3 = \left(k_9 C_E - \dfrac{k_{10} k_{12} C_C C_F}{k_{11} C_B} \right)$
2	$r_1 = \left(k_1 C_A - k_2 \dfrac{k_4 C_C C_D}{k_3 C_B} \right)$	$r_2 = \left(k_5 C_D - \dfrac{k_6 k_8 C_C C_E}{k_7 C_B} \right)$	$r_3 = \left(\dfrac{k_{11} k_9 C_E C_B}{k_{10}} - k_{12} C_C C_F \right)$
3	$r_1 = \left(k_1 C_A - k_2 \dfrac{k_4 C_C C_D}{k_3 C_B} \right)$	$r_2 = \left(\dfrac{k_5 k_7 C_D C_B}{k_6} - k_8 C_C C_E \right)$	$r_3 = \left(k_9 C_E - \dfrac{k_{10} k_{12} C_C C_F}{k_{11} C_B} \right)$
4	$r_1 = \left(k_1 C_A - k_2 \dfrac{k_4 C_C C_D}{k_3 C_B} \right)$	$r_2 = \left(\dfrac{k_5 k_7 C_D C_B}{k_6} - k_8 C_C C_E \right)$	$r_3 = \left(\dfrac{k_{11} k_9 C_E C_B}{k_{10}} - k_{12} C_C C_F \right)$
5	$r_1 = \left(\dfrac{k_1 k_3 C_A C_B}{k_2} - k_4 C_C C_D \right)$	$r_2 = \left(k_5 C_D - \dfrac{k_6 k_8 C_C C_E}{k_7 C_B} \right)$	$r_3 = \left(k_9 C_E - \dfrac{k_{10} k_{12} C_C C_F}{k_{11} C_B} \right)$
6	$r_1 = \left(\dfrac{k_1 k_3 C_A C_B}{k_2} - k_4 C_C C_D \right)$	$r_2 = \left(k_5 C_D - \dfrac{k_6 k_8 C_C C_E}{k_7 C_B} \right)$	$r_3 = \left(\dfrac{k_{11} k_9 C_E C_B}{k_{10}} - k_{12} C_C C_F \right)$
7	$r_1 = \left(\dfrac{k_1 k_3 C_A C_B}{k_2} - k_4 C_C C_D \right)$	$r_2 = \left(\dfrac{k_5 k_7 C_D C_B}{k_6} - k_8 C_C C_E \right)$	$r_3 = \left(k_9 C_E - \dfrac{k_{10} k_{12} C_C C_F}{k_{11} C_B} \right)$
8	$r_1 = \left(\dfrac{k_1 k_3 C_A C_B}{k_2} - k_4 C_C C_D \right)$	$r_2 = \left(k_5 C_D - \dfrac{k_6 k_8 C_C C_E}{k_7 C_B} \right)$	$r_3 = \left(\dfrac{k_{11} k_9 C_E C_B}{k_{10}} - k_{12} C_C C_F \right)$

3.7.2 Heterogeneous Systems

Different heterogeneous systems have been solved in Sections 3.6.3 and 3.6.4, and a complete case is presented in Chapter 4. However, here a complex reaction system will be presented, considered, and solved accordingly.

Consider the following set of reactions:

$$A + B \leftrightarrows C + D \quad (3.194)$$

$$D + B \leftrightharpoons C + E \tag{3.195}$$

We will assume the following reaction steps taking place:

$$A + S \leftrightharpoons AS \tag{3.196}$$
$$AS + B \leftrightharpoons C + DS \tag{3.197}$$
$$DS + B \leftrightharpoons ES \tag{3.198}$$
$$ES \leftrightharpoons E + S \tag{3.199}$$

We can present a reaction rate for each reaction step involve:

$$r_1 = (k_1 C_A C_S - k_2 C_{AS}) \tag{3.200}$$
$$r_2 = (k_3 C_{AS} C_B - k_4 C_C C_{DS}) \tag{3.201}$$
$$r_3 = (k_5 C_{DS} C_B - k_6 C_{ES}) \tag{3.202}$$
$$r_4 = (k_7 C_{ES} - k_8 C_E C_S) \tag{3.203}$$

Therefore, there are four reactions rates for four different reactions, in order to solve for all scenarios it needs to be assumed that all possible steps could be the controlling step.

If step 1 is controlling, then using the equilibrium method, all the remaining steps are in equilibrium; this is Equations 3.201, 3.202, and 3.203, giving:

$$0 = (k_3 C_{AS} C_B - k_4 C_C C_{DS}) \tag{3.204}$$
$$0 = (k_5 C_{DS} C_B - k_6 C_{ES}) \tag{3.205}$$
$$0 = (k_7 C_{ES} - k_8 C_E C_S) \tag{3.206}$$

We can know solve Equation 3.206 to obtain C_{ES}:

$$C_{ES} = \frac{k_8 C_E C_S}{k_7} \tag{3.207}$$

Similarly, for Equation 3.205, we can get C_{DS}:

$$C_{DS} = \frac{k_6 C_{ES}}{k_5 C_B} \tag{3.208}$$

Substituting Equation 3.207 into Equation 3.208, we obtain:

$$C_{DS} = \frac{k_6}{k_5 C_B} \frac{k_8 C_E C_S}{k_7} \tag{3.209}$$

Reaction Kinetics

Analogously, we can solve Equation 3.204 for C_{AS} and replacing 3.209 we obtain:

$$C_{AS} = \frac{k_4 C_C C_{DS}}{k_3 C_B} = \frac{k_4 C_C}{k_3 C_B} \frac{k_6}{k_5 C_B} \frac{k_8 C_E C_S}{k_7} \qquad (3.210)$$

Substituting into Equation 3.200, we obtain:

$$r_1 = \left(k_1 C_A C_S - k_2 \frac{k_4 k_6 k_8 C_C C_E C_S}{k_3 k_5 k_7 C_B^2} \right) \qquad (3.211)$$

Finally, a catalyst site balance must be done to obtain C_S as a function of known concentrations. Generic balance for this problem is:

$$C_S^0 = C_S + C_{AS} + C_{DS} + C_{ES} \qquad (3.212)$$

Substituting Equations 3.207, 3.209, and 3.210 into 3.212, we obtain:

$$C_S^0 = C_S + \frac{k_4 k_6 k_8 C_C C_E C_S}{k_3 k_5 k_7 C_B^2} + \frac{k_6 k_8 C_E C_S}{k_5 k_7 C_B} + \frac{k_8 C_E C_S}{k_7} \qquad (3.213)$$

Rearranging it:

$$C_S = \frac{C_S^0}{\left(1 + \dfrac{k_4 k_6 k_8 C_C C_E}{k_3 k_5 k_7 C_B^2} + \dfrac{k_6 k_8 C_E}{k_5 k_7 C_B} + \dfrac{k_8 C_E}{k_7} \right)} \qquad (3.214)$$

Substituting Equation 3.214 into 3.211, we obtain:

$$r_1 = \left(k_1 C_A - k_2 \frac{k_4 k_6 k_8 C_C C_E}{k_3 k_5 k_7 C_B^2} \right) \frac{C_S^0}{\left(1 + \dfrac{k_4 k_6 k_8 C_C C_E}{k_3 k_5 k_7 C_B^2} + \dfrac{k_6 k_8 C_E}{k_5 k_7 C_B} + \dfrac{k_8 C_E}{k_7} \right)} \qquad (3.215)$$

If step 2 is controlling, then using the equilibrium method, all the remaining steps are in equilibrium; this is 3.200, 3.202, and 3.203, giving:

$$0 = \left(k_1 C_A C_S - k_2 C_{AS} \right) \qquad (3.216)$$

$$0 = \left(k_5 C_{DS} C_B - k_6 C_{ES} \right) \qquad (3.217)$$

$$0 = \left(k_7 C_{ES} - k_8 C_E C_S \right) \qquad (3.218)$$

We can know solve Equation 3.218 to obtain C_{ES}:

$$C_{ES} = \frac{k_8 C_E C_S}{k_7} \qquad (3.219)$$

Similarly, for 3.217 we can get C_{DS}:

$$C_{DS} = \frac{k_6 C_{ES}}{k_5 C_B} \qquad (3.220)$$

Substituting 3.219 into 3.220, we obtain:

$$C_{DS} = \frac{k_6}{k_5 C_B} \frac{k_8 C_E C_S}{k_7} \qquad (3.221)$$

Analogously, we can solve Equation 3.216 for C_{AS}:

$$C_{AS} = \frac{k_1 C_A C_S}{k_2} \qquad (3.222)$$

Substituting Equations 3.221 and 3.222 into Equation 3.201, we obtain:

$$r_2 = \left(\frac{k_1 k_3 C_A C_B C_S}{k_2} - \frac{k_4 k_6 k_8 C_E C_C C_S}{k_5 k_7 C_B} \right) \qquad (3.223)$$

Finally, a catalyst site balance must be done to obtain C_S as a function of known concentrations. Replacing Equations 3.219, 3.221, 3.222 into 3.212, we obtain generic balance for this problem:

$$C_S^0 = C_S + \frac{k_1 C_A C_S}{k_2} + \frac{k_6 k_8 C_E C_S}{k_5 k_7 C_B} + \frac{k_8 C_E C_S}{k_7} \qquad (3.224)$$

Rearranging it:

$$C_S = \frac{C_S^0}{\left(1 + \dfrac{k_1 C_A}{k_2} + \dfrac{k_6 k_8 C_E}{k_5 k_7 C_B} + \dfrac{k_8 C_E}{k_7} \right)} \qquad (3.225)$$

Substituting Equation 3.225 into 3.223, we obtain:

$$r_2 = \left(\frac{k_1 k_3 C_A C_B}{k_2} - \frac{k_4 k_6 k_8 C_E C_C}{k_5 k_7 C_B} \right) \frac{C_S^0}{\left(1 + \dfrac{k_1 C_A}{k_2} + \dfrac{k_6 k_8 C_E}{k_5 k_7 C_B} + \dfrac{k_8 C_E}{k_7} \right)} \qquad (3.226)$$

If step 3 is controlling, then using the equilibrium method, all the remaining steps are in equilibrium; this is 3.200, 3.201, and 3.203, giving:

$$0 = \left(k_1 C_A C_S - k_2 C_{AS} \right) \qquad (3.227)$$

Reaction Kinetics

$$0 = \left(k_3 C_{AS} C_B - k_4 C_C C_{DS}\right) \tag{3.228}$$

$$0 = \left(k_7 C_{ES} - k_8 C_E C_S\right) \tag{3.229}$$

We can know solve Equation 3.229 to obtain C_{ES}

$$C_{ES} = \frac{k_8 C_E C_S}{k_7} \tag{3.230}$$

Similarly, for 3.227, we can get C_{AS}

$$C_{AS} = \frac{k_1 C_A C_S}{k_2} \tag{3.231}$$

Solving for Equation 3.228, we can obtain C_{DS}:

$$C_{DS} = \frac{k_3 C_{AS} C_B}{k_4 C_C} \tag{3.232}$$

Substituting Equation 3.231 into 3.232, we obtain:

$$C_{DS} = \frac{k_3 C_B}{k_4 C_C} \frac{k_1 C_A C_S}{k_2} \tag{3.233}$$

We can now substitute Equations 3.230, 3.231, and 3.233 into 3.202:

$$r_3 = \left(\frac{k_1 k_3 k_5 C_A C_B^2 C_S}{k_2 k_4 C_C} - \frac{k_6 k_8 C_E C_S}{k_7}\right) \tag{3.234}$$

Finally, a catalyst site balance must be done to obtain C_S as a function of known concentrations. Replacing Equations 3.230, 3.231, and 3.233 into 3.212, we obtain generic balance for this problem:

$$C_S^0 = C_S + \frac{k_1 C_A C_S}{k_2} + \frac{k_1 k_3 C_B C_A C_S}{k_2 k_4 C_C} + \frac{k_8 C_E C_S}{k_7} \tag{3.235}$$

Rearranging it:

$$C_S = \frac{C_S^0}{\left(1 + \frac{k_1 C_A}{k_2} + \frac{k_1 k_3 C_B C_A}{k_2 k_4 C_C} + \frac{k_8 C_E}{k_7}\right)} \tag{3.236}$$

Substituting Equation 3.236 into 3.234, we obtain:

$$r_3 = \left(\frac{k_1 k_3 k_5 C_A C_B^2}{k_2 k_4 C_C} - \frac{k_6 k_8 C_E}{k_7} \right) \frac{C_S^0}{\left(1 + \dfrac{k_1 C_A}{k_2} + \dfrac{k_1 k_3 C_B C_A}{k_2 k_4 C_C} + \dfrac{k_8 C_E}{k_7}\right)} \qquad (3.237)$$

If step 4 is controlling, then using the equilibrium method, all the remaining steps are in equilibrium; this is 3.200, 3.201, and 3.202, giving:

$$0 = (k_1 C_A C_S - k_2 C_{AS}) \qquad (3.238)$$

$$0 = (k_3 C_{AS} C_B - k_4 C_C C_{DS}) \qquad (3.239)$$

$$0 = (k_5 C_B C_{DS} - k_6 C_{ES}) \qquad (3.240)$$

We can know solve equation 3.238 to obtain C_{AS}:

$$C_{AS} = \frac{k_1 C_A C_S}{k_2} \qquad (3.241)$$

Similarly, for 3.239, we can get C_{DS}:

$$C_{DS} = \frac{k_3 C_{AS} C_B}{k_4 C_C} \qquad (3.242)$$

Substituting Equation 3.241 into 3.242, we obtain:

$$C_{DS} = \frac{k_1 k_3 C_B C_A C_S}{k_2 k_4 C_C} \qquad (3.243)$$

Solving for equation 3.240, we can obtain C_{ES}:

$$C_{ES} = \frac{k_5 C_{DS} C_B}{k_6} \qquad (3.244)$$

Substituting Equation 3.243 into 3.244, we obtain:

$$C_{ES} = \frac{k_1 k_3 k_5 C_B^2 C_A C_S}{k_2 k_4 k_6 C_C} \qquad (3.245)$$

We can now substitute Equations 3.241, 3.243, and 3.245 into 3.203:

$$r_4 = \left(\frac{k_1 k_3 k_5 k_7 C_B^2 C_A C_S}{k_2 k_4 k_6 C_C} - k_8 C_E C_S \right) \qquad (3.246)$$

Reaction Kinetics

Finally, a catalyst site balance must be done to obtain C_S as a function of known concentrations. Replacing Equations 3.241, 3.243, 3.245 into 3.212 and we obtain generic balance for this problem:

$$C_S^0 = C_S + \frac{k_1 C_A C_S}{k_2} + \frac{k_1 k_3 C_B C_A C_S}{k_2 k_4 C_C} + \frac{k_1 k_3 k_5 C_B^2 C_A C_S}{k_2 k_4 k_6 C_C} \qquad (3.247)$$

Rearranging it:

$$C_S = \frac{C_S^0}{\left(1 + \dfrac{k_1 C_A}{k_2} + \dfrac{k_1 k_3 C_B C_A}{k_2 k_4 C_C} + \dfrac{k_1 k_3 k_5 C_B^2 C_A}{k_2 k_4 k_6 C_C}\right)} \qquad (3.248)$$

Substituting Equation 3.248 into 3.246, we obtain:

$$r_4 = \left(\frac{k_1 k_3 k_5 k_7 C_B^2 C_A}{k_2 k_4 k_6 C_C} - k_8 C_E\right) \frac{C_S^0}{\left(1 + \dfrac{k_1 C_A}{k_2} + \dfrac{k_1 k_3 C_B C_A}{k_2 k_4 C_C} + \dfrac{k_1 k_3 k_5 C_B^2 C_A}{k_2 k_4 k_6 C_C}\right)} \qquad (3.249)$$

In order to know which reaction is controlling, if any, we need to solve the system for each case and compare the outcome with experimental data. We encourage the reader to look into Chapter 4 for a very detailed problem with a full solution.

3.8 CATALYST DEACTIVATION

We have been trying to obtain the reaction rate for various, simple and slightly more complex reactions. In all cases, the reaction rate, with different homogeneous and/or heterogeneous catalyst, has been introduced as independent of the performance of the catalyst. This means that the catalyst has been considered as equally good during its whole service life and has not suffer from any modifications due to the reaction. Unfortunately, this is not true for most of the catalytic reactions; catalytic materials suffer from various physical phenomena that make them less active toward the reactions they should catalyze. The catalyst can get poisoned and loose activity, different materials can get deposited over them leading to fewer active sites; the catalyst can agglomerate, and the surface area and/or porous amount can be considerably reduced, reducing the active material available.

The deactivation of a catalyst can be thought of the ratio of the real reaction rate and the ideal reaction rate. Mathematically,

$$\delta = Deactivation = \frac{r_r}{r_i} \qquad (3.250)$$

where r_r is the real reaction rate, measured in the system, and r_i is the ideal reaction rate when the deactivation is nonexistent. The real reaction rate is time dependent and therefore, the deactivation, from now on called as δ, is also a function of time. More specifically, it goes from 1 to 0, where 1 corresponds to the mathematical value for a zero deactivation. This approach presented here allows us to separate the two processes, the reaction rate itself, and the deactivation of the catalyst. We will only focus on the latter one.

In order to solve this type of problem, when deactivation occurs, we need to know the dependence of δ on time and based on that, we can obtain the reaction rate for the catalyst deactivation r_δ and if it is dependent on the concentration of any reactant, product, or the catalyst concentration itself. Therefore, we can define:

$$\delta = f(t) \tag{3.251}$$

Therefore:

$$r_\delta = -\frac{d\delta}{dt} = f\left(\delta(t), k_\delta(T), C_i\right) \tag{3.252}$$

where the reaction rate for the deactivation decay can be a function of time, of the deactivation profile, of the deactivation reaction rate constant, and of the concentration (either products and/or raw materials). In order to solve this type of problem, we need to know δ(t). Several proposals have been done by different authors; we strongly recommend reading this section from Folger's book for a table with some options on the profiles for δ(t). We will just solve one here to show the procedure.

Let's considered the reaction A → B, an elementary irreversible reaction that takes place over a solid catalyst. Assuming that Equation 3.252 is only dependent on time, we can then integrate this separately from the reaction rate for our problem:

$$r_\delta = -\frac{d\delta}{dt} = k_\delta(T) \tag{3.253}$$

We can then integrate this expression:

$$\int_1^\delta -d\delta = \int_0^t k_\delta(T)dt \tag{3.254}$$

It is very important to see here that the integration limits for δ are from 1 to an arbitrarily value that will be smaller than 1, while time goes from a starting point at zero to a time that is greater than zero. This is crucial as the number of active sites can only decrease.

Solving Equation 3.198, we obtain:

$$1 - \delta = k_\delta(T)t \tag{3.255}$$

Reorganizing it:

$$\delta = 1 - k_\delta(T)t \tag{3.256}$$

Reaction Kinetics

With this expression, having an Arrhenius expression for k_δ, we can then substitute it into the reaction rate expression and solve our problem.

For our case, the reaction rate will be the following, knowing that $r_A = \delta * r_{Aideal} = \delta * k_A * C_A$:

$$\frac{dC_A}{dt} = r_A \rho = -k_A C_A \delta \rho \tag{3.257}$$

Substituting Equation 3.256 into 3.257, we get:

$$\frac{dC_A}{dt} = -k_A C_A \left(1 - k_\delta(T) t\right) \rho \tag{3.258}$$

Therefore, we can reorganize this equation and solve it analytically for an isothermal case:

$$\int_{C_A^0}^{C_A} \frac{dC_A}{C_A} = \int_0^t -k_A \rho \left(1 - k_\delta(T) t\right) dt \tag{3.259}$$

Resulting in:

$$\ln\left(\frac{C_A}{C_A^0}\right) = -k_A \rho t + \frac{k_A \rho k_\delta(T) t^2}{2} \tag{3.260}$$

If $k_\delta = 0$, and $k_A \rho = k$, then, we can reduce Equation 3.260:

$$\ln\left(\frac{C_A}{C_A^0}\right) = -kt \tag{3.261}$$

Expression 3.261 is the typical mole balance result for the reaction under consideration when there is no deactivation. The deactivation effect can be seen in Figure 3.11 where we compare the result for Equations 3.260 and 3.261.

It can be seen in Figure 3.11 that the concentration of reactant A is decreasing faster when there is no catalyst deactivation; this also means that the production of chemical B is faster, or for the same time, we have produced more amount. For the plot, we have used $k_A = 2 \left[\frac{l}{kg * min}\right]$ and $k_\delta = 0.009 \left[\frac{1}{min}\right]$.

The previous case that we looked into was a case where the deactivation was only a function of time. We will now take a look to a second case where the deactivation rate is a function of time as well as the concentration of one of the reactants. In this case, we are facing a system of differential equations which need to be solved simultaneously.

Let's consider the reaction of isomerization of A to B with the reaction rate for chemical A being a second-order irreversible reaction. If the deactivation of the catalyst follows a linear behavior with the concentration of reactant A as well as for δ with $k_\delta = 0.01$ [l/(mol*min)]:

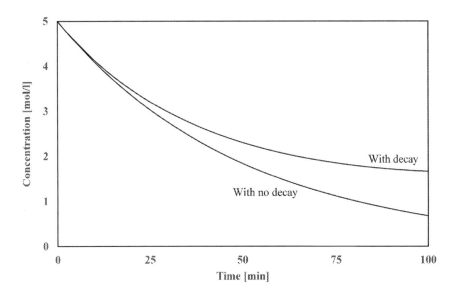

FIGURE 3.11 Concentration profiles with and without catalyst deactivation.

$$-\frac{d\delta}{dt} = k_\delta \delta C_A \tag{3.262}$$

Since the deactivation is a function of C_A as well, we need to solve the mole balance simultaneously; this can be done for the different types of reactors, like batch, CSTR, and PF. Here, the results for the batch reactor are presented, for which we need to solve Equation 3.255 simultaneously with the following equation and using the given numerical values for the reactions and deactivation constants.

$$r_A = \frac{dC_A}{dt} = -0.2\,\delta\,C_A^{\,2} \tag{3.263}$$

Knowing the initial conditions for each differential equation as well as the range for the independent variable (time), we can then solve the problem using Polymath for example. Figure 3.12 shows the variation of the concentration of chemical A as well as the deactivation decay profile.

3.9 MASS TRANSFER LIMITATIONS

We have been deducing kinetics models based on experimental data. For the catalytic systems, firstly, we assumed that the catalyst was ideal in the sense that it has always the same performance. Secondly, we remove that assumption and we looked at systems where the catalyst suffers deactivation. In this section, we will take into consideration that the reaction rate equations are based on the concentration values where the reaction takes place; however, when taking a sample from a reactor, we are almost always far away from that location. Figure 3.13 shows this situation.

Reaction Kinetics

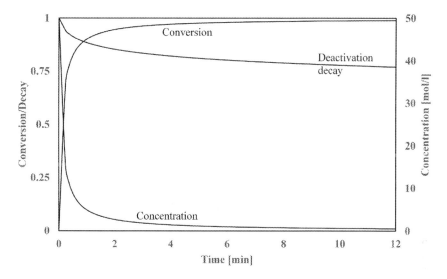

FIGURE 3.12 Concentration, conversion, and decay profiles for a batch system.

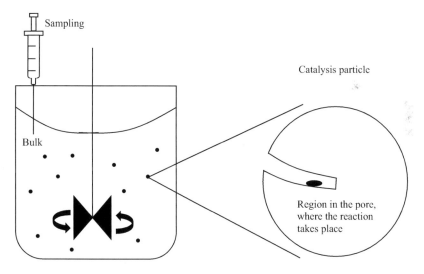

FIGURE 3.13 Schematic of a reactive system.

As it can easily be seen from Figure 3.13, when sampling from our reaction system, we are not taking a sample exactly where the reaction takes place. Therefore, we can be facing mass transfer limitations, external to the catalyst or internally in the catalyst.

For heterogeneous reactions, there are several steps in order to transform the reactants into products; these steps are:

1. The raw materials need to be moved from the bulk of the reactor to the surface of the catalyst.

2. From the surface of the catalyst, if the catalyst is porous, the reactants now have to diffuse internally in the material.
3. Once the reactants have reached the active site, one of more of them might be adsorbed on the catalytic surface.
4. The adsorbed/non-adsorbed chemicals will react.
5. The products need to be desorbed from the catalytic surface.
6. The product needs to diffuse through the porosity of the catalyst to allow new materials to diffuse in.
7. Once the product has reached the surface of the catalyst, they need to be removed from the surface of the catalyst and into the bulk.

Steps 1, 2, 6 and 7 are limited by mass transfer while 3, 4 and 5 are related to adsorption, reaction, and desorption of reactants products.

We have been focusing on steps 3, 4, and 5 as limiting steps to obtain a reaction rate mechanism and a reaction rate expression. However, as we mentioned, those equations assume that the concentration used is the same as the one we can measure. Figure 3.14a–d shows the concentration profile when there are (a) no limitations of any kinds (the scenario we have been solving so far) as well as when we have (b) external, (c) internal, and (d) external and internal limitations.

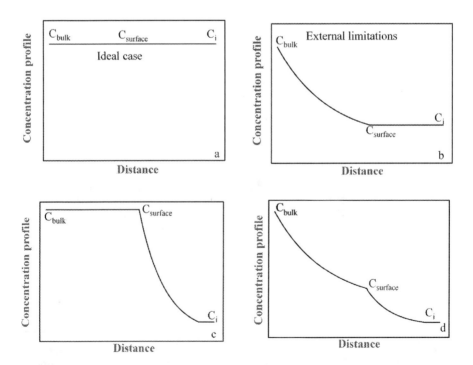

FIGURE 3.14 Concentration profiles for the different mass transfer limitation cases: (a) no limitations, (b) external limitations, (c) internal limitations, and d) external and internal limitations (C_{bulk} is the concentration in the bulk, $C_{surface}$ is the concentration on the surface of the catalyst, and C_i is the concentration that is used for the kinetics expression).

Reaction Kinetics

It is important to notice that in Figure 3.14b–d, the slope of the concentration profile has an inflection point; this is the limit between the external and internal parts of the system, at which point the raw materials enter the pores of the solid catalyst. It is important to note that the flat line at the end is not technically correct; otherwise, it can be misinterpreted as there is another profile inside; this line should be a point, where the reactants are adsorbed and the reaction occurs. However, for a visual effect, it has been presented as a line, but it is always constant and is not a new resistance or profile for mass limitations. It is very important to distinguish between the following: C_{bulk} is the concentration in the bulk, $C_{surface}$ is the concentration on the surface of the catalyst, and C_i is the concentration that is used for the kinetics expression.

So far, we have been looking at the case where $C_{bulk} = C_{surface} = C_i$. Therefore, we have not been having problems with the different concentration values. This is the case when all the mass transfer limitations are avoided.

We will be focusing now on the internal mass transfer limitations and the overall mass transfer limitations (internal and externals). The external mass transfer limitations, except some cases in particular are mainly mass transfer balances and not reaction engineering balances. The case where the reaction engineering might come in handy is when the catalyst is solid with no pores and therefore, the boundary conditions for the mass balance is the reaction itself. We will not focus on these problems as they should be addressed in a mass transfer course.

3.9.1 Internal Mass Transfer Limitations

For this scenario, we will assume that from the bulk of the reaction medium to the surface of the catalyst, there is no mass transfer limitation. Therefore, the problems could appear due to the catalyst, small porous, viscous liquid, large molecules, or a combination of all. In any case, we have a problem where the concentration profile in the catalyst follows that presented in Figure 3.14c.

We know that $C_{bulk} = C_{surface}$; however, for our reaction rate expression (generically):

$$r_i = -k\, C_i \tag{3.264}$$

Therefore, we need to find the relationship between C_i and $C_{surface}$; so that while we measure $C_{surface}$, we can predict the real concentration for our model C_i.

The relation between the concentration on the surface of the catalyst and the real concentration where the reaction takes place is given by the Thiele's modulus. This modulus considers the ratio:

$$\phi^2 = \frac{reaction\ rate}{diffusion\ rate} \tag{3.265}$$

It is obvious that the dependence of the modulus on the reaction rate and the diffusion rate depends on the order of the reaction and on the geometry of the porous/material/catalyst. Based on these two parameters, different Thiele moduli for cylindrical cases, for different reaction orders, can be obtained. Plotting the ratio of the concentration on the reaction site and the concentration on the surface of the materials under study as a function of the Thiele modulus will provide quite a lot of information of the mass transfer situation on the case under study.

Figure 3.15 shows a generic representation of how the Thiele modulus value can provide valuable information about the mass transfer limitations that are in our system, even if we do not have the mathematical formulation (we will come back to that).

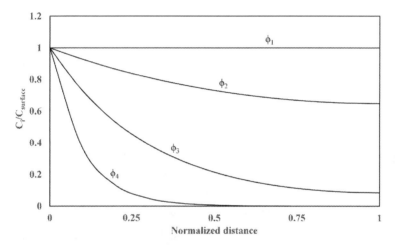

FIGURE 3.15 Variation of the ratio of concentration in the reaction site and concentration on the surface as a function of the Thiele modulus for an arbitrary normalized distance, where $\phi_4 > \phi_3 > \phi_2 > \phi_1$.

It is important to see that as the Thiele modulus increases, the diffusion rate needs to decrease in comparison with the reaction rate. This means that the diffusion rate becomes more and more relevant and therefore, limits the process. This means that the concentration in the reaction site is considerably different from the one on the surface. On the opposite end, if the reaction rate is infinitely smaller in comparison with the diffusion rate (or the diffusion rate is significantly larger), then, there is no limitation to the transfer of mass and therefore, the controlling step is the reaction itself. This will give us the scenario we have been working up to this point where the concentration of the surface is the concentration we use to solve our reactor mole balance equations ($C_i = C_{surface}$).

Solving the mass balances (including the reaction term) for different geometries for the porous on the catalytic materials (spherical, cylindrical), will allow us to develop the mathematical equations to represent the Thiele modulus and, eventually the effectiveness factor. Different derivations can be found in the literature depending on if we solve for a cylindrical porous, or if the catalytic material is a sphere, and if we have a first-order reaction or a higher order reaction. These derivations can be found in the literature [1–6] and therefore, we encourage the reader to study a more detailed representation in the cited books. We will focus here on the equation itself, it uses, and its physical meaning.

3.9.1.1 Cylindrical Pore

For a cylinder, we need to consider the scenario that is presented in Figure 3.16 which shows that the reactant is located at the surface (entrance) of the catalyst pore, and needs to diffuse or move, within the cylinder.

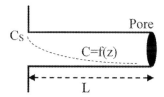

FIGURE 3.16 Cylindrical catalyst pore representation.

Figure 3.16 also presents the possible concentration profile that might occur within the catalyst pore; this profile is arbitrarily presented and can be steeper, or not, depending on the value of reaction and diffusion rates. For this case, the mass balance with reaction, considering a first-order irreversible system, to be solved is:

$$\frac{d^2C_A}{dz^2} = \frac{k}{D} * C_A \qquad (3.266)$$

where k is the reaction rate (volumetric based) and D is the effective diffusivity. This expression can be solved using the conditions that:

$$z = 0 \to C_A = C_{AS} \qquad (3.267)$$

$$z = L \to \frac{dC_A}{dz} = 0 \qquad (3.268)$$

Plus, we need to remember the assumption that at the end of the pore, at $z = L$, there is no reaction taking place; the reaction only happens on the side of the pore.

The solution to Equation 3.266 with the boundary conditions of 3.267 and 3.268 will be:

$$\frac{C_A}{C_{As}} = \frac{\cosh\left(\sqrt{\frac{k}{D}}(L-z)\right)}{\cosh\left(\sqrt{\frac{k}{D}}L\right)} \qquad (3.269)$$

We can then calculate the effectiveness factor for this case; this is calculated as the ratio of the actual reaction rates within the pore over the reaction rate as if there are no mass limitations. In other words, we are comparing the actual reaction rate (with a concentration profile) with the reaction rate considering the concentration, temperature, and other properties as constants, with values equal to those at the surface, symbolically, $\eta = \dfrac{-r_{A\,real}}{-r_{A\,surface}}$. For the first-order kinetics, we can then integrate the real reaction rate over the whole cylinder since this value changes with the distance in the pore:

$$\eta = \frac{\tanh\left(\sqrt{\frac{k}{D}}L\right)}{\sqrt{\frac{k}{D}}L} \qquad (3.270)$$

We can define the Thiele modulus as, for a first-order reaction:

$$\phi_1 = \sqrt{\frac{k}{D}}L \qquad (3.271)$$

where the subscript is the order of the reaction in consideration.

Substituting Equations 3.271 into 3.270, we obtain:

$$\eta = \frac{\tanh(\phi_1)}{\phi_1} \qquad (3.272)$$

where Equation 3.272 is the effectiveness factor as a function of the Thiele modulus for a first-order irreversible reaction with mass transfer limitations.

If we plot the effectiveness factor as a function of the Thiele modulus, we obtain Figure 3.17.

Besides the calculations that are presented here for a first-order irreversible reaction, a similar analysis can be done for different reactions orders and the plots will have a similar tendency but slightly shifted. It is important to notice that the calculations have been done for an isothermal system.

Reaction Kinetics 235

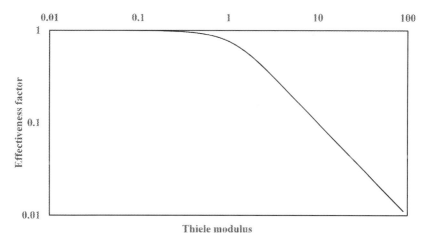

FIGURE 3.17 The effectiveness factor as a function of the Thiele modulus.

3.9.1.2 Spherical Pore

Besides the study of the diffusion in a cylindrical pore, we can study the diffusion and reactions in a spherical catalyst. The derivation of the expression is based on the fact that the steady-state balance takes place in a shell of radius r, where the chemicals under consideration need to flow in, flow out, and react. Figure 3.18 shows a schematic representation of the problem.

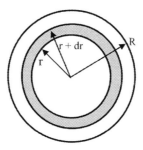

FIGURE 3.18 Spherical catalytic pellet scheme.

Doing the desired balance on the dr and reorganizing the terms, we obtain:

$$\frac{d^2C_A}{dr^2} + \frac{2}{r}\frac{dC_A}{dr} = \frac{k_n}{D} * C_A^n \qquad (3.273)$$

Equation 3.273 is a generic expression for a reaction rate that is irreversible but of order n. Therefore, the result obtained in this case is slightly more generic than in the previous scenario. It is important to point out that k_n has the units $\left[\left(\frac{Volume}{Mol}\right)^{(n-1)} \frac{1}{time}\right]$;

however, it can be calculated based on different information from the pellet and its properties.

k_n, as presented before, can be calculated as $k_n = k_s * S * \rho$, where k_s is the reaction rate per unit of surface area [m/s], S is the surface area per unit of mass of catalysts [m²/kg] and ρ is the density of the catalysts [kg/volume].

In order to solve this problem, we need boundary conditions:

$$r = R \rightarrow C_A = C_{AS} \tag{3.274}$$

$$r = 0 \rightarrow \frac{dC_A}{dr} = 0 \tag{3.275}$$

The procedure to solve this is to express Equation 3.273 as a dimensionless expression, while doing this, Equation 3.273 will take a new form which will be used to present the Thiele modulus, and the generic expression will take the form:

$$\Phi_n^2 = \frac{k_n R^2 C_{As}^{n-1}}{D} \tag{3.276}$$

We can, as for the cylindrical pore, solve this problem for the first-order irreversible reaction. In this case:

$$\Phi_1^2 = \frac{k_n R^2}{D} \tag{3.277}$$

Or:

$$\Phi_1 = R \sqrt{\frac{k_n}{D}} \tag{3.278}$$

Then, we can solve the dimensionless differential equation and we obtain:

$$\frac{C_A}{C_{As}} = \frac{1}{\gamma} \left(\frac{\sinh(\gamma \Phi_1)}{\sinh(\Phi_1)} \right) \tag{3.279}$$

where γ is the dimensionless number related to r/R. Equation 3.279 can be used to predict the variation of concentration of chemical A along the radius of the pellet, for a first-order irreversible reaction. Even though this is very useful information, for our reaction rate analysis, it is more relevant to relate the actual reaction rate with the reaction rate as if the reactions is taking place with the concentration equal to that on the catalysts surface. The link among these reactions rates, as presented before, is the effectiveness factor; therefore, we can obtain Equation 3.280 which is the effectiveness factor as a function of the Thiele modulus for a first-order irreversible reaction with mass transfer limitations in a spherical porous.

Reaction Kinetics

$$\eta = \frac{3}{\Phi_1}\left(\frac{1}{\tan h(\Phi_1)} - \frac{1}{\Phi_1}\right) \quad (3.280)$$

We can plot this expression as a function of the Thiele modulus and obtain Figure 3.19

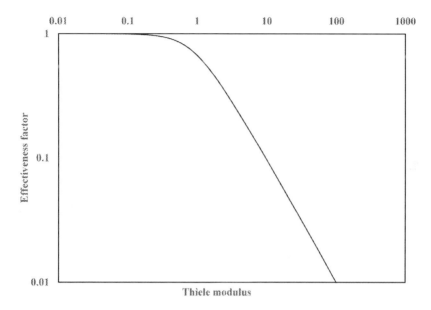

FIGURE 3.19 Effectiveness factor for a spherical catalyst.

We have obtained the Thiele modulus and the effectiveness factor for the first-order kinetics. However, for the spherical catalyst, we have a generic expression for this modulus, Equation 3.280. Then, we can apply this equation to different reaction orders like zero, one, and three, and plot the different effectiveness factor for each case. We strongly recommend reading Satterfield's book [7] and Bischoff work on the topic [8].

We can then plot the effectiveness factor as a function of the Thiele modulus for each case and we get Figure 3.20.

It is important to notice that there are different definitions for the "characteristic length" of penetration within the catalyst. This definition is important since it will determine the mathematical expression of the effectiveness factor; however, the outcome will be the same. It is strongly recommended to read Levenspiel book to know more on the different possible definitions for "characteristic length."

As mentioned, this is done for isothermal cases, while non-isothermal scenarios will have a small variations from the results presented here. We will just mention these cases and mention that the effectiveness factor can be much higher than 1. This goes against our reasoning since the mass transfer limitations is a resistance and

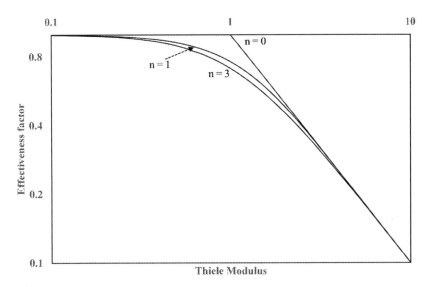

FIGURE 3.20 Effectiveness factor for a spherical catalyst.

therefore, values should decrease, but here we have the combination of different effects and the thermal one is larger than the mass transfer one.

3.9.2 Overall Mass Transfer Limitations

We have been working with the internal mass transfer limitations, meaning that from the bulk to the surface of the catalyst, the concentration remains constant. We will add here what happens if that last-mentioned concentration is not constant. We have omitted the case where only external limitations exist since that is a case for the mass transfer courses.

The overall factor considers the limitations that appear from things not moving fast enough as well as the limitations for diffusion problems within the pore. We will define the overall effectiveness factor as the ratio between the overall actual reaction rate and the reaction rate as if everything was constant at the bulk conditions (C_{Ab}, T_b, etc.)

With this in mind:

$$-r_A = k C_A \tag{3.281}$$

However, this concentration is in the reaction site inside the pores and for this case, there are internal and external limitations; therefore, we need to relate this concentration to the one in the bulk; so the equation can be rewritten as:

$$-r_A = \Theta k C_{Ab} \tag{3.282}$$

where Θ is the overall effectiveness factor. This factor can be defined for a first-order reaction as:

$$\Theta = \frac{\eta}{\left(1 + \dfrac{\eta k_r}{k_m a_c}\right)} \tag{3.283}$$

where
 η is the internal effectiveness factor
 k_r is the reaction rate constant
 k_m is the external mass transfer coefficient
 a_c is the ratio of external surface area per reactor volume unit.

where $k_r = \rho_b * S * k_r^{II}$ with ρ_b being the bulk density and related to the porosity as $\rho_b = \rho_c (1 - \chi)$ where χ is the porosity and ρ_c is the density of the catalytic pellet. S is the surface area and k_r^{II} is the reaction rate constant expressed in units of catalyst surface area.

It is important to see that in this new effectiveness factor, we not only have the reaction rate equation, but also the internal effectiveness factor as well as the overall mass transfer coefficient. This is due to the fact that now we are considering the mass transfer issues from the bulk to the surface of the catalyst, in addition to the mass transfer limitations inside the pellet. Once both of these issues have been overcome, the reaction rate is then the third kind of resistance involved, which can't be avoided.

We can see from Equation 3.284, that if the overall mass transfer coefficient is large, $k_c \rightarrow \infty$ then, the denominator gets closer to 1 and the overall effectiveness factor is equal to the internal effectiveness factor. This implies that the external mass diffusion problems have been overcome, which is the case when increasing steering in a reactor and therefore, the mass coefficient increases. Similarly, if the internal effectiveness factor is 1 (no limitations inside the particle) there could still be some diffusion limitations due to the external mass transfer (unless the steering is large enough).

We can now substitute Equation 3.284 into Equation 3.282 to obtain the reaction rate equation based on the concentration of the bulk, the one we can actually measure, but considering all limitations so that we can predict the concentration in the catalytic location, the one we need for the reaction rate expression:

$$-r_A = \frac{\eta}{\left(1 + \dfrac{\eta k_r}{k_m a_c}\right)} k_r C_{Ab} \tag{3.284}$$

Figure 3.21 shows a comparison of the variation of the reaction rate as a function of C_{Ab} for different regimes, we will present the reaction rate when there is no mass transfer limitations, when the internal diffusion is 50% but no external, when the external limitation is relevant but there is no internal mass limitations, as well as when both cases are relevant.

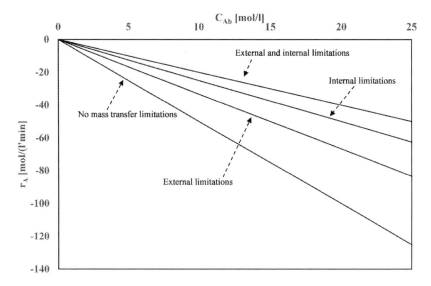

FIGURE 3.21 Variation of the reaction rate for different mass transfer limitation scenarios.

There are two criteria, Mears for external diffusion and Weisz–Prater for internal diffusion, that are useful for a rough estimation if there are any external and or internal mass transfer limitations.

For internal limitations, we have the Weisz–Prater criteria, this is:

$$C_{WP} = \eta \Phi_1^2 = 3\left(\Phi_1 \coth(\Phi_1) - 1\right) \tag{3.285}$$

which can be reorganized as:

$$C_{WP} = \eta \Phi_1^2 = \frac{-r_{Aobs} * \rho_c * R^2}{D * C_{As}} \tag{3.286}$$

where
- $-r_{Aobs}$ is the reaction rate that can be measured.
- ρ_c is the density of the catalyst
- R is the pellet radius
- D is the diffusivity
- C_{As} is the concentration of A at the surface

If the C_{WP} is much lower than 1, then there is no internal diffusion, and if C_{WP} is much larger than 1, then, we have mass transfer issues and limitations.

Similarly, Mears proposed the following criteria:

$$MR = \frac{-r_{Aobs} * \rho_b * R * n}{k_c * C_{Ab}} \tag{3.287}$$

where
- $-r_{Aobs}$ is the reaction rate that can be measured.

Reaction Kinetics

ρ_b is the bulk density
R is the pellet radius
n is the reaction order
k_c is the mass transfer coefficient.
C_{As} is the concentration of A at the surface

If MR is below 0.15, then the external mass transfer limitations can be neglected.
Now that we have introduced the theory behind this effectiveness factor, let's solve an example.

Example 3.1

Consider the irreversible first-order reaction in liquid phase where chemical A is transformed into B, A → B. The reaction is being carried out in a packed bed reactor where a solid catalyst is being used. Consider three scenarios: (a) no mass transfer limitations, (b) internal mass transfer limitations, (c) internal and external mass transfer limitations.

Plot the concentration of reactant A as a function of the amount of catalyst in the reactor for each of the previous cases and discuss the results.

In addition, if the radius of the pellets decreases to half, when having only internal mass transfer limitations, what will be the effect of such reduction? Plot the concentration of A for both cases, compare the results and discuss.

Finally, if the external mass transfer coefficient is increased by a factor of 10, for the overall mass limitations scenario, what is the effect of such modification on the concentration of A? Plot concentrations, compare, and discuss.

Solution

The reaction under evaluation is a first-order irreversible reaction; therefore, the reaction rate can be given by:

$$-r_A^w = k_1^w C_A \tag{3.288}$$

It is important to notice that the superscript w indicates that the reaction rate constant is based on the amount of catalyst employed; therefore, the reaction rate will be also given in weight of catalyst instead of volume. If we consider the relationship among k_1 and k_1^w, it is easy to see that this is:

$$\rho_{catalyst} * k_1^w = k_1 \tag{3.289}$$

Since this problem has mass transfer limitations, we will express the reaction rate in a more generic expression by adding the overall effectiveness factor. This gives:

$$-r_A^w = \Theta k_1^w C_A \tag{3.290}$$

With Equation 3.290 for the reaction rate, we can now try to solve the mole balance in order to obtain the desired profiles. The mole balance to solve:

$$\frac{dF_A}{dW} = r_A^w = -\Theta k_1^w C_A \tag{3.291}$$

We can now solve Equation 3.291 for all the desired scenarios, in order to obtain the C_A profiles.

For the first scenario, when there are no mass transfer limitations, we can set $\Theta = 1$. The expression we obtain is the typical expression for a packed bed reactor:

$$\frac{dF_A}{dW} = k_1^w C_A \tag{3.292}$$

In order to solve it, we need the additional data that this is a liquid phase reaction so that $F_A = v*C_A$, as well as values for the reaction rate constants.

For this case, the script to be solved is:

Script:

```
#Mole Balance
    d(Ca)/d(W) = ra/v
    Ca(0) = 10
#Reaction rate
    ra = -k1*Ca
#Auxiliary equations
    k1 = k100*EXP(-Ea/(R*T))
#Constants
    Ea = 100
    k100 = 1
    R = 8.3143
    T = 298
    v = 2
#Independent variable
    W(0) = 0
    W(f) = 25
```

The result for C_A as a function of W is presented in Figure 3.22.

Secondly, we need to solve the mole balance, when there are no external mass transfer limitations. If we take a look at Equation 3.275, for the overall effectiveness factor, we can see that if k_m tends to infinity, then there are no mass transfer limitations and $\Theta = \eta$. Therefore, we can rewrite the mole balance:

$$\frac{dF_A}{dW} = -\eta k_1^w C_A \tag{3.293}$$

In order to solve this problem, we need to calculate the value of the internal effectiveness factor η, which can be calculated as:

$$\eta = \frac{3}{\Phi_1}\left(\frac{1}{\tanh(\Phi_1)} - \frac{1}{\Phi_1}\right) \tag{3.294}$$

Reaction Kinetics

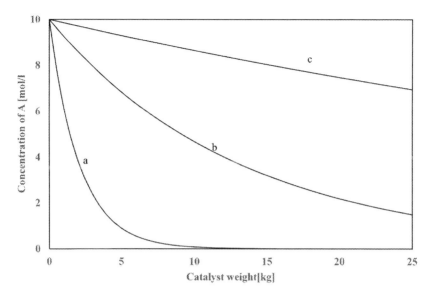

FIGURE 3.22 Variation of C_A as a function of the amount of catalyst for: (a) no mass transfer limitations; (b) internal mass transfer limitations; (c) internal and external mass transfer limitations.

where

$$\Phi_1 = R\sqrt{\frac{k_1}{D}} \qquad (3.295)$$

Equation 3.295 shows that we need k_1; however, to solve the mole balance, we need k_1^W, we can then use Equation 3.290 to relate them; this leads to:

$$\Phi_1 = R\sqrt{\frac{\rho_{cat} k_1^W}{D}} \qquad (3.296)$$

Substituting Equation 3.296 into 3.295 and further into Equation 3.294, we can solve our problem and obtain the concentration profile. The script used in Polymath for solving the problem is:

Script:

```
#Mole balance
    d(Ca)/d(W) = ra/v
    Ca(0) = 10
#Reaction rate
    ra = -k1*Ca*Nu
#Auxiliary equations
    k1 = k100*EXP(-Ea/(R*T))
```

```
Nu = (3/(Phi^2))*((Phi*coth(Phi))−1)
Phi = R2*SQRT(k2/De)
k2 = k1*Rho
#Constants
    Ea = 100
    k100 = 1
    R = 8.3143
    T = 298
    v = 2
    De = 0.0000003
    R2 = 0.01
    Rho = 1
#Independent variable
    W(0) = 0
    W(f) = 25
```

The variation of C_A is also presented in Figure 3.22 in comparison with the previous profile.

Finally, we will now try to solve Equation 3.292. In this case, we have internal and external mass transfer limitations; therefore, we need to find the overall effectiveness factor which is given by Equation 3.294.

Our mole balance to be solved is:

$$\frac{dF_A}{dW} = r_A^w = -\Theta k_1^w C_A \tag{3.297}$$

The overall effectiveness factor has the mathematical expression:

$$\Theta = \frac{\eta}{\left(1 + \dfrac{\eta k_1}{k_m a_c}\right)} \tag{3.298}$$

where

$$\eta = \frac{3}{\Phi_1}\left(\frac{1}{\tanh(\Phi_1)} - \frac{1}{\Phi_1}\right) \tag{3.299}$$

$$\Phi_1 = R\sqrt{\frac{k_1}{D}} \tag{3.300}$$

Equations 3.299 and 3.300 shows that we need k_1, however, to solve the mole balance, we need k_1^w, we can then use Equation 3.290 to relate them; this leads to:

Reaction Kinetics

$$\Theta = \frac{\eta}{\left(1 + \dfrac{\eta \rho_{cat} k_1^w}{k_m a_c}\right)} \qquad (3.301)$$

$$\Phi_1 = R\sqrt{\frac{\rho_{cat} k_1^w}{D}} \qquad (3.302)$$

In addition, k_m is the overall mass transfer coefficient, set to the value given in the problem and a_c is the ratio of surface area and volume, for a spherical case $a_c = 3/R$.

For this case, the script is:

Script:

```
#Mole balance
    d(Ca)/d(W) = ra/v
    Ca(0) = 10
#Reaction rate
    ra = -k1*Ca*RO
#Auxiliary equations
    k1 = k100*EXP(-Ea/(R*T))
    RO = Nu/(1+((Nu*k2)/(km*ac)))
    Nu = (3/(Phi^2))*((Phi*coth(Phi))-1)
    Phi = R2*SQRT(k2/De)
    k2 = k1*Rho
    ac = 3/R
#Constants
    Ea = 100
    k100 = 1
    R = 8.3143
    T = 298
    v = 2
    Rho=1
    De = 0.0000003
    R2 = 0.01
    km = 0.1
#Independent variable
    W(0) = 0
    W(f) = 25
```

The variation of C_A for this case is also presented in Figure 3.22.

Figure 3.22 shows the variations of the concentration of A for the three desired scenarios. It can be seen that case (a) presents the most abrupt tendency; this is due to the fact that the reaction takes place quite fast and the reactant is being consumed. We can compare this scenario with case (b) where there are some internal mass transfer limitations, and in this case, it can be seen that the concentration of A decreases in a much slower pace. This is because, due to the internal mass transfer limitations, reactant A is not available to be consumed at all times not at the same pace as in case (a). Some time and energy are being consumed for reactant A to reach the reactive site, a situation which in case (a) was more *instantaneous*. If we take a look at case (c), we can see that the tendency is even slower, this means that having internal mass transfer limitations (reactant A has a more difficult time to reach the site) in addition to external mass transfer limitations (reactant A cannot reach the outside of the catalytic pellet as easy as in the previous two cases) makes the availability of A much less limited for the reaction to take place.

The second part of this problem involves studying the concentration profile when the radius of the pellet is decreased to half its size. For this case, we need to solve the case (b) script one more time with the radius being divided by a factor of 2. Figure 3.23 shows the variation of the concentration of A for both cases. As shown, a reduction of the catalyst radius to half its size, increases the amount of A being consumed. This is because having a smaller particle makes the internal mass limitations to be considerably reduced; the distance that the reactant needs to travel inside the particle is smaller and therefore, can reach faster the active site to react.

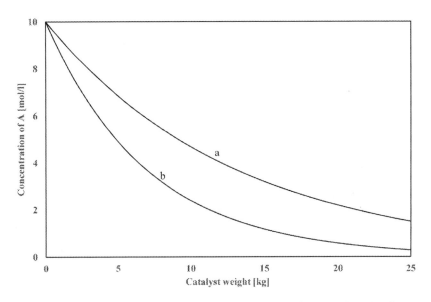

FIGURE 3.23 Variation of C_A as a function of the amount of catalyst for: (a) radius as before, (b) radius decreased in half.

It is important to mention that we reduced the particle size (smaller spherical pellets); however, there has been no modification to the porous diameter.

Finally, the effect of increasing the mass transfer coefficient by a factor of 10, is shown in Figure 3.24.

Figure 3.24 shows that an increase in the value of the mass transfer coefficient produces a shift in the consumption of chemical A to a faster degree. The reason

Reaction Kinetics

is that a higher coefficient means that the diffusion is improved and therefore, is less controlling than in previous scenarios. If k_m increases indefinitely, it will reach the same tendency as it will have if there are no external mass transfer issues.

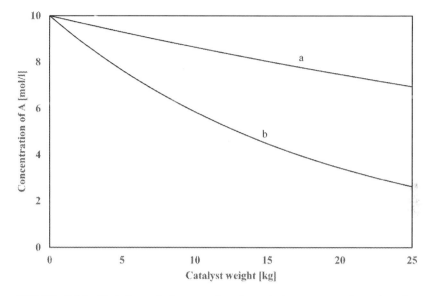

FIGURE 3.24 Variation of C_A as a function of the amount of catalyst for: (a) $k_m = 0.1$ [m/s] and (b) $k_m = 1$ [m/s].

Extra:

We will include, in addition to this analysis, the variation of the concentration of A if there is some catalyst deactivation. For this case, we will assume that the deactivation will follow:

$$-\frac{d\delta}{dt} = k_\delta(T) * \delta \qquad (3.303)$$

We can rewrite this as a function of the catalyst weight; if we have M [mass/time] flowing at a given location in the reactor, we can express:

$$t = \frac{W}{M} \qquad (3.304)$$

which gives:

$$dt = \frac{dW}{M} \qquad (3.305)$$

Substituting into Equation 3.295:

$$\frac{d\delta}{dW} = \frac{-k_\delta(T)}{M} * \delta \qquad (3.306)$$

M is mass per time, but we know the volumetric flow; therefore, if the density of the liquid does not change during the reaction, which is generally is the case for liquid reactions but not necessarily for gas phase processes, then, we can use that:

$$M = v * density_{liquid} \qquad (3.307)$$

Substituting it:

$$\frac{d\delta}{dW} = \frac{-k_\delta(T)}{v * density_{liquid}} * \delta \qquad (3.308)$$

We can then add this expression to our script and solve the problem for when there is a decay due to catalyst deactivation.

The new script to be solved is:

Script:

```
#Mole balance
    d(Ca)/d(W) = (deact*ra)/v
    Ca(0) = 10
#Deactivation balance
    d(deact)/d(W) = (-kd*deact)/(v*density)
    deact(0) = 1
#Reaction rate
    ra = -k1*Ca*RO
#Auxiliary equations
    k1 = k100*EXP(-Ea/(R*T))
    RO = Nu/(1+((Nu*k2)/(km*ac)))
    Nu = (3/(Phi^2))*((Phi*coth(Phi))-1)
    Phi = R2*SQRT(k2/De)
    k2 = k1*rho
    ac = 3/R
# Constants
    Ea = 100
    k100 = 1
    R = 8.3143
    T = 298
    v = 2
    De = 0.0000003
    R2 = 0.01
    km = 0.1
    rho =1
    kd=0.1
```

Reaction Kinetics

```
    density =1
# Independent variable
    W(0) = 0
    W(f) = 25
```

Figure 3.25 presents the comparison for the process with and without catalyst deactivation. In all cases, temperature has remained constant, so the process is isothermal.

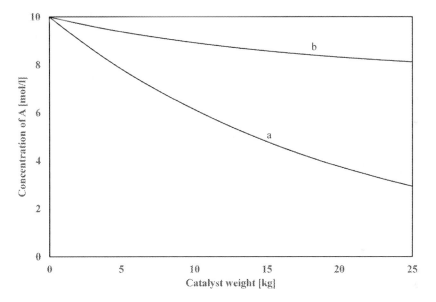

FIGURE 3.25 Variation of C_A as a function of the amount of catalyst for: (a) non-catalyst deactivation and (b) catalyst deactivation.

It can be seen that when there is deactivation of the catalyst, the final concentration of A is higher than when the catalyst is active; meaning that less product is being produced. If the deactivation is too large, the final concentration of A might be constant; meaning that there is no more possibility of reaction and this can be due to various reasons like poisoning in the catalyst, agglomeration, sintering, or coke deposition.

NOTE

1 Obviously, this can be applied for equilibrium reactions.

REFERENCES

1. Westerterp, K.R., van Swain, W.P.M., Beenackers, A.A.C.M. *Chemical Reactor Design and Operation*. 2001. John Wiley & Sons. ISBN: 0471917303.
2. Fogler, H.S. *Elements of Chemical Reaction Engineering*. 5th edition. 2016. Pearson Education Inc. ISBN: 9780133887518.

3. Levenspiel, O. *Chemical Reaction Engineering*. 3rd edition. 1999. John Wiley & Sons. ISBN: 9780471254249.
4. Froment, G.F., Bischoff, K.B., De Wilde, J. *Chemical Reactor Analysis and Design*. 3rd edition. 2011. John Wiley & Sons. ISBN: 9780470565414.
5. Welty, J.R., Wicks, C.E., Wilson, R.E., Rorrer, G.L. *Fundamentals of Momentum, Heat, and Mass Transfer*. 5th edition. 2008. John Wiley & Sons. ISBN: 9780470128688.
6. Bird, R.B., Stewart, W.E., Lightfoot, E.N. *Transport Phenomena*. 2nd edition. 2002. John Wiley & Sons. ISBN: 0471410772.
7. Satterfield, C.N. *Mass Transfer in Heterogeneous Catalysis*. 1970. MIT Press. ISBN: 262190621.
8. Bischoff, K.B. "Effectiveness Factors for General Reaction Rate Forms". *A.I.Ch.E. Journal*. 11(2), (1965), 351–355.

PROBLEMS

PROBLEM 1
Determine if the following reactions are elementary reactions or not.

a. $A + 2B \rightleftarrows C$; $r_A = -k_1 C_A C_B^2 + k_2 C_C$

b. $A + B \rightleftarrows C$; $r_A = -k_1 C_A C_B^2 + k_2 C_C$

c. $2A + B \rightleftarrows 2C + 3D$; $r_A = -k_1 C_A^2 C_B + k_2 C_C^2 C_D^2$

d. $2A + B \rightleftarrows 2C + 3D$; $r_A = -k_1 C_A^2 C_B + k_2 C_C^2 C_D^3$

PROBLEM 2
For the four different plots of energy versus reaction coordinate presented here, determine if the reactions will be exothermic or endothermic.

a.

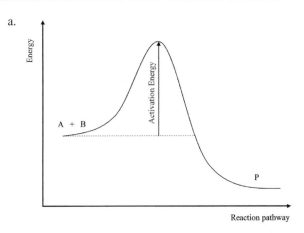

Reaction Kinetics

b.

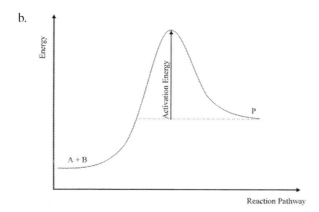

c.

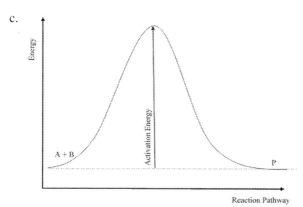

d.

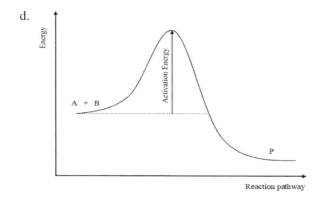

PROBLEM 3
For the experimental data presented here, obtain the order of reaction and the value of the Arrhenius constant (k), using the integration approach.

Time [min]	Concentration of A [mol/l]
0.00	1.850
0.01	1.760
0.10	1.250
0.50	0.700
1.00	0.500
3.00	0.300
5.00	0.250
7.00	0.200
9.00	0.170
13.00	0.145
15.00	0.135
20.00	0.110
25.00	0.100
30.00	0.097
40.00	0.080

PROBLEM 4
Solve Problem 3 but using the differential method instead.

PROBLEM 5
When using the differential method for obtaining the order of reaction, the data can be interpolated with different mathematical functions. In this case, for the data provided below, solve and obtain the reaction order and the Arrhenius constant when the data are interpolated with:

a. An exponential function.
b. A degree 4 polynomial function.
c. Solve this problem again, using the integration methodology.

Time [min]	Concentration of A [mol/l]
0.00	1.850
0.01	1.830
0.10	1.819
0.50	1.700
1.00	1.600
3.00	1.200
5.00	0.880
7.00	0.650
9.00	0.500
13.00	0.250
15.00	0.200
20.00	0.100
25.00	0.050

PROBLEM 6
For the following chemical reactions, considering them as elementary reactions, write their reaction rates. In addition, write the relationship among the reactions rates for all the reactants or products present in each reaction.

Reaction Kinetics

$$A \rightarrow B$$
$$A \rightleftarrows B$$
$$A + B \rightarrow C$$
$$A + B \rightleftarrows C + 2D$$
$$2A + B \rightarrow 3C + \tfrac{1}{2}D$$
$$\tfrac{1}{2}A + \tfrac{3}{2}B \rightleftarrows \tfrac{1}{4}C + D$$
$$2A + B \rightleftarrows 2C + 3D$$
$$2A + B + 2C \rightarrow 2D + 3E$$
$$2A + 2B + 3E \rightleftarrows 2F + 2Y + 3P$$
$$A + B \rightleftarrows 2E + C + 2F$$

PROBLEM 7

The transformation of two chemicals takes place following the reaction A + B → D. This reaction is not elementary but irreversible. Therefore, in order to obtain the reaction rate, it is imperative to know the reaction steps.

a. Propose a mechanism where A has a reaction intermediate and obtain the reaction rate for r_D.
b. One of your colleagues found out that the reaction can be simplified to A → D. However, it involved an intermediate for chemical A (A*) which transforms into a second intermediate (B*), which eventually transforms into D. For this new approach, propose a mechanism, solve it, and compare it with case (a).
c. Another group, from another university, found experimental information that shows a new possible mechanism, which is as follows:

$$A \rightarrow A*$$
$$A* + B \rightarrow C*$$
$$B + C* \rightarrow D*$$
$$D* \rightarrow D$$

Solve this mechanism and obtain r_D.

PROBLEM 8

The transformation of A and B into C is a reversible non-elementary reaction. Since the reaction is non-elementary, different mechanisms can be proposed to solve and obtain the reaction rate. The following mechanism was proposed, where an active intermediate is produced from the reactant A, to then have a surface reaction with B, to produce the activate intermediate from C which can be transformed into C.

Mechanism:

$$\text{Step } 1: A \rightleftarrows A*$$

$$\text{Step } 2: A* + B \rightleftarrows C*$$

$$\text{Step } 3: C* \rightleftarrows C$$

For the above-mentioned mechanism:

a. Solve the problem using the equilibrium approach. Solve this for cases when:
   i Step 1: is controlling
   ii Step 2: is controlling
   iii Step 3: is controlling.
b. Solve the mechanism using the PSSH methodology.

PROBLEM 9

The reaction where A and B are transformed into a desired product P and a by-product C takes place via a reversible non-elementary reaction. The main reaction is $A + B \rightleftarrows C + P$. The proposed mechanism obtained from experimental information is:

Mechanism:

$$\text{Step } 1: A \rightleftarrows A*$$

$$\text{Step } 2: A* + B \rightleftarrows C + D*$$

$$\text{Step } 3: A* + D* \rightleftarrows E + P*$$

$$\text{Step } 4: P* \rightleftarrows P$$

For the above-mentioned mechanism:

a. Solve the mechanism when step 4 is controlling.
b. Show the equations when using the PSSH; do not solve it, just present the equations.

PROBLEM 10

Work has been done over the heterogeneous reactions for the transformation of chemicals A and B into two products C and D, this reaction is an irreversible non-elementary reaction. Using the PSSH methodology, acquire the reaction rate expression for the product C.

Experimentally, it has been found that A is adsorbed while D is being desorbed. The main reaction is $A + B \longrightarrow C + D$, while the proposed mechanism is as follows:

Reaction Kinetics

Mechanism:

$$Step\ 1: A + S \rightarrow AS$$

$$Step\ 2: AS + B \rightarrow C + DS$$

$$Step\ 3: DS \rightarrow D + S$$

PROBLEM 11
For the following proposed mechanisms, develop the reaction rate expression for each case considering that all reactions could be the controlling steps. It is advised to use the equilibrium approach to do this problem.

a. *Main reaction:*

$$A \rightleftarrows C$$

Proposed mechanism:

$$Step\ 1: A + S \rightarrow AS$$
$$Step\ 2: AS \rightleftarrows C + S$$

b. *Main reaction:*

$$A + B \rightleftarrows C$$

Proposed mechanism:

$$Step\ 1: A + S \rightleftarrows AS$$
$$Step\ 2: AS + B \rightleftarrows CS$$
$$Step\ 3: CS \rightleftarrows C + S$$

c. *Main reaction:*

$$A + B \rightleftarrows C$$

Proposed mechanism:

$$Step\ 1: A + S \rightleftarrows AS$$
$$Step\ 2: B + S \rightleftarrows BS$$
$$Step\ 3: AS + BS \rightleftarrows C + 2S$$

d. *Main reaction:*

$$A + B \rightleftarrows C$$

Proposed mechanism:

$$Step\ 1: A + S \rightleftarrows AS$$
$$Step\ 2: B + S \rightleftarrows BS$$
$$Step\ 3: AS + BS \rightleftarrows CS + S$$
$$Step\ 4: CS \rightleftarrows C + S$$

PROBLEM 12

A plug flow reactor is being used to carry out the following reactions, taking place simultaneously in liquid phase:

Reaction 1: $A + B \rightleftarrows C + D$
Reaction 2: $A + C \rightleftarrows E$
Reaction 3: $B + D \rightleftarrows F + G$
Reaction 4: $E + F \rightleftarrows H$

a. Write the mole balances for all the components involved in this complex multiple reaction system.
b. Present an estimation of the evolution of the different flows along the volume of the reactor.

PROBLEM 13

The production of a product P is being done following these reactions:

Reaction 1: $A + B \rightleftarrows C + D$
Reaction 2: $B + D \rightleftarrows C + E$
Reaction 3: $E \rightleftarrows P$

However, these reactions are not elementary steps and therefore, experimental work has shown that the mechanism involved is as follows:

Reaction 1.α: $A \rightleftarrows A*$
Reaction 1.β: $A* + B \rightleftarrows C + D$
Reaction 2.α: $D + B \rightleftarrows C + E$
Reaction 3.α: $E \rightleftarrows E*$
Reaction 3.β: $E* \rightleftarrows P$

a. Present the reaction rate for all the five reactions mentioned above.

Reaction Kinetics

b. Reactions 1 and 3 have two steps each; therefore, obtain the reaction rate for step 1, 2, and 3 (one for each), when all possible steps are being considered. Present all possible combinations for all three reactions. (Hint: Obtain one for each reaction and use the other one as a reaction in equilibrium) (Hint 2: the total amount of combination is 4)

PROBLEM 14

Catalyst deactivation is very important and relevant since it determines the useful life of a material. The deactivation of catalysts, as seen before, can be due to different factors that will lead to different deactivations rates and therefore, different physical causes, like coke deposition, poisoning of a catalyst, or inhibitors. The deactivation rate can be expressed as:

$$\frac{d\delta}{dt} = -r_\delta$$

Obtain the expression for the deactivation, δ, as a function of time when r_δ is:

$$r_\delta = k$$
$$r_\delta = k\delta$$
$$r_\delta = k\delta^2$$

PROBLEM 15

For all the three deactivation expressions obtained from Problem 14, solve the mole balances for the irreversible reaction of A to B when:

a. Reaction order is 1 with respect to A. ($r_{Aideal} = k_1 * C_A$)
b. Reaction order is 2 with respect to A. ($r_{Aideal} = k_1 * C^2_A$)
c. Make a comparative plot for cases from (a) and another for cases from (b).

PROBLEM 16

A heterogeneous reaction where A and B are being transformed takes place in a plug flow reactor. The catalyst is a porous material with a cylindrical shape.

a. The reaction has an internal effectiveness factor of 0.85; plot the variation of L as a function of k/D. (Refer to Equations 3.125 and 3.216)
b. How does this plot change if $\eta = 0.5$ instead?
c. If the Thiele modulus for a first-order reaction in a spherical catalysts $\phi = 0.9$, what is the radius of such spherical catalysts when $k = 52.3$ [1/min] and $D = 3.486*10^{-4}$ [m²/s]?
d. For case (c), plot the variation of C_A as a function of r. (Eventually, it can be done as a function of r/R; this will give a smaller domain.)
e. For the same case (c), plot the variation of the internal effectiveness factor as a function of the radius (r).
f. Present in a comparative plot, the variation of C_A when $\phi = 0, 0.25, 0.5,$ and 1.

PROBLEM 17
Chemical A is being consumed following a first-order irreversible reaction. A researcher in your group managed to get to the lab and carry on some experimental work to obtain the variation of C_A as a function of time. Since this is a heterogeneous reaction that takes place, there are some concerns regarding internal mass transfer limitations. Therefore:

a. Plot the data.
b. Plot the kinetic model when $C_A^0 = 2.35$ [mol/l] and $k = 3.78$ [1/min]
c. Calculate the value of the internal effectiveness factor (η).
d. Plot the new model with the real value of the internal effectiveness factor and compare with the previous case.
e. For the obtained value of η, what will be the value of ϕ_1?
f. Based on ϕ_1, considering $k = 3.78$ [1/min] and $D = 2*10^{-6}$ [m²/s], what is the radius [m] of the spherical catalysts?
g. If ϕ_1 is increased by a factor of 10, what will be the new radius of the catalysts and the new internal effectiveness factor?

Time [min]	Concentration of A [mol/l]
0.00	2.960
0.10	2.000
0.20	1.400
0.30	0.900
0.40	0.647
0.50	0.444
0.60	0.304
0.70	0.205
0.80	0.143
0.90	0.098
1.00	0.065

PROBLEM 18
After discussing the results with some of your colleagues, you find out that the reaction that has been considered in Problem 17 (A → B) has been wrongly studied. The mass transfer limitations that were found were the overall mass transfer limitations and not only the internal. Therefore, you need to study the reaction with an $\Omega = 0.8$. For this new scenario:

a. What is the new real value for η? What is the value for ϕ_1? Consider that $k_c = 0.105$[m/min] and $a_C = 3/R$ [1/m] with $R = 0.0115$ [m]
b. With the new value of η, plot the new model for P17 and discuss the quality of fitting?
c. What happened to the values η and Ω if the value of k_1 is reduced to half?
d. What happened to the values η and Ω if the value of k_c is increased by a factor of 10?
e. What happened to the values η and Ω if the value of a_c is increased by a factor of 10?

4 Completely Solved Example

4.1 INTRODUCTION

Within this chapter I would like to present a problem and to solve it from start to finish, as complete as possible for the level of this book. I will show all the steps needed to go from the description of the problem, the equipment used, how to obtain the data in the laboratory as well as processing such data, to obtained reactions kinetics expression. Finally we will use those expressions in a plug flow reactor, or in a packed bed reactor, and we will see and study the effects of each different scenario.

This problem will be tackled in the following steps:

Section 4.2: Description of the Problem.
Section 4.3: Laboratory Equipment Employed.
Section 4.4: Experimental Procedure.
Section 4.5: Sample Analysis and Errors.
Section 4.6: Data Evaluation.
Section 4.7: Mathematical Model 1.
Section 4.8: Comparison of Data and Model 1.
Section 4.9: Mathematical Model 2.
Section 4.10: Comparison of Data and Model 2.
Section 4.11: Final Expression for Kinetics.
Section 4.12: Simulation of an Isothermal Plug Flow Reactor Using Kinetics from 4.11.
Section 4.13: Simulation of an Adiabatic Plug Flow Reactor with the Kinetics from 4.11.
Section 4.14: Simulation of a Constant Heat Transfer Plug Flow Reactor with Kinetics from 4.11.
Section 4.15: Simulation of a Co-current Heat Transfer Flow in a Plug Flow Reactor with Kinetics from 4.11.
Section 4.16: Simulation of a Counter-Current Heat Transfer Flow in a Plug Flow Reactor with Kinetics from 4.11.
Section 4.17: Comparison For A Gas Phase System with Pressure Drop.

While some readers may not use the same equipment mentioned here, we would like to highlight that the procedure presented, and the algorithm used herein, are fundamental and basic to all reaction systems and can be adapted for modeling other kinds of systems.

4.2 DESCRIPTION OF THE PROBLEM

The first step is to identify the problem, the information available, and what results are needed. In this case, we will start with a simple situation and then add information and assumptions to cover different scenarios.

Here, we deal with an equilibrium solid catalyst-based reaction; following is the generic reaction:

$$\alpha A + \beta B \leftrightarrows \delta C + \Omega D \tag{4.1}$$

For the purpose of this problem, *i.e.*, to demonstrate how information is extracted from experiments, we choose to avoid complicated mathematics by selecting every stoichiometric constant to be equal to 1. Therefore, our equation is simplified to:

$$A + B \leftrightarrows C + D \tag{4.2}$$

This reaction is not yet assigned to be an elementary reaction, to maintain its generality; this will be addressed in Section 4.6.

Based on this reaction, and that it is a solid catalyst-based reaction, *i.e.*, the reaction takes place over the surface of a catalyst, the type of reaction model to be solved can then be selected accordingly.

4.3 LABORATORY EQUIPMENT EMPLOYED

In order to find out the kinetics expression, a batch reactor is the main equipment to be used; more information about this equipment and it uses and how to get the expressions from it can be found in Chapter 2. In summary, a batch reactor does not have a flow in or out of any component; therefore, only changes with respect to time are studied since it is normally run at a constant temperature. This variation with time is extremely important for the development of a kinetics expression.

Regularly, we said that we used a batch reactor for the developed of the kinetics expressions. However, the experimental equipment that is used in the lab has a few more parts than those mentioned in Chapter 2. An experimental reaction set up can be seen in Figure 4.1, which shows a glass reactor so that the flow can be easily seen and recorded. This helps study the effect of mixing and identify any zones of dead volume and to know if the RPM needs to be modified. It also has a condenser on the top, to not allow any of the liquids involved (this is used for a liquid/liquid/solid reactions), to evaporate and leak out of the reaction set up. It has an external water bath in order to have a better control on the reaction temperature. However, a thermocouple is required to know the temperature inside the reaction mixture properly. The reactor has a glass jacket where the water goes around in order to warm up the system as uniformly as possible. The mixing can be done with an external engine from the top, or as in this case, with a magnetic steering from below. The steering needs to be strong enough to assure perfect mixing so that there are no external mass transfer limitations or gradients of temperature within the reactor.

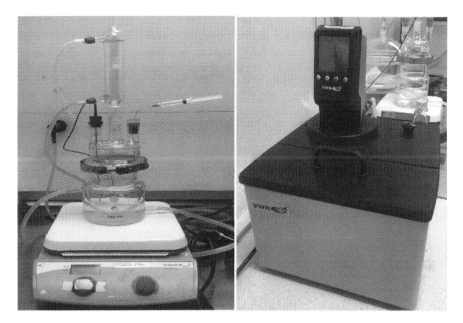

FIGURE 4.1 Batch experimental setup with water bath.

4.4 EXPERIMENTAL PROCEDURE

It is very important to mention here that this is just a suggestion of a procedure, and there are many differences in how to do a procedure depending on the system, reactants, type of reaction, and catalyst involved. For the cases presented here, using the equipment mentioned above for kinetics, it is generally considered better to add the reactants, start the steering and wait until the desired temperature is reached and is stable. This procedure is suitable if the reactants do not react with each other without a catalyst. However, if they do react with each other, it is better to warm them up separately and then mix them together when the reaction process starts. When the desired temperature is reached, the catalyst should be added and the reaction clock started.

For reaction under study, there is no information about the pressure; therefore, a proper assumption is that it is under atmospheric pressure. However, the same reaction could be improved under high pressure and temperature (we saw that effect within the production of biodiesel from jojoba oil [1–3]). If that is the case, a Parr reactor will be used. In these cases, all components should be added at the same time and sampling should be done before the temperature is set. When the desired temperature is reached and the reaction clock has started, a new sample should be withdrawn; this will allow you to know the degree of reaction that takes place in the warming up part of the process.

4.5 SAMPLE ANALYSIS AND ERRORS

In order to be able to have proper data for kinetics modeling, sufficient data should be taken to be able to cover the whole spectrum of time over which the reaction happens to avoid missing parts of the reaction that are crucial for the kinetics. Taking too many samples will lead to a reduction of the volume of the reaction set up and in the case of a heterogeneous system, that should be reduced to its minimum, since a batch reactor has the assumption of constant volume.

The withdrawing time needs to be established beforehand, based on information from the literature. The number of samples, and the amount, to be removed will also be a condition for the initial concentration and quantity of each reactant. Each withdrawn sample should be as small as possible, but large enough to allow you to test it twice in order to evaluate the reproducibility of the measurement. So, for example, if we want to remove 1 [mL] per withdrawal and 20 samples total; in order to not produce a significant change in the volume of the reactor, the total reaction volume could be 2 [liters] or 2,000 [mL]; so the total removed volume is 1%. There is disagreement on this percentage; some researchers will allow 2–4% of the volume; however, the amount should not be too large, or the ideal batch reactor model will no longer apply.

Once the samples have been withdrawn, analysis of the samples should be carried out as soon as possible. An online measuring system will be optimal, but that is not the case in many laboratories or applications. Therefore, when possible, the sample should be prepared and tested immediately in order to know the degree of reaction and conversion of the main component. If direct quantification is not an alternative, proper storage and condition should be considered in order to avoid degradation of the samples.

When preparing a sample for testing, the preparation of each sample should be accurate, and each sample should be prepared and measured twice when possible; this will allow the reproducibility of the result and the repeatability of the preparation and analytical methods employed. Following well-developed and accepted techniques and methods will minimize the errors that can be encountered when preparing the samples. However, it is important to remind the reader that each technique will be associated to some error. Therefore, having the appropriate equipment is crucial to have accurate results.

Errors in the samples can be minimized by the use of computer-based tools and robots such as an auto-sampler, workbenches, or auto-mixing. This will considerably reduce the random errors and will allow you to spot systematic errors that occur from the use of the machine and can then be taken into consideration for the calculations, if required. Quantification of the amount of error in each equipment has to be established based on each equipment individually; this cannot be comprehensively covered in this book since each technique/equipment has an intrinsic error.

Completely Solved Example 263

4.6 DATA EVALUATION

Once the experiment is prepared and is running, samples are withdrawn as presented earlier in this chapter. Once samples have been extracted, analyzing them is crucial in order to know the composition and concentration [mol/l or wt.%] of each component in the sample. This is an indication if the reaction has moved forward, as desired, or if the conditions we are using are not suitable.

Analyzing the data is a challenge in itself; usually, there are techniques that can be used for very different applications. For instance, Gas Chromatography (GC) is a very versatile tool that can be used to measure the composition of the reactants and the products in the reaction. This technique is complex in comparison with others and requires more time. However, coupled to an Mass Spectrometry (MS), it will not only give you the amount of each chemical, but also what chemicals are being produced or consumed, allowing the user to have a much better understanding of the process.

Other techniques are as good as GC for other chemicals; titration is a simple and reliable technique to measure the acidity or basicity of a solution. Therefore, it can be used to measure fatty acid, for example, and its production or consumption in a simple and effective way.

High Performance Liquid Chromatography (HPLC) can also be used for measuring the degree of reaction of a given process by doing liquid samples analysis for processes like sugar production and composition, or methane production from a biogas. Karl Fischer water titration can be used to determine the amount of water in a sample; if water is being produced, this is an indirect indication of the degree of advancement of the reaction.

In other cases, indirect measurements can be done as well. In a process where gas is being produced, even as a byproduct, a measure in the pressure of the system will give an indication of the production of the component. Therefore, an estimation of the conversion achieved, and the amount of product generated, can be obtained. It is important to notice that this technique should be complimentary to a more definitive measurement since the production of any gas will give a reading which is not necessarily an indication that the desired product was produced.

Besides the reactants and the products, in several cases when using heterogeneous catalyst, it is also important to see the degradation (if any) of the catalyst due to the reaction. Therefore, we will refer you to Chapter 1 for catalyst preparation and characterization for more detailed information on the matter and only give a summary here. Table 4.1 has some of the most common preparation techniques and what they are used for.

As an example, in the work of Avhad *et al.* [4], we studied the production of biodiesel using a heterogeneous catalyst which was characterized using XRD and TGA. Figure 4.2 shows the XRD pattern for the glycerol enriched CaO catalysts employed. The catalyst was calcinated at 800 [°C], the result from the calcination was analyzed in comparison to fresh catalysts, as shown in the XRD results, which confirm the presence of glycerol-enriched CaO.

Furthermore, a TGA analysis was carried out. Figure 4.3 presents the variation of the weight for each case as a function of the temperature. The TGA profiles for the synthesized MC and GMC are shown in Figure 4.3a and b, respectively.

TABLE 4.1
Summary of Characterization Techniques and Their Uses

Acronym	Full name	Uses
XRD	X-Ray diffraction	Provides the structure and active phases of the catalysts. Also understanding the particle size is possible.
FTIR	Fourier transform infrared spectroscopy	It helps in understanding the bonding between the atoms and their degree of strength.
TGA	Thermogravimetric analysis	This technique helps in understanding the thermal and mechanical stability of the catalyst.
BET	Specific surface area	This technique is important to understand the exposed surface area of the catalyst, pore volume, and pore diameter.
SEM	Scanning electron microscopy	This is an imaging technique where we can see the looks of the materials.
TEM	Transmission electron microscopy	This technique is generally required if you have a catalyst supported on another material in order to see the pores and the particle size.

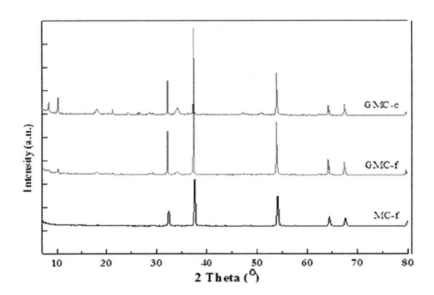

FIGURE 4.2 XRD patterns for CaO and glycerol-enriched CaO. MC-f mussel shells derived CaO-fresh, GMC-f glycerol-enriched CaO-fresh, GMC-c glycerol-enriched CaO collected. (*Reproduced with permission* [4])

For MC, a weight loss of 3.1% was observed after heating the sample from 25 to 800 [°C]. When looking at fresh GMC, a total weight loss of 17.8% was obtained. The high weight loss indicated the formation of the CaO–glycerol complex. As shown, there are three main points where weight loss occurs and for each of them, a different process is involved. For example, the weight lost between 600 and 700 [°C] is attributed to the decarbonation process to give CaO. The TGA profiles of the collected GMC is shown in Figure 4.3c. The weight loss in the case of the collected GMC increased from 17.8% to 25.9%.

Completely Solved Example

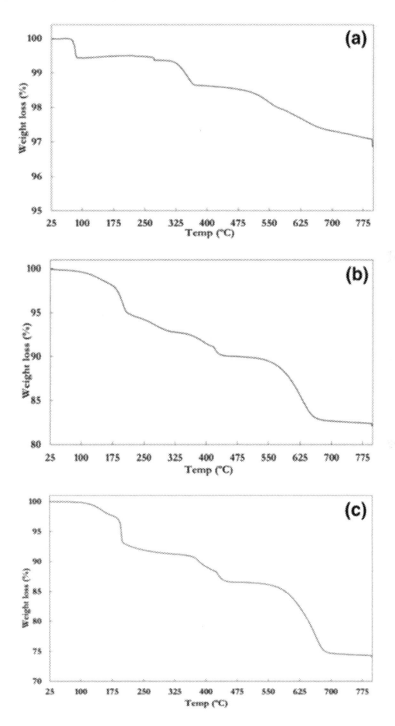

FIGURE 4.3 Thermal analysis for Cao and glycerol-enriched CaO. (a) Mussel shells derived CaO-fresh, (b) glycerol-enriched CaO-fresh, (c) glycerol-enriched CaO collected. (*Reproduced with permission* [4])

From our example problem, we have the reaction of A and B to give C and D. As mentioned before, here, we will present the plots of the experimental results that were obtained by following the reaction procedure, the data sampling, the data analysis, and errors. The errors are represented for just one of the figures because the error is small. For the rest of the figures, only average values are presented.

Figure 4.4 shows the experimental value for conversion (with errors) for one temperature, and Figure 4.5 shows the variation of concentration of all the components for the same temperature. Figure 4.6a–e presents the variation of all the components for three different experimental temperatures and their effect on the reaction. It is evident that a higher conversion and concentration of products is achieved with higher temperatures.

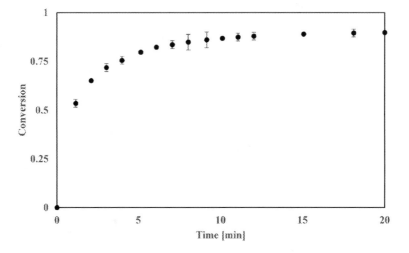

FIGURE 4.4 Conversion of component A as a function of time with error bars for $T = 300$ K.

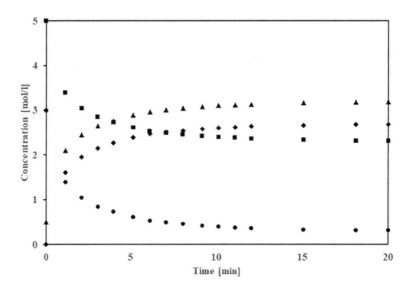

FIGURE 4.5 Variation of the concentration of all reactants and products as a function of time.

Completely Solved Example

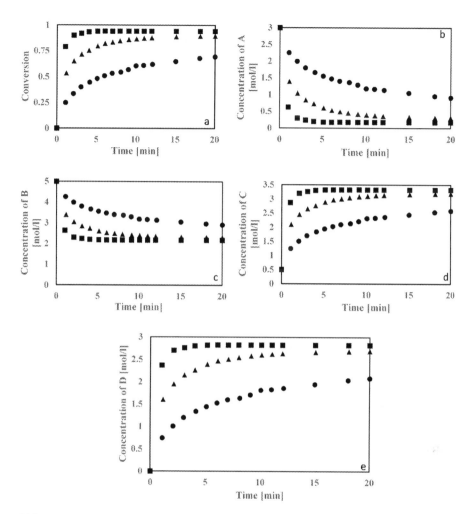

FIGURE 4.6 Variation of conversion (a), concentration of component A (b), component B (c), component C (d), and component D (e), as a function of time for all three tested temperatures. (●) $T = 275$ K, (▲) $T = 300$ K, and (■) $T = 325$ K.

Before we can start with more advanced models, it is important to try to see if the reaction we are considering is not a one-step elementary reaction. In this case, it means that the reaction, as presented in Equation 4.2, is the reaction we need to study.

Based on Equation 4.2, we can obtain the kinetics expression that will be:

$$r = k_1 C_A C_B - k_2 C_C C_d \tag{4.3}$$

Is important to notice that in the reaction expression that we just obtained, there is no evaluation for the amount of catalyst used. This means that any modification in the catalyst amount should not produce any change in the reaction itself. Even though this is not impossible, it is very unlikely that this is the case since higher amount of catalyst has the tendency to increase the initial reaction rate, making the reaction go faster to the equilibrium values. However, due to the fact that we do not know the behavior of this reaction in particular, we need to evaluate it in order to fit the data and see the goodness of fit.

After fitting the data for all three temperatures, and trying to optimize the kinetics parameters, we obtain the fitting presented in Figure 4.7.

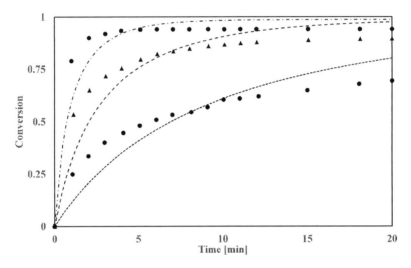

FIGURE 4.7 Variation of conversion of component A as a function of time. (●) Experimental data and (··) model for $T = 275K$, (▲) experimental data and (- -) model for $T = 300$ K, (■) experimental data and (-.-) model for $T = 325K$.

The fit of experimental and model data, in Figure 4.7 is satisfactory, but not good enough. It is evident that the model either underestimates or overestimates the values for all three temperatures. This is a good indicator that this reaction follows a mechanism. Due to this indication, the upcoming sections are used to present different mechanisms and their fitting against the data.

In order to understand the goodness of the fit, some statistical numbers can be calculated, such as the model selection criteria (MSC) [5], which means that the higher the value, the better the fitting. The MSC formula is as follows:

$$\text{MSC} = \ln\left[\frac{\sum(\beta_{exp,i}-\beta_{exp,a})^2}{\sum(\beta_{exp,i}-\beta_{pre,i})^2}\right] - \frac{2p}{n} \qquad (4.4)$$

where $\beta_{exp,i}$ is the experimental data at the ith reaction time; $\beta_{exp,a}$ is an average of the experimental data; $\beta_{pre,i}$ is the predicted values at the ith reaction time; p is the number of the parameter involved in the model; and n is the number of experimental data.

For this previous case, the MSC value is 3.16. The value for MSC should be as large as possible, and even thought 3.16 is large, we are in the search of values that could be above 7 or 8. However, visual inspection of the fitting is crucial to not misinterpret the data. A lower MSC might be better if the fitting (result, or mathematical expression employed) has more physical meaning.

4.7 MATHEMATICAL MODEL 1

Based on the experimental information collected, we can develop a mathematical model to fit the data and that can be used for further simulations in other types of reactors. Our goal is to present a model and to fit the model against the data.

It is important to note that there are vast possibilities of models that can be proposed and different techniques to find the reaction rate, as explained before in Chapter 3. Here, we will propose one model; we will use the pseudo-equilibrium approach to obtain the kinetics expression and use a software to solve the mole balances and compare the model against the experimental data (by minimizing the square differences) to see the goodness of the fitting.

The model proposed here is simple; it is a three-step mechanism where component A is adsorbed over the catalysis surface (Step 1), it reacts with component B, that is not adsorbed, to produce component C, which is adsorbed over the catalyst and component D, which is not absorbed (Step 2), and finally component C is desorbed from the surface (Step 3).

The three steps can be seen as follows:

Step 1: Adsorption of reactant

$$A + S \underset{k_2}{\overset{k_1}{\rightleftarrows}} AS$$

Step 2: Surface reaction

$$AS + B \underset{k_4}{\overset{k_3}{\rightleftarrows}} CS + D$$

Step 3: Desorption reaction

$$CS \underset{k_6}{\overset{k_5}{\rightleftarrows}} C + S$$

where S represents the site over the catalyst surface.

In order to obtain the possible kinetics expression, we need to determine the controlling step. In this case, since we have three elementary steps, we need to study the possibility that each of them might be the controlling step. By doing that, we will end up with three kinetics expressions that will be then tested against the experimental data to evaluate the goodness of the fit.

In order to establish the kinetics expression, we will use the pseudo-equilibrium approach. As previously mentioned, this approach considers one of the reactions to be the limiting step and all the others are considered to be in equilibrium, meaning $r_i = 0$.

For our case, we can develop three reactions equations, one for each step. The subscripts for r represent the step being referring to:

$$\text{Step 1)} \quad r_1 = k_1 C_A C_S - k_2 C_{AS} \quad (4.5)$$

$$\text{Step 2)} \quad r_2 = k_3 C_{AS} C_B - k_4 C_{CS} C_D \quad (4.6)$$

$$\text{Step 3)} \quad r_3 = k_5 C_{CS} - k_6 C_C C_S \quad (4.7)$$

As it was mentioned, we have three equations and therefore, we have three possible candidates for controlling step; we will solve for each of the cases to obtain the final expression for the kinetics.

Step 1 is controlling

If step 1 is controlling, this means that $r_2 = r_3 = 0$. Therefore:

$$0 = r_2 = k_3 C_{AS} C_B - k_4 C_{CS} C_D \quad (4.8)$$

$$0 = r_3 = k_5 C_{CS} - k_6 C_C C_S \quad (4.9)$$

Solving Equation 4.9 first, we can obtain that:

$$C_{CS} = \frac{k_6 C_C C_S}{k_5} \quad (4.10)$$

Now, we can substitute 4.10 into 4.8 and solved to obtain C_{AS}; doing that will lead to:

$$C_{AS} = \frac{k_4 C_D}{k_3 C_B} \frac{k_6 C_C C_S}{k_5} \quad (4.11)$$

We substitute Equation 4.11 into $r_1 = k_1 C_A C_S - k_2 C_{AS}$ in order to obtain the expression of r_1.

$$r_1 = k_1 C_A C_S - k_2 \frac{k_4 C_D}{k_3 C_B} \frac{k_6 C_C C_S}{k_5} \quad (4.12)$$

Completely Solved Example

This expression is a function of all the concentration of products and reactants and the concentration of the catalyst. In order to find a solution for the concentration of C_s, we need to do a catalysts site balance. This takes into account that all the available catalyst that was added to the reactor (C_S^0) is separated between the amount that has not reacted yet, C_S, plus the amount that is used up by the reactants or the products, (C_{AS}, C_{CS}), given:

$$C_S^0 = C_S + C_{AS} + C_{CS} \tag{4.13}$$

Substituting Equations 4.11 and 4.12 into 4.13, we can obtain:

$$C_S^0 = C_S + \frac{k_4 C_D}{k_3 C_B}\frac{k_6 C_C C_S}{k_5} + \frac{k_6 C_C C_S}{k_5} \tag{4.14}$$

We can obtain a common factor (C_s), leading to:

$$C_S^0 = C_S \left(1 + \frac{k_4 C_D}{k_3 C_B}\frac{k_6 C_C}{k_5} + \frac{k_6 C_C}{k_5}\right) \tag{4.15}$$

This expression can be rearranged in order to obtain C_s as a function of the concentration of each component involved, the reaction rate constant, and the initial values of the concentration of the catalyst. The final expression for C_s is:

$$C_S = \frac{C_S^0}{\left(1 + \dfrac{k_4 C_D}{k_3 C_B}\dfrac{k_6 C_C}{k_5} + \dfrac{k_6 C_C}{k_5}\right)} \tag{4.16}$$

Equation 4.16 is the site balance for the catalyst. Then, this expression can be substituted into Equation 4.13 in order to get the final reaction rate expression for the case when step 1 is the controlling step:

$$r_1 = \left(k_1 C_A - k_2 \frac{k_4 C_D}{k_3 C_B}\frac{k_6 C_C}{k_5}\right)\frac{C_S^0}{\left(1 + \dfrac{k_4 C_D}{k_3 C_B}\dfrac{k_6 C_C}{k_5} + \dfrac{k_6 C_C}{k_5}\right)} \tag{4.17}$$

This is the final expression for r_1 that needs to be put into the batch reactor mole balance in order to fit the experimental data. This fitting will be carried out in the upcoming section.

Step 2 is controlling

If step 2 is controlling, this means that $r_1 = r_3 = 0$. Therefore:

$$0 = r_1 = k_1 C_A C_S - k_2 C_{AS} \tag{4.18}$$

$$0 = r_3 = k_5 C_{CS} - k_6 C_C C_S \qquad (4.19)$$

Solving Equation 4.18 first, we can obtain that:

$$C_{AS} = \frac{k_1 C_A C_S}{k_2} \qquad (4.20)$$

Solving Equation 4.19 first, we can obtain that:

$$C_{CS} = \frac{k_6 C_C C_S}{k_5} \qquad (4.21)$$

We substitute Equations 4.20 and 4.21 into $r_2 = k_3 C_{AS} C_B - k_4 C_{CS} C_D$ in order to obtain the expression of r_3.

$$r_2 = \frac{k_1 k_3 C_A C_b C_S}{k_2} - \frac{k_4 k_6 C_D C_C C_S}{k_5} \qquad (4.22)$$

This expression is a function of the concentrations of all products and reactants and the concentration of the catalyst. In order to find a solution for the concentration of C_s, we need to do a site balance. This takes into account that all the available catalyst that was added to the reactor (C_S^0) is separated between the amount that has not reacted yet, C_S, plus the amount that is used by the reactants or the products, (C_{AS}, C_{CS}), giving:

$$C_S^0 = C_S + C_{AS} + C_{CS} \qquad (4.23)$$

Substituting Equations 4.20 and 4.21 into 4.23, we can obtain:

$$C_S^0 = C_S + \frac{k_1 C_A C_S}{k_2} + \frac{k_6 C_C C_S}{k_5} \qquad (4.24)$$

We can obtain a common factor (C_s), leading to:

$$C_S^0 = C_S \left(1 + \frac{k_1 C_A}{k_2} + \frac{k_6 C_C}{k_5}\right) \qquad (4.25)$$

This expression can be rearranged in order to obtain C_s, as a function of the concentration of each component involved, the reaction rate constant, and the initial values of the concentration of the catalyst. The final expression for C_s is:

$$C_S = \frac{C_S^0}{\left(1 + \dfrac{k_1 C_A}{k_2} + \dfrac{k_6 C_C}{k_5}\right)} \qquad (4.26)$$

Completely Solved Example

Equation 4.26 is the site balance for the catalyst. Then, this expression can be substituted into Equation 4.22 in order to get the final reaction rate expression for the case when step 2 is the controlling step:

$$r_2 = \left(\frac{k_1 k_3 C_A C_b}{k_2} - \frac{k_4 k_6 C_D C_C}{k_5} \right) \left(\frac{C_S^0}{1 + \frac{k_1 C_A}{k_2} + \frac{k_6 C_C}{k_5}} \right) \quad (4.27)$$

This is the final expression for r_2 that needs to be put into the batch reactor mole balance in order to fit the experimental data. This fitting will be carried out in the upcoming section.

Step 3 is controlling

If step 3 is controlling, this means that $r_1 = r_2 = 0$. Therefore:

$$0 = r_1 = k_1 C_A C_S - k_2 C_{AS} \quad (4.28)$$

$$0 = r_2 = k_3 C_{AS} C_B - k_4 C_{CS} C_D \quad (4.29)$$

Solving Equation 4.28 first, we can obtain that:

$$C_{AS} = \frac{k_1 C_A C_S}{k_2} \quad (4.30)$$

We can then substitute Equation 4.30 into 4.29 and solve for C_{CS}, getting the following expression:

$$C_{CS} = \frac{k_1 k_3 C_A C_B C_S}{k_2 k_4 C_D} \quad (4.31)$$

We substitute Equation 4.31 into step (3), $r_3 = k_5 C_{CS} - k_6 C_C C_S$ in order to obtain r_3.

$$r_3 = \frac{k_1 k_3 k_5 C_A C_B C_S}{k_2 k_4 C_D} - k_6 C_C C_S \quad (4.32)$$

This expression is a function of all the concentrations of products and reactants and the concentration of the catalyst. In order to find a solution for the concentration of C_s, we need to do a site balance. This takes into account that all the available catalyst that was added to the reactor (C_S^0) is separated between the amount that has not reacted yet, C_S, plus the amount that is used by the reactants or the products, (C_{AS}, C_{CS}), giving:

$$C_S^0 = C_S + C_{AS} + C_{CS} \quad (4.33)$$

Substituting Equations 4.30 and 4.31 into 4.33, we can obtain:

$$C_S^0 = C_S + \frac{k_1 C_A C_S}{k_2} + \frac{k_1 k_3 C_A C_B C_S}{k_2 k_4 C_D} \quad (4.34)$$

We can obtain a common factor (C_s), leading to:

$$C_S^0 = C_S \left(1 + \frac{k_1 C_A}{k_2} + \frac{k_1 k_3 C_A C_B}{k_2 k_4 C_D} \right) \quad (4.35)$$

This expression can be rearranged in order to obtain C_s as a function of the concentration of each component involved, the reaction rate constant, and the initial values of the concentration of the catalyst. The final expression for C_s is:

$$C_S = \frac{C_S^0}{\left(1 + \dfrac{k_1 C_A}{k_2} + \dfrac{k_1 k_3 C_A C_B}{k_2 k_4 C_D} \right)} \quad (4.36)$$

Equation 4.36 is the site balance for the catalyst which can be substituted into Equation 4.32 to get the final reaction rate expression when step 3 is the controlling step:

$$r_3 = \left(\frac{k_1 k_3 k_5 C_A C_B}{k_2 k_4 C_D} - k_6 C_C \right) \left(\frac{C_S^0}{1 + \dfrac{k_1 C_A}{k_2} + \dfrac{k_1 k_3 C_A C_B}{k_2 k_4 C_D}} \right) \quad (4.37)$$

This is the final expression for r_3 that needs to be put into the batch reactor mole balance in order to fit the experimental data. This fitting will be carried out in the upcoming section.

The models presented here are based on the mechanism that was selected for this example. There are vast possibilities and in order to determine which one works the best; different options should be evaluated against the experimental data. This is one of the simplest mechanisms, which is very similar to the Eley–Rideal mechanism which has been developed for one of the components to remain in the gas phase.

4.8 COMPARISON OF DATA AND MODEL 1

It is important to know if the model that has been proposed fits the experimental data correctly and if it is an accurate representation of the physical world under study. A fitting regression must be done over all the possible values obtained experimentally using the different expressions for the reaction rate for cases where different controlling steps are considered.

Completely Solved Example

To solve the problem, we need to solve the mole balance for the conversion of component A as a function of time for all three different temperatures and compare the values from the model with the experimental information. It is important for the kinetics parameters to be assigned with values that have physical meaning and not just mathematical numbers.

Using a software like Polymath or Aspen, the values for the kinetics parameters can be estimated while minimizing the difference between the experimental data and the model for all the experimental information simultaneously.

When step 1 is the controlling step, we solve the mathematical model using Polymath, the script used with the optimal values for the kinetics parameters can be seen below (this is only for one temperature):

Script:

```
#Mole balance
    d(x)/d(t) = −r1/Ca0
    x(0) = 0

#Reaction rate
    r1 = −Cs*(k1*Ca−(k2*k4*k6*Cc*Cd/(k3*k5*Cb)))
    Cs = (Cs0)/(1+(k4*k6*Cd*Cc/(k3*k5*Cb))+(k6*Cc/k5))

#Auxiliary equations
    k1 = k1oo*EXP(−Ea1/(8.3143*T0))
    k2 = k2oo*EXP(−Ea2/(8.3143*T0))
    k3 = k3oo*EXP(−Ea3/(8.3143*T0))
    k4 = k4oo*EXP(−Ea4/(8.3143*T0))
    k5 = k5oo*EXP(−Ea5/(8.3143*T0))
    k6 = k6oo*EXP(−Ea6/(8.3143*T0))
    Ca = Ca0*(1−x)
    Cb = Cb0−Ca0*x
    Cc = 0.5+Ca0*x
    Cd = Ca0*x

#Constants and initial values
    Ca0 = 3
    Cb0 = 5
    Cs0 = 10
    T0 = 300
    k1oo = 25,000
```

Ea1 = 33,000
k2oo = 12,000
Ea2 = 29,000
k3oo = 11,000
Ea3 = 35,000
k4oo = 9,500
Ea4 = 31,000
k5oo = 95,000
Ea5 = 30,000
k6oo = 1,800
Ea6 = 32,000

#Independent variable
t(0) = 0
t(f) = 20

The script does not allow the notation of units; therefore, units have to be consistent. In our case, we are using the SI units; however, some units might be different. For all these models, the units are listed in Table 4.2.

TABLE 4.2
Units for the Different Parameters Used in the Modeling

Variable	Units
C_i	mol/l
Temperature	K
Time	min
k_{i00}	$l^{(\alpha-1)}/(t*mol^{(\alpha-1)})^a$
E_i	J/mol
8.3143 (universal gas constant)	J/(mol*K)

[a] α is the order of the reaction, this applies for the forward and the backward reactions and it is important to point out that they can be different from each other and for the different elementary steps as well.

When using these values with the units as presented in Table 4.2, the final plot of conversion as a function of time, data, and model can be seen in Figure 4.8.

It is clear that assuming step 1 to be the controlling step does not result in a good fit with the data. It is important to notice that all the software programs that use regression to calculate the kinetics parameters are heavily depend on the initial value of the parameters and therefore, a better fitting than the one presented here might be achieved with a new set of initial values.

Completely Solved Example

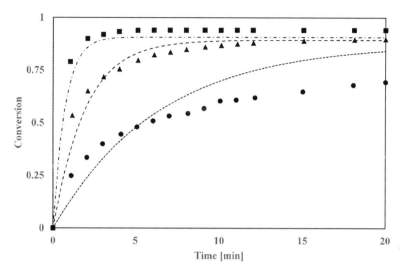

FIGURE 4.8 Variation of conversion of component A as a function of time. (●) Experimental data and (--) model for $T = 275$ [K], (▲) experimental data and (– –) model for $T = 300$ [K], (■) experimental data and (-•-) model for $T = 325$ [K].

The MSC calculated for the fitting presented in Figure 4.8 gives a value of 3.89.

While an MSC value of 3.89 is high, it is very important not only to look at this value alone, and also pay close attention to the plot itself and see that the errors are not being compensated even if the MSC predicts an "*ok*" fit of the data.

Since this model was not completely satisfactory with the data, assuming step 2 as the controlling step was considered and the kinetics parameters where fitted accordingly.

The result of the fitting can be seen in Figure 4.9.

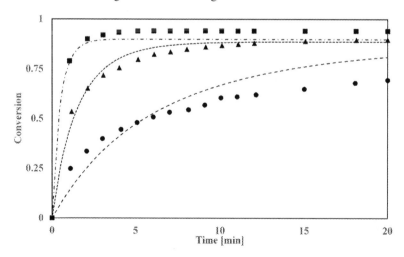

FIGURE 4.9 Variation of conversion of component A as a function of time. (●) Experimental data and (– –) model for $T = 275$[K], (▲) experimental data and (--) model for $T = 300$ [K], (■) experimental data and (-•-) model for $T = 325$ [K].

Even though the plots in Figures 4.8 and 4.9 seem the same, especially in regard to the fitting, they are not exactly the same.

The script to plot the optimized kinetics parameters condition from Polymath can be seen below, for a given temperature.

Script:

#Mole balance
d(x)/d(t) = −r2/Ca0
x(0) = 0

#Reaction rates
r2 = −Cs*((k1*k3*Ca*Cb/k2)−(k4*k6*Cc*Cd/k5))
Cs = (Cs0)/(1+(k1*Ca/k2)+(k6*Cc/k5))

#Auxiliary equations
Ca = Ca0*(1−x)
Cb = Cb0−Ca0*x
Cc = 0.5+Ca0*x
Cd = Ca0*x

#Constants and initial values
Ca0 = 3
Cb0 = 5
Cs0 = 10
T0 = 300
k1 = k1oo*EXP(−Ea1/(8.3143*T0))
k2 = k2oo*EXP(−Ea2/(8.3143*T0))
k3 = k3oo*EXP(−Ea3/(8.3143*T0))
k4 = k4oo*EXP(−Ea4/(8.3143*T0))
k5 = k5oo*EXP(−Ea5/(8.3143*T0))
k6 = k6oo*EXP(−Ea6/(8.3143*T0))
k1oo = 100,000
Ea1 = 40,000
k2oo = 2,000
Ea2 = 25,000
k3oo = 11,000
Ea3 = 28,000
k4oo = 9,500
Ea4 = 31,000
k5oo = 65,000

Completely Solved Example

       Ea5 = 25,000
       k6oo = 48,000
       Ea6 = 32,000

    #Independent variable
       t(0) = 0
       t(f) = 20

For this case as well, the MSC value was calculated based on Equation 4.4 and the value for this model adjustment of the data obtained is 4.064. This is an improvement from the previous fitting but is still not good enough; the ideal value for MSC should be as high as possible.

In order to seek for an improvement in the fitting, we need to test the fitting when the step 3 is the controlling step. For that purpose, we fit the data with the proposed model from Equation 4.37 and evaluate the smoothness of the fit against the data.

The results for all three evaluated temperatures and how well the fittings are can be seen in Figure 4.10.

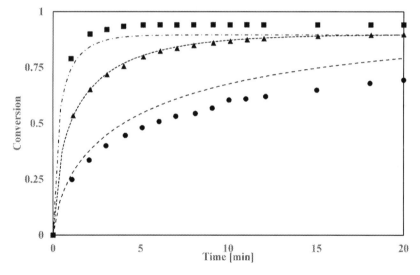

FIGURE 4.10 Variation of conversion of component A as a function of time. (●) Experimental data and (– –) model for $T = 275$ [K], (▲) experimental data and (--) model for $T = 300$ [K], (■) experimental data and (-•-) model for $T = 325$ [K].

Evidently, the model for controlling step 3 is very suitable for the middle range temperature and less accurate for the low and high temperature values. The higher temperature profile is completely underpredicted, and the model indicates that the reaction has reached the equilibrium at a lower equilibrium conversion. The MSC for this case was calculated to be 6.89. Even though this value is much better than the value previously obtained, the visual representation is clearly suggesting that a better fitting is required.

The optimized model used in Polymath can be seen in the script presented below. Note that the initial conversion is not zero, but 0.00001, since mathematical division by zero is impossible therefore, a very small number, close to zero, was assigned as the initial value.

Script:

```
#Mole balance
    d(x) d(t) = −r3/Ca0
    x(0) = 0.00001

#Reaction rates
    r3 = −Cs*(((k1*k3*k5*Ca*Cb)/(k2*k4*Cd))−(k6*Cc))
    Cs = (Cs0)/(1+(k1*Ca/k2)+((k1*k3*Ca*Cb)/(k2*k4*Cd)))

#Auxiliary equations
    Ca = Ca0*(1−x)
    Cb = Cb0−Ca0*x
    Cc = 0.5+Ca0*x
    Cd = Ca0*x

#Constants and initial values
    Ca0 = 3
    Cb0 = 5
    Cs0 = 10
    T0 = 300
    k1 = k1oo*EXP(−Ea1/(8.3143*T0))
    k2 = k2oo*EXP(−Ea2/(8.3143*T0))
    k3 = k3oo*EXP(−Ea3/(8.3143*T0))
    k4 = k4oo*EXP(−Ea4/(8.3143*T0))
    k5 = k5oo*EXP(−Ea5/(8.3143*T0))
    k6 = k6oo*EXP(−Ea6/(8.3143*T0))
    k1oo = 45,000
    Ea1 = 40,000
    k2oo = 16,000
    Ea2 = 25,000
    k3oo = 11,000
    Ea3 = 19,000
```

Completely Solved Example 281

```
k4oo = 7,000
Ea4 = 22,000
k5oo = 65,000
Ea5 = 29,000
k6oo = 35,000
Ea6 = 42,000
```

#Independent variable
```
t(0) = 0
t(f) = 20
```

All three previous models have had some deficiencies and could not perfectly fit all the data for all the tested temperatures. This is a good indication that a new model might need to be proposed, developed, and evaluated.

4.9 MATHEMATICAL MODEL 2

Based on the previous results from the modeling and the comparison with the data, a new model, with different controlling steps, needs to be assumed and tested in order to obtain a good fitting with the data.

For that purpose, a second model, with more elementary steps, is being proposed and the kinetics expression can be derived from it.

Within this model, both reactants get adsorbed over the surface of the catalytic materials. Then, a surface reaction of the adsorbed reactants giving adsorbed products takes place. Finally, both products are desorbed from the catalyst surface.

The five steps can be seen as follow:

Step 1: Adsorption of reactant A

$$A + S \underset{k_2}{\overset{k_1}{\rightleftarrows}} AS$$

Step 2: Adsorption of reactant B

$$B + S \underset{k_4}{\overset{k_3}{\rightleftarrows}} AS$$

Step 3: Surface reaction

$$AS + BS \underset{k_6}{\overset{k_5}{\rightleftarrows}} CS + DS$$

Step 4: Desorption reaction for component C

$$CS \underset{k_8}{\overset{k_7}{\rightleftarrows}} C + S$$

Step 5: Desorption reaction for component D

$$DS \underset{k_{10}}{\overset{k_9}{\rightleftarrows}} D + S$$

where S represents the site over the catalyst surface.

In order to obtain the possible kinetics expression, we need to determine the controlling step. In this case, we have five elementary steps that need to be considered as the controlling step. Thus, we will end up with five kinetics expressions that will then be tested against the experimental data to evaluate the goodness of such fitness.

For our case, we can develop five reaction equations, one for each step. The sub-index for the r indicate the step being referred to:

Step 1) $\quad r_1 = k_1 C_A C_S - k_2 C_{AS}$ \hfill (4.38)

Step 2) $\quad r_2 = k_3 C_B C_S - k_4 C_{BS}$ \hfill (4.39)

Step 3) $\quad r_3 = k_5 C_{AS} C_{BS} - k_6 C_{CS} C_{DS}$ \hfill (4.40)

Step 4) $\quad r_4 = k_7 C_{CS} - k_8 C_C C_S$ \hfill (4.41)

Step 5) $\quad r_5 = k_9 C_{DS} - k_{10} C_D C_S$ \hfill (4.42)

As it was done for the previous mechanism, we will evaluate all possible scenarios, *i.e.*, all the steps can be considered as the controlling step while the others are in equilibrium.

Step 1 is controlling

If step 1 is controlling, this means that $r_2 = r_3 = r_4 = r_5 = 0$. Therefore:

$$0 = r_2 = k_3 C_B C_S - k_4 C_{BS} \qquad (4.43)$$

$$0 = r_3 = k_5 C_{AS} C_{BS} - k_6 C_{CS} C_{DS} \qquad (4.44)$$

$$0 = r_4 = k_7 C_{CS} - k_8 C_C C_S \qquad (4.45)$$

$$0 = r_5 = k_9 C_{DS} - k_{10} C_D C_S \qquad (4.46)$$

Solving Equation 4.43, we can obtain a relationship for C_{BS}:

$$C_{BS} = \frac{k_3 C_B C_S}{k_4} \qquad (4.47)$$

Completely Solved Example

From Equation 4.44, we can obtain a relationship for C_{AS}:

$$C_{AS} = \frac{k_6 C_{CS} C_{DS}}{k_5 C_{BS}} \quad (4.48)$$

From Equation 4.45, we can obtain a relationship for C_{CS}:

$$C_{CS} = \frac{k_8 C_C C_S}{k_7} \quad (4.49)$$

From Equation 4.43, we can obtain a relationship for C_{DS}:

$$C_{DS} = \frac{k_{10} C_D C_S}{k_9} \quad (4.50)$$

We can now substitute Equations 4.47, 4.49, and 4.50 into Equation 4.48 in order to obtain an expression for C_{AS}, that can be substituted into $r_1 = k_1 C_A C_S - k_2 C_{AS}$ in order to obtain the expression of r_1. Therefore, for C_{AS} we obtained:

$$C_{AS} = \frac{k_4 k_6 k_8 k_{10} C_C C_D C_S}{k_3 k_5 k_7 k_9 C_B} \quad (4.51)$$

We can now substitute C_{AS} into step 1 and obtain and expression for r_1 as follows:

$$r_1 = k_1 C_A C_S - k_2 \frac{k_4 k_6 k_8 k_{10} C_C C_D C_S}{k_3 k_5 k_7 k_9 C_B} \quad (4.52)$$

This expression is a function of the concentration of all products and reactants and the concentration of the catalyst. In order to find a solution for the concentration of C_s, we need to do the site balance. For this case, the total site balance is:

$$C_S^0 = C_S + C_{AS} + C_{BS} + C_{CS} + C_{DS} \quad (4.53)$$

Substituting Equations 4.47, 4.49, 4.50, and 4.51 into Equation 4.53, we obtain:

$$C_S^0 = C_S + \frac{k_4 k_6 k_8 k_{10} C_C C_D C_S}{k_3 k_5 k_7 k_9 C_B} + \frac{k_3 C_B C_S}{k_4} + \frac{k_8 C_C C_S}{k_7} + \frac{k_{10} C_D C_S}{k_9} \quad (4.54)$$

We can obtain a common factor (C_s), leading to:

$$C_S^0 = C_S \left(1 + \frac{k_4 k_6 k_8 k_{10} C_C C_D}{k_3 k_5 k_7 k_9 C_B} + \frac{k_3 C_B}{k_4} + \frac{k_8 C_C}{k_7} + \frac{k_{10} C_D}{k_9} \right) \quad (4.55)$$

This expression can be rearranged in order to obtain C_s as a function of the concentration of each component involved, the reaction rate constant, and the initial values of the concentration of the catalyst. The final expression for C_s is:

$$C_S = \frac{C_S^0}{\left(1 + \dfrac{k_4 k_6 k_8 k_{10} C_C C_D}{k_3 k_5 k_7 k_9 C_B} + \dfrac{k_3 C_B}{k_4} + \dfrac{k_8 C_C}{k_7} + \dfrac{k_{10} C_D}{k_9}\right)} \tag{4.56}$$

Equation 4.56 is the site balance for the catalyst. Then, this expression can be substituted into Equation 4.52 in order to get the final reaction rate expression when step 1 is the controlling step:

$$r_1 = \left(k_1 C_A - k_2 \frac{k_4 k_6 k_8 k_{10} C_C C_D}{k_3 k_5 k_7 k_9 C_B}\right) \frac{C_S^0}{\left(1 + \dfrac{k_4 k_6 k_8 k_{10} C_C C_D}{k_3 k_5 k_7 k_9 C_B} + \dfrac{k_3 C_B}{k_4} + \dfrac{k_8 C_C}{k_7} + \dfrac{k_{10} C_D}{k_9}\right)} \tag{4.57}$$

This is the final expression for r_1 that needs to be put into the batch reactor mole balance in order to fit the experimental data. This fitting will be carried out in the upcoming section.

Here, we move from a three-step to five-step mechanism, i.e., in the previous system it was one adsorption and one desorption step; here, we have two adsorption and two desorption steps. This brings a difference in the terms in the denominator of the site balance, as seen in Equation 4.57.

Step 2 is controlling
If step 2 is controlling, this means that $r_1 = r_3 = r_4 = r_5 = 0$. Therefore:

$$0 = r_1 = k_1 C_A C_S - k_2 C_{AS} \tag{4.58}$$

$$0 = r_3 = k_5 C_{AS} C_{BS} - k_6 C_{CS} C_{DS} \tag{4.59}$$

$$0 = r_4 = k_7 C_{CS} - k_8 C_C C_S \tag{4.60}$$

$$0 = r_5 = k_9 C_{DS} - k_{10} C_D C_S \tag{4.61}$$

Solving Equation 4.58, we can obtain a relationship for C_{AS}:

$$C_{AS} = \frac{k_1 C_A C_S}{k_2} \tag{4.62}$$

From Equation 4.59, we can obtain a relationship for C_{AS}:

$$C_{BS} = \frac{k_6 C_{CS} C_{DS}}{k_5 C_{AS}} \tag{4.63}$$

Completely Solved Example

From Equation 4.60, we can obtain a relationship for C_{CS}:

$$C_{CS} = \frac{k_8 C_C C_S}{k_7} \tag{4.64}$$

From Equation 4.61, we can obtain a relationship for C_{DS}:

$$C_{DS} = \frac{k_{10} C_D C_S}{k_9} \tag{4.65}$$

We can now substitute Equations 4.62, 4.64, and 4.65 into Equation 4.63 in order to obtain an expression for C_{BS}, that can be substituted into $r_2 = k_3 C_B C_S - k_4 C_{BS}$ in order to obtain the expression of r_2. Therefore, for C_{BS}, we obtain:

$$C_{BS} = \frac{k_2 k_6 k_8 k_{10} C_C C_D C_S}{k_1 k_5 k_7 k_9 C_A} \tag{4.66}$$

We can now substitute C_{BS} into step 2 and obtain and expression for r_2 as follows:

$$r_2 = k_3 C_B C_S - k_2 \frac{k_4 k_6 k_8 k_{10} C_C C_D C_S}{k_1 k_5 k_7 k_9 C_A} \tag{4.67}$$

This expression is a function of all the concentration of products and reactants and the concentration of the catalyst. In order to find a solution for the concentration of C_s, we need to do the site balance. For this case, the total site balance is:

$$C_S^0 = C_S + C_{AS} + C_{BS} + C_{CS} + C_{DS} \tag{4.68}$$

Substituting Equations 4.62, 4.64, 4.65, and 4.66 into Equation 4.68, we obtain:

$$C_S^0 = C_S + \frac{k_1 C_A C_S}{k_2} + \frac{k_2 k_6 k_8 k_{10} C_C C_D C_S}{k_1 k_5 k_7 k_9 C_A} + \frac{k_8 C_C C_S}{k_7} + \frac{k_{10} C_D C_S}{k_9} \tag{4.69}$$

We can obtain a common factor (C_s), leading to:

$$C_S^0 = C_S \left(1 + \frac{k_1 C_A}{k_2} + \frac{k_2 k_6 k_8 k_{10} C_C C_D}{k_1 k_5 k_7 k_9 C_A} + \frac{k_8 C_C}{k_7} + \frac{k_{10} C_D}{k_9} \right) \tag{4.70}$$

This expression can be rearranged in order to obtain C_s as a function of the concentration of each component involved, the reaction rate constant, and the initial values of the concentration of the catalyst. The final expression for C_s is:

$$C_S = \frac{C_S^0}{\left(1 + \frac{k_1 C_A}{k_2} + \frac{k_2 k_6 k_8 k_{10} C_C C_D}{k_1 k_5 k_7 k_9 C_A} + \frac{k_8 C_C}{k_7} + \frac{k_{10} C_D}{k_9} \right)} \tag{4.71}$$

Equation 4.71 is the site balance for the catalyst. Then, this expression can be substituted into Equation 4.67 in order to get the final reaction rate expression when step 2 is the controlling step:

$$r_2 = \left(k_3 C_B - k_2 \frac{k_4 k_6 k_8 k_{10} C_C C_D}{k_1 k_5 k_7 k_9 C_A}\right) \frac{C_S^0}{\left(1 + \frac{k_1 C_A}{k_2} + \frac{k_2 k_6 k_8 k_{10} C_C C_D}{k_1 k_5 k_7 k_9 C_A} + \frac{k_8 C_C}{k_7} + \frac{k_{10} C_D}{k_9}\right)} \quad (4.72)$$

This is the final expression for r_2 that needs to be used with the mole balance for the batch reactor in order to predict the data and to evaluate the goodness of fit.

Step 3 is controlling

If step 3 is controlling, this means that $r_1 = r_2 = r_4 = r_5 = 0$. Therefore:

$$0 = r_1 = k_1 C_A C_S - k_2 C_{AS} \quad (4.73)$$

$$0 = r_2 = k_3 C_B C_S - k_4 C_{BS} \quad (4.74)$$

$$0 = r_4 = k_7 C_{CS} - k_8 C_C C_S \quad (4.75)$$

$$0 = r_5 = k_9 C_{DS} - k_{10} C_D C_S \quad (4.76)$$

Solving Equation 4.73, we can obtain a relationship for C_{AS}:

$$C_{AS} = \frac{k_1 C_A C_S}{k_2} \quad (4.77)$$

From Equation 4.74, we can obtain a relationship for C_{BS}:

$$C_{BS} = \frac{k_3 C_B C_S}{k_4} \quad (4.78)$$

From Equation 4.75, we can obtain a relationship for C_{CS}:

$$C_{CS} = \frac{k_8 C_C C_S}{k_7} \quad (4.79)$$

From Equation 4.76, we can obtain a relationship for C_{DS}:

$$C_{DS} = \frac{k_{10} C_D C_S}{k_9} \quad 0 = r_1 = k_1 C_A C_S - k_2 C_{AS} \quad (4.80)$$

With Equations 4.77, 4.78, 4.79, and 4.80 we can now substitute into $r_3 = k_5 C_{AS} C_{BS} - k_6 C_{CS} C_{DS}$, giving:

$$r_3 = C_S^2 \left(\frac{k_1 k_3 k_5 C_A C_B}{k_2 k_4} - \frac{k_6 k_8 k_{10} C_C C_D}{k_7 k_9}\right) \quad (4.81)$$

Completely Solved Example

This expression is a function of the concentrations of all products and reactants and the concentration of the catalyst. In order to find a solution for the concentration of C_s, we need to do the site balance. For this case, the total site balance is:

$$C_S^0 = C_S + C_{AS} + C_{BS} + C_{CS} + C_{DS} \tag{4.82}$$

Substituting Equations 4.77, 4.78, 4.79, and 4.80 into Equation 4.82, we obtain:

$$C_S^0 = C_S + \frac{k_1 C_A C_S}{k_2} + \frac{k_3 C_B C_S}{k_4} + \frac{k_8 C_C C_S}{k_7} + \frac{k_{10} C_D C_S}{k_9} \tag{4.83}$$

We can obtain a common factor (C_s), leading to:

$$C_S^0 = C_S \left(1 + \frac{k_1 C_A}{k_2} + \frac{k_3 C_B}{k_4} + \frac{k_8 C_C}{k_7} + \frac{k_{10} C_D}{k_9} \right) \tag{4.84}$$

This expression can be rearranged in order to obtain C_s as a function of the concentration of each component involved, the reaction rate constant, and the initial values of the concentration of the catalyst. The final expression for C_s is:

$$C_S = \frac{C_S^0}{\left(1 + \frac{k_1 C_A}{k_2} + \frac{k_3 C_B}{k_4} + \frac{k_8 C_C}{k_7} + \frac{k_{10} C_D}{k_9} \right)} \tag{4.85}$$

Equation 4.85 is the site balance for the catalyst. Then, this expression can be substituted into Equation 4.81 in order to get the final reaction rate expression when step 3 is considered the controlling step:

$$r_3 = \left(\frac{k_1 k_3 k_5 C_A C_B}{k_2 k_4} - \frac{k_6 k_8 k_{10} C_C C_D}{k_7 k_9} \right) \left(\frac{C_S^0}{\left(1 + \frac{k_1 C_A}{k_2} + \frac{k_3 C_B}{k_4} + \frac{k_8 C_C}{k_7} + \frac{k_{10} C_D}{k_9} \right)} \right)^2 \tag{4.86}$$

This is the final expression for r_3 which needs to be used with the mole balance for the batch reactor in order to predict the data and to evaluate the goodness of fit.

Step 4 is controlling

If step 4 is controlling, this means that $r_1 = r_2 = r_3 = r_5 = 0$. Therefore:

$$0 = r_1 = k_1 C_A C_S - k_2 C_{AS} \tag{4.87}$$

$$0 = r_2 = k_3 C_B C_S - k_4 C_{BS} \tag{4.88}$$

$$0 = r_3 = k_5 C_{AS} C_{BS} - k_6 C_{CS} C_{DS} \tag{4.89}$$

$$0 = r_5 = k_9 C_{DS} - k_{10} C_D C_S \qquad (4.90)$$

Solving Equation 4.87, we can obtain a relationship for C_{AS}:

$$C_{AS} = \frac{k_1 C_A C_S}{k_2} \qquad (4.91)$$

From Equation 4.88, we can obtain a relationship for C_{BS}:

$$C_{BS} = \frac{k_3 C_B C_S}{k_4} \qquad (4.92)$$

From Equation 4.89, we can obtain a relationship for C_{CS}:

$$C_{CS} = \frac{k_5 C_{AS} C_{BS}}{k_6 C_{DS}} \qquad (4.93)$$

From Equation 4.90, we can obtain a relationship for C_{DS}:

$$C_{DS} = \frac{k_{10} C_D C_S}{k_9} \qquad (4.94)$$

Substituting Equations 4.91, 4.92, and 4.94 into Equation 4.93, we can obtain the expression for C_{CS} giving:

$$C_{CS} = \frac{k_1 k_3 k_5 k_9 C_A C_B C_S}{k_2 k_4 k_6 k_{10} C_D} \qquad (4.95)$$

Substituting 4.95 into $r_4 = k_7 C_{CS} - k_8 C_C C_S$, we get:

$$r_4 = \left(\frac{k_1 k_3 k_5 k_7 k_9 C_A C_B C_S}{k_2 k_4 k_6 k_{10} C_D} - k_8 C_C C_S \right) \qquad (4.96)$$

This expression is a function of the concentration of products and reactants and the concentration of the catalyst. In order to find a solution for the concentration of C_s, we need to do the catalyst balance. For this case, the total balance is:

$$C_S^0 = C_S + C_{AS} + C_{BS} + C_{CS} + C_{DS} \qquad (4.97)$$

Substituting Equations 4.91, 4.92, 4.94, and 4.95 into Equation 4.97, we obtain:

$$C_S^0 = C_S + \frac{k_1 C_A C_S}{k_2} + \frac{k_3 C_B C_S}{k_4} + \frac{k_1 k_3 k_5 k_9 C_A C_B C_S}{k_2 k_4 k_6 k_{10} C_D} + \frac{k_{10} C_D C_S}{k_9} \qquad (4.98)$$

Completely Solved Example

We can obtain a common factor (C_s), leading to:

$$C_S^0 = C_S\left(1 + \frac{k_1 C_A}{k_2} + \frac{k_3 C_B}{k_4} + \frac{k_1 k_3 k_5 k_9 C_A C_B}{k_2 k_4 k_6 k_{10} C_D} + \frac{k_{10} C_D}{k_9}\right) \quad (4.99)$$

This expression can be rearranged in order to obtain C_s as a function of the concentration of each component involved, the reaction rate constant, and the initial values of the concentration of the catalyst. The final expression for C_s is:

$$C_S = \frac{C_S^0}{\left(1 + \dfrac{k_1 C_A}{k_2} + \dfrac{k_3 C_B}{k_4} + \dfrac{k_1 k_3 k_5 k_9 C_A C_B}{k_2 k_4 k_6 k_{10} C_D} + \dfrac{k_{10} C_D}{k_9}\right)} \quad (4.100)$$

Equation 4.100 is the site balance for the catalyst. Then, this expression can be substituted into Equation 4.96 in order to get the final reaction rate expression for the case when step 4 is the controlling step:

$$r_4 = \left(\frac{k_1 k_3 k_5 k_7 k_9 C_A C_B}{k_2 k_4 k_6 k_{10} C_D} - k_8 C_C\right) \frac{C_S^0}{\left(1 + \dfrac{k_1 C_A}{k_2} + \dfrac{k_3 C_B}{k_4} + \dfrac{k_1 k_3 k_5 k_9 C_A C_B}{k_2 k_4 k_6 k_{10} C_D} + \dfrac{k_{10} C_D}{k_9}\right)} \quad (4.101)$$

When step 4 is controlling, the final expression that needs to be used in the mole balance in order to obtain the final kinetics expression, can be seen in Equation 4.101.

Step 5 is controlling

If step 5 is controlling, this means that $r_1 = r_2 = r_3 = r_4 = 0$. Therefore:

$$0 = r_1 = k_1 C_A C_S - k_2 C_{AS} \quad (4.102)$$

$$0 = r_2 = k_3 C_B C_S - k_4 C_{BS} \quad (4.103)$$

$$0 = r_3 = k_5 C_{AS} C_{BS} - k_6 C_{CS} C_{DS} \quad (4.104)$$

$$0 = r_4 = k_7 C_{CS} - k_8 C_C C_S \quad (4.105)$$

Solving Equation 4.102, we can obtain a relationship for C_{AS}:

$$C_{AS} = \frac{k_1 C_A C_S}{k_2} \quad (4.106)$$

From Equation 4.103, we can obtain a relationship for C_{BS}:

$$C_{BS} = \frac{k_3 C_B C_S}{k_4} \quad (4.107)$$

From Equation 4.104, we can obtain a relationship for C_{DS}:

$$C_{DS} = \frac{k_5 C_{AS} C_{BS}}{k_6 C_{CS}} \qquad (4.108)$$

From Equation 4.105, we can obtain a relationship for C_{CS}:

$$C_{CS} = \frac{k_8 C_C C_S}{k_7} \qquad (4.109)$$

Substituting Equations 4.106, 4.107, and 4.109 into Equation 4.108, we can obtain the expression for C_{CS} giving:

$$C_{DS} = \frac{k_1 k_3 k_5 k_7 C_A C_B C_S}{k_2 k_4 k_6 k_8 C_C} \qquad (4.110)$$

Substituting 4.110 into $r_5 = k_9 C_{DS} - k_{10} C_D C_S$, we get:

$$r_5 = \left(\frac{k_1 k_3 k_5 k_7 k_9 C_A C_B C_S}{k_2 k_4 k_6 k_8 C_C} - k_{10} C_D C_S \right) \qquad (4.111)$$

This expression is a function of the concentrations of products and reactants and the concentration of the catalyst. In order to find a solution for the concentration of C_s, we need to do the site balance. For this case, the total balance is:

$$C_S^0 = C_S + C_{AS} + C_{BS} + C_{CS} + C_{DS} \qquad (4.112)$$

Substituting Equations 4.106, 4.107, 4.109, and 4.110 into Equation 4.112, we obtain:

$$C_S^0 = C_S + \frac{k_1 C_A C_S}{k_2} + \frac{k_3 C_B C_S}{k_4} + \frac{k_8 C_C C_S}{k_7} + \frac{k_1 k_3 k_5 k_7 C_A C_B C_S}{k_2 k_4 k_6 k_8 C_C} \qquad (4.113)$$

We can obtain a common factor (C_s), leading to:

$$C_S^0 = C_S \left(1 + \frac{k_1 C_A}{k_2} + \frac{k_3 C_B}{k_4} + \frac{k_8 C_C}{k_7} + \frac{k_1 k_3 k_5 k_7 C_A C_B}{k_2 k_4 k_6 k_8 C_C} \right) \qquad (4.114)$$

This expression can be rearranged in order to obtain C_s as a function of the concentration of each component involved, the reaction rate constant, and the initial values of the concentration of the catalyst. The final expression for C_s is:

$$C_S = \frac{C_S^0}{\left(1 + \dfrac{k_1 C_A}{k_2} + \dfrac{k_3 C_B}{k_4} + \dfrac{k_8 C_C}{k_7} + \dfrac{k_1 k_3 k_5 k_7 C_A C_B}{k_2 k_4 k_6 k_8 C_C} \right)} \qquad (4.115)$$

Equation 4.115 is the site balance for the catalyst. Then, this expression can be substituted into Equation 4.111 in order to get the final reaction rate expression for the case when step 5 is considered the controlling reaction:

$$r_5 = \left(\frac{k_1 k_3 k_5 k_7 k_9 C_A C_B}{k_2 k_4 k_6 k_8 C_C} - k_{10} C_D \right) \frac{C_S^0}{\left(1 + \frac{k_1 C_A}{k_2} + \frac{k_3 C_B}{k_4} + \frac{k_8 C_C}{k_7} + \frac{k_1 k_3 k_5 k_7 C_A C_B}{k_2 k_4 k_6 k_8 C_C}\right)} \quad (4.116)$$

When step 5 is the controlling step, the equation for the reaction rate can be seen in Equation 4.116; this equation will be substituted in the mole balance to see if the model fits the experimental data.

Now that all the five-reaction kinetics expressions have been developed, based on each step being the controlling one, we need to fit the data with these new expressions to see if the fitting is better than that when considering the global reaction as elementary and/or with the simple mechanism that has been proposed and evaluated.

4.10 COMPARISON OF DATA AND MODEL 2

A fitting regression must be done over all the possible values obtained in the lab using the different expressions for the reaction rate for the different controlling steps.

Based on the reaction rate expression, we can now solve the mole balance as we have done previously. We will rewrite the mole balance as a function of conversion of component A in order to reduce the amount of differential equations. This is solved for all three temperatures and the kinetics parameters are optimized to the best of our ability. It is important to mention once more that the software and mathematical tools used to solve the differential equations are dependent on the initial value. It is important that the kinetics parameters are assigned values that have physical meaning and are not just mathematical numbers.

We will solve the problem for all five controlling steps. We will show the best fitting for each case and the respectively optimized kinetics parameters.

For the case when step one is controlling the script (for Polymath) to plot the best fitting with the best estimated kinetics is as follows (this is only for one temperature, i.e., 300 [K]).

Script:

```
#Mole balance
    d(x)/d(t) = -r1/Ca0
    x(0) = 0

#Reaction rates
    r1 = -((k1*Ca)-((k2*k4*k6*k8*k10*Cc*Cd)/(k3*k5*k7*k9*Cb)))*Cs
    Cs = (Cs0/(1+((k4*k6*k8*k10*Cc*Cd)/(k3*k5*k7*k9*Cb))+(k3*Cb/
        k4)+(k8*Cc/k7)+(k10*Cd/k9)))
```

#Auxiliary equations
Ca = Ca0*(1−x)
Cb = Cb0−Ca0*x
Cc = 0.5+Ca0*x
Cd = Ca0*x

#Constants and initial values
Ca0 = 3
Cb0 = 5
Cs0 = 10
T0 = 300
k1 = k1oo*EXP(−Ea1/(8.3143*T0))
k2 = k2oo*EXP(−Ea2/(8.3143*T0))
k3 = k3oo*EXP(−Ea3/(8.3143*T0))
k4 = k4oo*EXP(−Ea4/(8.3143*T0))
k5 = k5oo*EXP(−Ea5/(8.3143*T0))
k6 = k6oo*EXP(−Ea6/(8.3143*T0))
k7 = k7oo*EXP(−Ea7/(8.3143*T0))
k8 = k8oo*EXP(−Ea8/(8.3143*T0))
k9 = k9oo*EXP(−Ea9/(8.3143*T0))
k10 = k10oo*EXP(−Ea10/(8.3143*T0))
k1oo = 55,000
Ea1 = 26,500
k2oo = 10,000
Ea2 = 23,000
k3oo = 9,000
Ea3 = 32,000
k4oo = 9,500
Ea4 = 32,000
k5oo = 55,000
Ea5 = 25,000
k6oo = 12,000
Ea6 = 32,000
k7oo = 20,000
Ea7 = 38,000
k8oo = 6,000
Ea8 = 35,000
k9oo = 10,000

Ea9 = 30,000
k10oo = 5,000
Ea10 = 22,000

#Independent variable
t(0) = 0
t(f) = 20

When using these values with the units as presented in Table 4.2, the final plot of conversion as a function of time, data, and model can be seen in Figure 4.11.

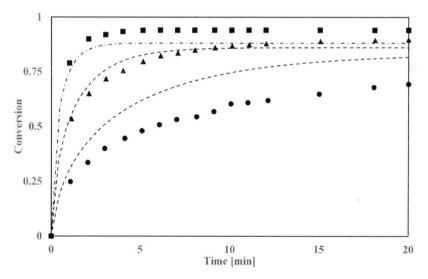

FIGURE 4.11 Variation of conversion of component A as a function of time. (●) Experimental data and (– –) model for $T = 275$ [K], (▲) experimental data and (--) model for $T = 300$ [K], (■) experimental data and (-•-) model for $T = 325$ [K].

From a visual interpretation, the fitting is not good, as it underestimates the values at higher temperatures and overestimates the values at low temperatures; the model does not capture the essence of the temperature effect properly.

Thus, as seen visually from the Figure 4.11, if step 1 is controlling, a good fitting with the data is not obtained. We also calculate the MSC, given a value of 3.65. As mentioned before, the value of MSC itself might be misleading the user. Therefore, it is important to also have a visual interpretation of the number, along with a value for variance or standard deviation.

It is very important to not only look at the MSC value alone. The value of 3.65 is high, but not very high, yet the user should try to avoid forming decision with that number alone. The plot should be carefully examined, and it should be noted that some errors are not compensated.

As a perfect fit of this model is not obtained when considering step one as controlling step, we move into assuming that step 2 as the controlling step. The Polymath script for this case can be seen below.

Script:

#Mole balance
   d(x)/d(t) = −r2/Ca0
   x(0) = 0

#Reaction rates
   r2 = −((k1*Ca)−((k2*k4*k6*k8*k10*Cc*Cd)/(k3*k5*k7*k9*Cb)))*Cs
   Cs = (Cs0/(1+(k1*Ca/k2)+((k2*k4*k6*k8*k10*Cc*Cd)/(k1*k5*k7*k9*Ca))+(k8*Cc/k7)+(k10*Cd/k9)))

#Auxiliary equations
   Ca = Ca0*(1−x)
   Cb = Cb0−Ca0*x
   Cc = 0.5+Ca0*x
   Cd = Ca0*x
   k1 = k1oo*EXP(−Ea1/(8.3143*T0))
   k2 = k2oo*EXP(−Ea2/(8.3143*T0))
   k3 = k3oo*EXP(−Ea3/(8.3143*T0))
   k4 = k4oo*EXP(−Ea4/(8.3143*T0))
   k5 = k5oo*EXP(−Ea5/(8.3143*T0))
   k6 = k6oo*EXP(−Ea6/(8.3143*T0))
   k7 = k7oo*EXP(−Ea7/(8.3143*T0))
   k8 = k8oo*EXP(−Ea8/(8.3143*T0))
   k9 = k9oo*EXP(−Ea9/(8.3143*T0))
   k10 = k10oo*EXP(−Ea10/(8.3143*T0))

#Constants and initial values
   Ca0 = 3
   Cb0 = 5
   Cs0 = 10
   T0 = 300
   k1oo = 45,000
   Ea1 = 26,500
   k2oo = 10,000
   Ea2 = 24,000
   k3oo = 9,000

Completely Solved Example

Ea3 = 32,000
k4oo = 9,500
Ea4 = 32,000
k5oo = 55,000
Ea5 = 25,000
k6oo = 12,000
Ea6 = 32,000
k7oo = 20,000
Ea7 = 38,000
k8oo = 6,000
Ea8 = 35,000
k9oo = 10,000
Ea9 = 30,000
k10oo = 5,000
Ea10 = 22,000

#Independent variable
t(0) = 0
t(f) = 20

The result of the fitting can be seen in Figure 4.12.

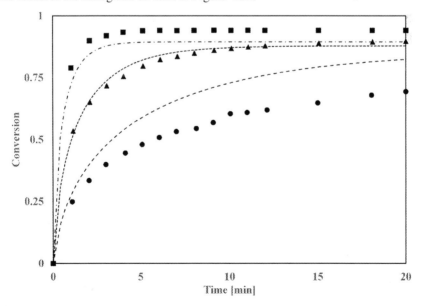

FIGURE 4.12 Variation of conversion of component A as a function of time. (●) Experimental data and (– –) model for $T = 275$ [K], (▲) experimental data and (--) model for $T = 300$ [K], (■) experimental data and (-•-) model for $T = 325$ [K].

This model presents the same extent of underestimation and overestimation as the previous model. This result is not unusual considering that in both cases, the controlling step was the adsorption of one of the reactants, and based on the stoichiometric values of the reaction, this outcome is among the possibilities.

From the visual analysis of the fitting, it is not a good match. These can also be seen in the value for the MSC, which is 4.14. Once more, the value seems to be a large number; however, values over 8 will give a high-quality fitting.

Due to the non-optimal fitting form the previous two models, a third option, with the reaction step (step 3) being the controlling step, has been proposed and compared with the experimental data. The script for this case can be seen below:

Script:

```
#Mole balance
    d(x)/d(t) = −r3/Ca0
    x(0) = 0

#Reaction rates
    r3 = −(((k1*k3*k5*Ca*Cb)/(k2*k4))−((k6*k8*k10*Cc*Cd)/(k7*k9)))*Cs
    Cs = ((Cs0/(1+(k1*Ca/k2)+(k3*Cb/k4)+(k8*Cc/k7)+(k10*Cd/k9)))^2)

#Auxiliary equations
    Ca = Ca0*(1−x)
    Cb = Cb0−Ca0*x
    Cc = 0.5+Ca0*x
    Cd = Ca0*x
    k1 = k1oo*EXP(−Ea1/(8.3143*T0))
    k2 = k2oo*EXP(−Ea2/(8.3143*T0))
    k3 = k3oo*EXP(−Ea3/(8.3143*T0))
    k4 = k4oo*EXP(−Ea4/(8.3143*T0))
    k5 = k5oo*EXP(−Ea5/(8.3143*T0))
    k6 = k6oo*EXP(−Ea6/(8.3143*T0))
    k7 = k7oo*EXP(−Ea7/(8.3143*T0))
    k8 = k8oo*EXP(−Ea8/(8.3143*T0))
    k9 = k9oo*EXP(−Ea9/(8.3143*T0))
    k10 = k10oo*EXP(−Ea10/(8.3143*T0))

#Constants and initial values
    Ca0 = 3
    Cb0 = 5
```

Completely Solved Example

```
Cs0 = 10
T0 = 300
k1oo = 95,000
Ea1 = 33,000
k2oo = 12,000
Ea2 = 29,000
k3oo = 11,000
Ea3 = 35,000
k4oo = 9,500
Ea4 = 31,000
k5oo = 95,000
Ea5 = 30,000
k6oo = 1,800
Ea6 = 32,000
k7oo = 10,000
Ea7 = 38,000
k8oo = 2,000
Ea8 = 35,000
k9oo = 12,000
Ea9 = 30,000
k10oo = 500
Ea10 = 18,000

#Independent variable
t(0) = 0
t(f) = 20
```

With this model, we plotted the experimental data against the model for all tested temperatures. The results can be seen in Figure 4.13.

The visual inspection of this model shows a great agreement between the data and the model. The results are very well represented by the model having the reaction step as the controlling step. The MSC value obtained is 8.91, which is large enough to assure a good convergence between the model and the data, as seen in Figure 4.13. One of the most common reasons for this model to be a better fit is that the reaction step is regularly the one that requires the most energy to occur. In the reaction step, some bonds are broken, and others are formed. These energy values are much greater than the typical adsorption values over a catalyst, which can be physisorption phenomena. Physisorption are weaker bonds among the elements, while chemisorption is stronger and with energy value comparable to the creation or rupture of chemical bonds. While typically, the reactions step is the controlling stage, it is not necessarily for all cases, and in some scenarios, adsorption has been proved to be the controlling step. We recommend looking at references [6,7].

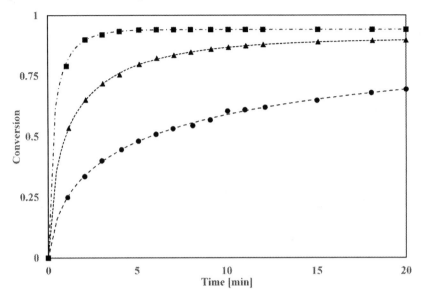

FIGURE 4.13 Variation of conversion of component A as a function of time. (●) Experimental data and (– –) model for $T = 275$ [K], (▲) experimental data and (--) model for $T = 300$ [K], (■) experimental data and (-•-) model for $T = 325$ [K].

Considering that the reaction step might not be the one with the highest energy dependency, it is important to test the other two models. This is when steps 4 and 5, i.e., desorption of products, are controlling as well.

When the desorption of product C, i.e., step 4, is controlling, the script in Polymath has the following formulation.

Script:

```
#Mole balance
    d(x)/d(t) = -r4/Ca0
    x(0) = 0.00001

#Reaction rates
    r4 = -(((k1*k3*k5*k7*k9*Ca*Cb)/(k2*k4*k6*k10*Cd))-(k8*Cc))*Cs
    Cs =  (Cs0/(1+(k1*Ca/k2)+(k3*Cb/k4)+((k1*k3*k5*k9*Ca*Cb)/
          (k2*k4*k6*k10*Cd))+(k10*Cd/k9)))

#Auxiliary equations
    Ca = Ca0*(1-x)
    Cb = Cb0-Ca0*x
```

Completely Solved Example

$Cc = 0.5+Ca0*x$
$Cd = Ca0*x$
$k1 = k1oo*EXP(-Ea1/(8.3143*T0))$
$k2 = k2oo*EXP(-Ea2/(8.3143*T0))$
$k3 = k3oo*EXP(-Ea3/(8.3143*T0))$
$k4 = k4oo*EXP(-Ea4/(8.3143*T0))$
$k5 = k5oo*EXP(-Ea5/(8.3143*T0))$
$k6 = k6oo*EXP(-Ea6/(8.3143*T0))$
$k7 = k7oo*EXP(-Ea7/(8.3143*T0))$
$k8 = k8oo*EXP(-Ea8/(8.3143*T0))$
$k9 = k9oo*EXP(-Ea9/(8.3143*T0))$
$k10 = k10oo*EXP(-Ea10/(8.3143*T0))$

#Constants and initial values
Ca0 = 3
Cb0 = 5
Cs0 = 10
T0 = 300
k1oo = 50,000
Ea1 = 34,000
k2oo = 1,000
Ea2 = 25,000
k3oo = 50,000
Ea3 = 31,000
k4oo = 10,000
Ea4 = 24,000
k5oo = 50,000
Ea5 = 31,500
k6oo = 1,000
Ea6 = 26,000
k7oo = 50,000
Ea7 = 30,000
k8oo = 1,000
Ea8 = 30,000
k9oo = 50,000
Ea9 = 35,000
k10oo = 1,000
Ea10 = 22,000

#Independent variable
t(0) = 0
t(f) = 20

With this script, the model versus the experimental data for all times and all temperatures can be plotted, as shown in Figure 4.14.

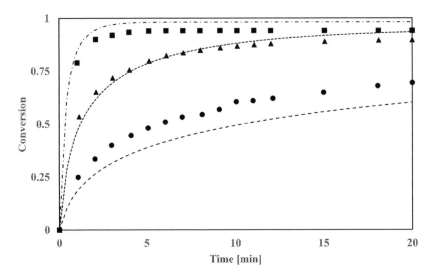

FIGURE 4.14 Variation of conversion of component A as a function of time. (●) Experimental data and (– –) model for $T = 275$ [K], (▲) experimental data and (--) model for $T = 300$ [K], (■) experimental data and (-•-) model for $T = 325$ [K].

In this case, the model underestimates the values at lower temperature and overestimates the value at higher temperatures, which is the opposite scenario from the situation when steps 1 and 2 were controlling. This analysis does not present a good fitting between the model and the data, which can also be seen in the MSC value of 3.64.

Finally, we will show the result when step 5 is controlling. For this case, the desorption of the second product is the controlling step. For this case, the script is as follows:

Script:
#Mole Balance
d(x)/d(t) = −r5/Ca0
x(0) = 0.00001

#Reaction rates
r5 = −(((k1*k3*k5*k7*k9*Ca*Cb)/(k2*k4*k6*k8*Cc))−(k10*Cd))*Cs

Completely Solved Example

$$Cs=(Cs0/(1+(k1*Ca/k2)+(k3*Cb/k4)+(k8*Cc/k7)+((k1*k3*k5*k7*Ca*Cb)/(k2*k4*k6*k8*Cc))))$$

#Auxiliary equations
Ca = Ca0*(1−x)
Cb = Cb0−Ca0*x
Cc = 0.5+Ca0*x
Cd = Ca0*x
k1 = k1oo*EXP(−Ea1/(8.3143*T0))
k2 = k2oo*EXP(−Ea2/(8.3143*T0))
k3 = k3oo*EXP(−Ea3/(8.3143*T0))
k4 = k4oo*EXP(−Ea4/(8.3143*T0))
k5 = k5oo*EXP(−Ea5/(8.3143*T0))
k6 = k6oo*EXP(−Ea6/(8.3143*T0))
k7 = k7oo*EXP(−Ea7/(8.3143*T0))
k8 = k8oo*EXP(−Ea8/(8.3143*T0))
k9 = k9oo*EXP(−Ea9/(8.3143*T0))
k10 = k10oo*EXP(−Ea10/(8.3143*T0))

#Constants and initial values
Ca0 = 3
Cb0 = 5
Cs0 = 10
T0 = 300
k1oo = 52,000
Ea1 = 35,000
k2oo = 2,000
Ea2 = 25,000
k3oo = 50,000
Ea3 = 27,000
k4oo = 8,000
Ea4 = 24,000
k5oo = 50,000
Ea5 = 29,500
k6oo = 8,000
Ea6 = 24,000
k7oo = 45,000
Ea7 = 35,000
k8oo = 1,000

Ea8 = 29,500
k9oo = 36,000
Ea9 = 30,500
k10oo = 1,000
Ea10 = 24,200

#Independent variable
t(0) = 0
t(f) = 20

Using this script for all evaluated temperatures, we managed to plot all models versus temperature. The results can be seen in Figure 4.15.

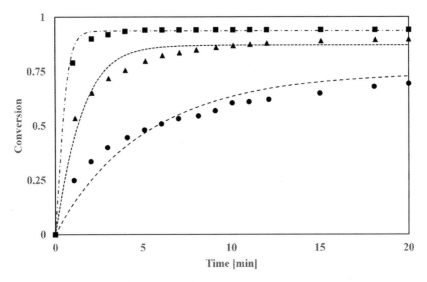

FIGURE 4.15 Variation of conversion of component A as a function of time. (●) Experimental data and (– –) model for $T = 275$ [K], (▲) experimental data and (--) model for $T = 300$ [K], (■) experimental data and (-•-) model for $T = 325$ [K].

It can be seen from Figure 4.15 that this model is much more accurate than the previous one; it does have a better fitting for higher temperatures, but it does not follow the data tendency that well for low and middle range temperatures. This can also be seen in the MSC value. This value is of 4.63.

As it was seen from all the five models presented and all the comparison with the data, the more complex model with the surface reaction as the controlling step was the one that gave the highest MSC value. This, together with the visual analysis of the fitting, shows that the mode accurately represents the experimental information. It is important to notice that this model is accurate under the conditions and assumptions for this problem; however, more complex models can also be obtained and could show better fitting than this one. Nevertheless, due to the goodness of the fitting we

4.11 FINAL EXPRESSION FOR KINETICS

Now that all the proposed models have been tested and compared their simulations with the experimental data and the kinetics parameters obtained accordingly, we would like to summarize the best result since it will be used in the next subsections. The reaction rate expression is:

$$r_3 = \left(\frac{k_1 k_3 k_5 C_A C_B}{k_2 k_4} - \frac{k_6 k_8 k_{10} C_C C_D}{k_7 k_9} \right) \left(\frac{C_S^0}{\left(1 + \frac{k_1 C_A}{k_2} + \frac{k_3 C_B}{k_4} + \frac{k_8 C_C}{k_7} + \frac{k_{10} C_D}{k_9}\right)} \right)^2 \quad (4.117)$$

For Equation 4.117, we also need Arrhenius equation ($k_i = k_{0i} e^{\left(\frac{-E_{ai}}{R*T}\right)}$) for $i = 1–10$, while the values for each term are presented in Table 4.3.

Now, based on the kinetics expression obtained, we will use it for the modeling of an isothermal plug flow reactor, and adiabatic plug flow reactor, a constant heat transfer plug flow reactor, a co-current heat transfer plug flow reactor, and a countercurrent plug flow reactor.

TABLE 4.3
Kinetics Values

Variable	Value	Units
k_{100}	95,000	1/(mol*min)
k_{200}	12,000	1/min
k_{300}	11,000	1/(mol*min)
k_{400}	9,500	1/min
k_{500}	95,000	1/(mol*min)
k_{600}	1,800	1/(mol*min)
k_{700}	10,000	1/min
k_{800}	2,000	1/(mol*min)
k_{900}	12,000	1/min
k_{1000}	500	1/(mol*min)
Ea_1	33,000	J/mol
Ea_2	29,000	J/mol
Ea_3	35,000	J/mol
Ea_4	31,000	J/mol
Ea_5	30,000	J/mol
Ea_6	32,000	J/mol
Ea_7	38,000	J/mol
Ea_8	35,000	J/mol
Ea_9	30,000	J/mol
Ea_{10}	18,000	J/mol

4.12 SIMULATION OF AN ISOTHERMAL PLUG FLOW REACTOR USING KINETICS FROM 4.11

This section is based on the development of the expression for an isothermal plug flow reactor, as in the previous chapter, and the kinetics expression obtained from the experimental information shown above. We will now use this information to model a plug flow reactor that operates in different temperature scenarios, in this case, isothermally. This means that the temperature does not change with the position of the reactor and remains constant for the entire problem. In this case, we will try to obtain the change in the molar flows of each of the components. We will make an assumption for the sake of simplification, that the process happens in liquid phase so that there is no change in the number of moles or pressure that could affect the mole balance. Therefore, the set of equations that we need to solve for this problem are:

$$\frac{dF_A}{dV} = r_A; \frac{dF_B}{dV} = r_B; \frac{dF_C}{dV} = r_C; \frac{dF_D}{dV} = r_D \qquad (4.118)$$

In order to obtain the kinetic expression for all components, we consider:

$$\frac{r_A}{-1} = \frac{r_B}{-1} = \frac{r_C}{1} = \frac{r_D}{1} \qquad (4.119)$$

We also use the expression to convert concentration to molar flow: $F_i = v_i * C_i$, as well as the definition for conversion, which is:

$$x = \frac{F_A^0 - F_A}{F_A^0} \qquad (4.120)$$

Some assumptions have been made such as that $T = 300$ [K], and as previously mentioned, there is no change in pressure or in the number of moles; therefore, $v_i = v$ for all components. The additional data required to solve this problem has been directly introduced into the script, the reader should extract this information as an exercise.

With this set of expressions, we can then write the script in Polymath.

Script:

#Mole balance
    d(Fa)/d(V) = ra
    Fa(0) = 3
    d(Fb)/d(V) = rb
    Fb(0) = 5
    d(Fc)/d(V) = rc
    Fc(0) = 0.5
    d(Fd)/d(V) = rd
    Fd(0) = 0

#Reaction rates
    ra = −(((k1*k3*k5*Ca*Cb)/(k2*k4))−((k6*k8*k10*Cc*Cd)/(k7*k9)))*Cs

Completely Solved Example

```
    Cs = ((Cs0/(1+(k1*Ca/k2)+(k3*Cb/k4)+(k8*Cc/k7)+(k10*Cd/k9)))^2)
    rb = ra
    rc = -ra
    rd = -ra

#Auxiliary equations
    x = (Fa0-Fa)/Fa0
    Ca = Fa/v
    Cb = Fb/v
    Cc = Fc/v
    Cd = Fd/v
    k1 = 95,000*EXP(-33,000/(8.3143*T0))
    k2 = 12,000*EXP(-29,000/(8.3143*T0))
    k3 = 11,000*EXP(-35,000/(8.3143*T0))
    k4 = 9,500*EXP(-31,000/(8.3143*T0))
    k5 = 95,000*EXP(-30,000/(8.3143*T0))
    k6 = 18,00*EXP(-32,000/(8.3143*T0))
    k7 = 10,000*EXP(-38,000/(8.3143*T0))
    k8 = 2,000*EXP(-35,000/(8.3143*T0))
    k9 = 12,000*EXP(-30,000/(8.3143*T0))
    k10 = 500*EXP(-18,000/(8.3143*T0))

#Constants and initial values
    Fa0 = 3
    v = 1
    Cs0 = 10
    T0 = 300

#Independent variable
    V(0) = 0
    V(f) = 20
```

By solving this set of equations, we can obtain the profile of all molar flows as a function of the volume of the reactor. These results, plus the conversion of A, can be seen in Figures 4.16 and 4.17, respectively.

As seen in Figure 4.16, the flow profile changes along the reactor tube. This result can help estimate the length of the reactor to prevent unused volume and obtain the desired conversion at the end of the process. Since it is an isothermal system, the profile of temperature is not presented in this case. In Figure 4.17, it can be seen that the conversion increases along the reaction, which is a good indicator of the length or volume of the system to achieve a desired yield. Conversion and heat transfer can be improved under an adiabatic reaction system.

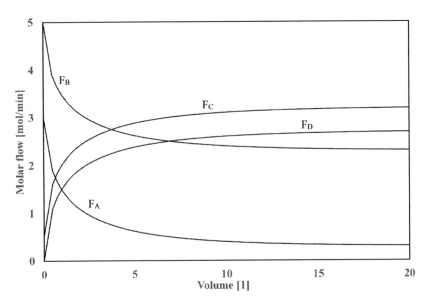

FIGURE 4.16 Variation of the molar flow as a function of the volume of the reactor.

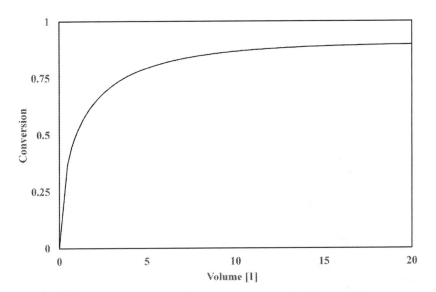

FIGURE 4.17 Variation of conversion as a function of the volume of the reactor.

Completely Solved Example

4.13 SIMULATION OF AN ADIABATIC PLUG FLOW REACTOR WITH THE KINETICS FROM 4.11

As mentioned, an adiabatic process implies that there is no heat transfer either in or out of the system ($Q = 0$). The expression for an adiabatic temperature profile as a function of conversion is:

$$T = \frac{-xF_A^0 \Delta H_{RX}(T_R) + \left(F_A^0 C_{P_A} + F_B^0 C_{P_B} + F_C^0 C_{P_C}\right)T_0}{\left(F_A^0 C_{P_A} + F_B^0 C_{P_B} + F_C^0 C_{P_C}\right)} \quad (4.121)$$

It is important to see that this equation was obtained based on Equation 2.228, where the simplification of heat of reaction contribution is considerably larger than the energy involved to heat the chemicals from the reference temperature to the reaction temperature. Example 2.11 is solved without this simplification; therefore, you have both cases to study and compare.

Using Equation 4.121 together with the mole balance and assumptions from the previous solution, we can then get the set of equations to be solved for an adiabatic scenario. For this case, we will need the value of Cp_i for each component plus the ΔH_{RX}.

We have, $Cp_A = 120 \left[\dfrac{J}{mol*K}\right]$, $Cp_B = 140 \left[\dfrac{J}{mol*K}\right]$, $Cp_C = 150 \left[\dfrac{J}{mol*K}\right]$, $Cp_D = 110 \left[\dfrac{J}{mol*K}\right]$, and $\Delta H_{RX} = -7,200 \left[\dfrac{J}{mol}\right]$.

With this information, we can then present the script to be used in Polymath.

Script:

```
#Mole balance
    d(Fa)/d(V) = ra
    Fa(0) = 3
    d(Fb)/d(V) = rb
    Fb(0) = 5
    d(Fc)/d(V) = rc
    Fc(0) = 0.5
    d(Fd)/d(V) = rd
    Fd(0) = 0

#Energy balance
    T = (Fa0*Cpa*T0+Fb0*Cpb*T0+Fc0*Cpc*T0–Fa0*x*DH)/
    (Fa0*Cpa+Fb0*Cpb+Fc0*Cpc)
```

#Reaction rates
ra = -(((k1*k3*k5*Ca*Cb)/(k2*k4))-((k6*k8*k10*Cc*Cd)/(k7*k9)))*Cs
Cs = ((Cs0/(1+(k1*Ca/k2)+(k3*Cb/k4)+(k8*Cc/k7)+(k10*Cd/k9)))^2)
rb = ra
rc = -ra
rd = -ra

#Auxiliary equations
x = (Fa0-Fa)/Fa0
Ca = Fa/v
Cb = Fb/v
Cc = Fc/v
Cd = Fd/v
k1 = 95,000*EXP(-33,000/(8.3143*T))
k2 = 12,000*EXP(-29,000/(8.3143*T))
k3 = 11,000*EXP(-35,000/(8.3143*T))
k4 = 9,500*EXP(-31,000/(8.3143*T))
k5 = 95,000*EXP(-30,000/(8.3143*T))
k6 = 1,800*EXP(-32,000/(8.3143*T))
k7 = 10,000*EXP(-38,000/(8.3143*T))
k8 = 2,000*EXP(-35,000/(8.3143*T))
k9 = 12,000*EXP(-30,000/(8.3143*T))
k10 = 500*EXP(-18,000/(8.3143*T))

#Constants and initial values
Fa0 = 3
Fb0 = 5
Fc0 = 0.5
v = 1
Cs0 = 10
T0 = 300
Cpa = 120
Cpb = 140
Cpc = 150
DH = -7,200

#Independent variable
V(0) = 0
V(f) = 20

Completely Solved Example

Based on the script presented above, we can obtain the profile for all the molar flows as a function of the volume. In this case, we can also see the conversion as well as the temperature profile as a function of the volume. The molar flows can be seen in Figure 4.18, the variations of conversion in Figure 4.19, and the temperature profile in Figure 4.20.

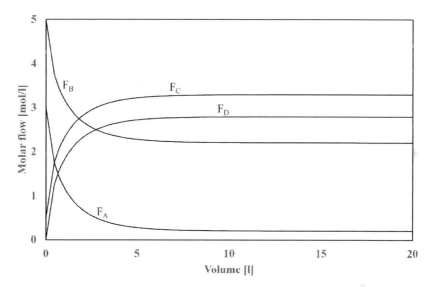

FIGURE 4.18 Variation of the molar flow as a function of the volume of the reactor.

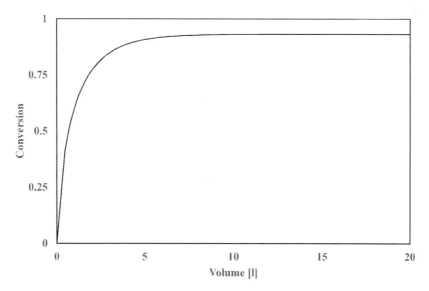

FIGURE 4.19 Variation of the conversion for an adiabatic scenario as a function of the volume.

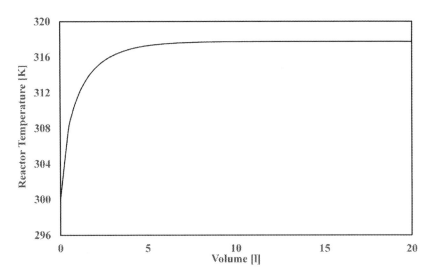

FIGURE 4.20 Temperature profile as a function of the volume for an adiabatic process.

As the process is adiabatic, the temperature of the process changes with the volume, which allows the process to reach a higher final conversion of 0.93, as compared to 0.89 in the isothermal scenario.

As shown in Figure 4.19, the temperature profile increases until it reaches a value of 318 [K], then it remains constant. It is important to consider that the final temperature that the process might reach when being driven adiabatically, should not overtake the maximum allowed temperature for the materials. This can lead to destruction of the reactants, a degradation of the products, or a breakage on the reactor itself. In order to avoid this type of problem, a cooling (or heating) flow is connected to the reactor. This flow can be run co-currently or counter-currently, depending on the needs and the process. If the flow of the cooling liquid is too high, it can be assumed that the heat transfer is constant and so is the temperature of the cooling liquid, which is the case we will tackle first.

4.14 SIMULATION OF A CONSTANT HEAT TRANSFER PLUG FLOW REACTOR WITH KINETICS FROM 4.11

From the energy balance, we know that the general expression for variation of temperature with volume is:

$$\frac{dT}{dV} = \frac{(r_A \Delta H_{RX}) - (U\varpi(T - T_f))}{\sum F_i Cp_i} \qquad (4.122)$$

In order to improve the reaction or to reduce the possibility of the temperature profile becoming too high, or low, the removal or addition of heat to the system can be considered. For this problem, since we are dealing with an exothermic reaction,

Completely Solved Example

heat will be removed from the system. However, the same expression can be used for adding heat to the reactor, but taking into consideration that the sign in the heat in the energy balance must be changed. In this case, we will first assume that the temperature of the external fluid is constant and equal to 280 [K], and the value of $U\varpi$ is 50 [W/(m*K)]. The units for the other variables agree with Tables 4.2 and 4.3.

For this case, the script is almost identical to the previous one.

Script:

```
#Mole balance
    d(Fa)/d(V) = ra
    Fa(0) = 3
    d(Fb)/d(V) = rb
    Fb(0) = 5
    d(Fc)/d(V) = rc
    Fc(0) = 0.5
    d(Fd)/d(V) = rd
    Fd(0) = 0

#Energy balance
    d(T)/d(V) = ((DH*ra)-(Ud*(T-Tf)))/(Fa*Cpa+Fb*Cpb+Fc*Cpc+Fd*Cpd)
    T(0) = 300

#Reaction rates
    ra = -(((k1*k3*k5*Ca*Cb)/(k2*k4))-((k6*k8*k10*Cc*Cd)/(k7*k9)))*Cs
    Cs = ((Cs0/(1+(k1*Ca/k2)+(k3*Cb/k4)+(k8*Cc/k7)+(k10*Cd/k9)))^2)
    rb = ra
    rc = -ra
    rd = -ra

#Auxiliary equations
    x = (Fa0-Fa)/Fa0
    Ca = Fa/v
    Cb = Fb/v
    Cc = Fc/v
    Cd = Fd/v
    k1 = 95,000*EXP(-33,000/(8.3143*T))
    k2 = 12,000*EXP(-29,000/(8.3143*T))
    k3 = 11,000*EXP(-35,000/(8.3143*T))
    k4 = 9,500*EXP(-31,000/(8.3143*T))
```

```
k5 = 95,000*EXP(-30,000/(8.3143*T))
k6 = 1,800*EXP(-32,000/(8.3143*T))
k7 = 10,000*EXP(-38,000/(8.3143*T))
k8 = 2,000*EXP(-35,000/(8.3143*T))
k9 = 12,000*EXP(-30,000/(8.3143*T))
k10 = 500*EXP(-18,000/(8.3143*T))

#Constants and initial values
    Fa0 = 3
    Fb0 = 5
    Fc0 = 0.5
    v = 1
    Cs0 = 10
    Tf = 280
    Ud = 50
    Cpa = 120
    Cpb = 140
    Cpc = 150
    Cpd = 110
    DH = -7,200

#Independent variable
    V(0) = 0
    V(f) = 20
```

Based on these assumptions included in this script, we can obtain the profile for the flows as well as for the temperature as a function of the volume. The results are plotted in Figures 4.21 to 4.23, the figures present the molar flow profiles, the conversion and the temperature profile, respectively (in the last one we have added T_f even though it is constant). The final conversion of component A has decreased a little, *i.e.*, to 0.90. However, it can also be seen that the temperature profile now goes through a maximum value, which is critical and needs to be known. Reactants and products can degenerate due to high temperatures and therefore, knowing this limit is important to assure a proper functioning of the process. If this value is too high, over the material or chemical component degradation limits, then a flow of cooling fluid must be added to the surroundings in order to keep the system below the desired temperature. This is presented in the upcoming subsections.

The variation of the conversion can be seen in Figure 4.22.
The variation of the temperature profile can be seen in Figure 4.23.

Completely Solved Example

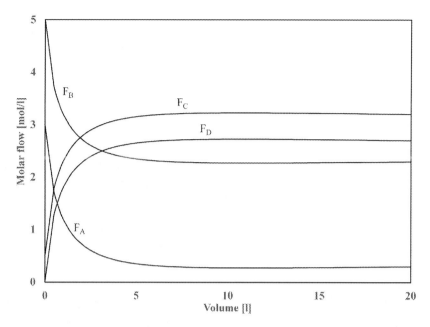

FIGURE 4.21 Variation of the molar flow as a function of the volume of the reactor.

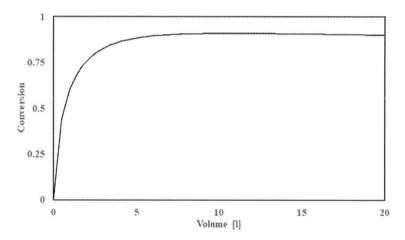

FIGURE 4.22 Variation of the conversion as function of the volume of the reactor.

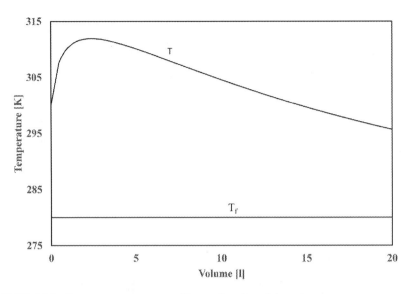

FIGURE 4.23 Reactor temperature profile as a function of the volume.

4.15 SIMULATION OF A CO-CURRENT HEAT TRANSFER FLOW IN A PLUG FLOW REACTOR WITH KINETICS FROM 4.11

In order to assure a lower temperature peak in the reactor, and that the temperature does not rise unconditionally to dangerous levels where the process is changed into producing undesired products; therefore, the use of a cooling liquid is very important. In this problem, we work with an exothermic reaction; therefore, we need to remove energy from the system as the reaction progresses. For this case, we need to add only the energy balance for the external liquid. The expression that needs to be added to the equation matrix is:

$$\frac{dT_f}{dV} = \frac{U\varpi(T - T_f)}{m_f C_{Pf}} \quad (4.123)$$

Adding this expression to the previous system, we can obtain a set of equations to be solved where the temperature of the external fluid varies.

Script:
```
#Mole balance
    d(Fa)/d(V) = ra
    Fa(0) = 3
    d(Fb)/d(V) = rb
    Fb(0) = 5
    d(Fc)/d(V) = rc
    Fc(0) = 0.5
```

Completely Solved Example

d(Fd)/d(V) = rd
Fd(0) = 0

#Energy balance
d(T)/d(V) = ((DH*ra)−(Ud*(T−Tf)))/(Fa*Cpa+Fb*Cpb+Fc*Cpc+Fd*Cpd)
T(0) = 300

#Energy balance external co-current flow
d(Tf)/d(V) = (Ud*(T−Tf))/(mf*Cpf)
Tf(0) = 280

#Reaction rates
ra = −(((k1*k3*k5*Ca*Cb)/(k2*k4))−((k6*k8*k10*Cc*Cd)/(k7*k9)))* Cs
Cs = ((Cs0/(1+(k1*Ca/k2)+(k3*Cb/k4)+(k8*Cc/k7)+(k10*Cd/k9)))^2)
rb = ra
rc = −ra
rd = −ra

#Auxiliary equations
x = (Fa0−Fa)/Fa0
Ca = Fa/v
Cb = Fb/v
Cc = Fc/v
Cd = Fd/v
k1 = 95,000*EXP(−33,000/(8.3143*T))
k2 = 12,000*EXP(−29,000/(8.3143*T))
k3 = 11,000*EXP(−35,000/(8.3143*T))
k4 = 9,500*EXP(−31,000/(8.3143*T))
k5 = 95,000*EXP(−30,000/(8.3143*T))
k6 = 1,800*EXP(−32,000/(8.3143*T))
k7 = 10,000*EXP(−38,000/(8.3143*T))
k8 = 2,000*EXP(−35,000/(8.3143*T))
k9 = 12,000*EXP(−30,000/(8.3143*T))
k10 = 500*EXP(−18,000/(8.3143*T))

#Constants and initial values
Fa0 = 3
Fb0 = 5
Fc0 = 0.5

```
v = 1
Cs0 = 10
mf = 5
Cpf = 52
Ud = 50
Cpa = 120
Cpb = 140
Cpc = 150
Cpd = 110
DH = -7,200

#Independent variable
V(0) = 0
V(f) = 20
```

This script will allow us to see the flow patterns for all the chemicals involved in the reaction, *i.e.*, the raw materials and products. The reaction temperature profile and the temperature profile for the cooling fluid.

The molar flows for each chemical can be seen in Figure 4.24. The conversion and therefore, the generation of the product increases to 0.92, shown in Figure 4.25. This conversion is as high as the conversion for the adiabatic scenario, but it has a much better temperature control. Having the external fluid allows us to optimize the process better within the reactor and to avoid empty and/or unused volume.

The system is running in co-current mode, and therefore, the temperature of the reactor and the temperature of the external fluid will tend to be equal to each other.

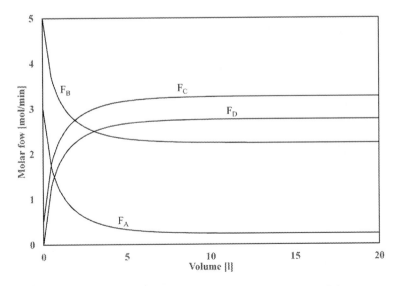

FIGURE 4.24 Variation of the molar flow as a function of the volume of the reactor.

Completely Solved Example

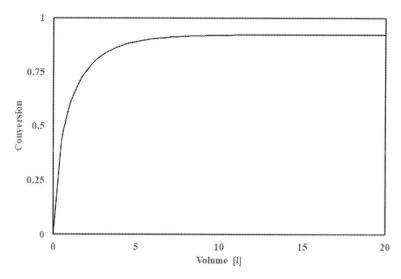

FIGURE 4.25 Variation of the conversion as a function of the volume of the reactor.

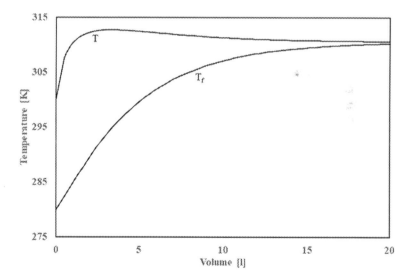

FIGURE 4.26 Temperature profiles as a function of the volume.

Mathematically, this is possible, for a practical perspective, it is important to know that a $\Delta T \geq 10$ K is a good rule of thumb to have good heat transfer from one source to the other. In this problem, we are not taking into consideration the rule of thumb in order to show that the software just solves mathematical equations. The physical meaning and practical constrains must be considered by the engineer.

The temperature profiles are provided in Figure 4.26.

From Figure 4.26, we can easily notice that the temperature peak is reached with a similar value as for the case at constant T_f. However, this value can easily change based on the value of the external flow of fluid that is used for the cooling device. We recommend the reader to run the script with several different values of m_c and to see the effect on the maximum value for the temperature profile.

The heat transfer takes place due to the difference in temperature, and in the co-current case, the temperature profile gets closer and closer together. This indicates that the value of heat transfer decreases constantly. A more practical approach is to have the flow in counter-current mode as it allows better heat transfer among the fluids involved.

4.16 SIMULATION OF A COUNTER-CURRENT HEAT TRANSFER FLOW IN A PLUG FLOW REACTOR WITH KINETICS FROM 4.11

The last process that will be simulated, in this example, is a plug flow with the external fluid in counter-current mode. This way, the heat transfer is more efficient than in co-current flow mode. The script is identical to the co-current flow, but to solve a set of differential equations in Polymath, we need to give initial values for all the variables. As we saw in the Section 2.11.4, for a counter-current flow, the initial value for the external fluid is unknown, but what we know its value when the volume is V. Therefore, here we need to apply the so-called shooting approach to solve these equations. As the name implies, we need to guess our initial value, and shoot that value through the equations to see if the final value is suitable. For this problem, several T_f values were proposed for $V = 0$ and the script was solved to get the value of T_f when $V = 20$ [m³]. We will only present one case, but we recommend that the student prepare a script and test it to see the effect of having counter-current flow.

Since the external flow is flowing in the opposite direction, it is important to change the sign for the evaluation of Taf with the volume; the equation will be written as:

$$\frac{dT_f}{dV} = \frac{-U\varpi(T-T_f)}{m_f C_{Pf}} \tag{4.124}$$

The script with this modification is:

Script:

```
#Mole balance
    d(Fa)/d(V) = ra
    Fa(0) = 3
    d(Fb)/d(V) = rb
    Fb(0) = 5
    d(Fc)/d(V) = rc
    Fc(0) = 0.5
    d(Fd)/d(V) = rd
    Fd(0) = 0
```

Completely Solved Example

#Energy balance
d(T)/d(V) = ((DH*ra)−(Ua*(T−Tf)))/(Fa*Cpa+Fb*Cpb+Fc*Cpc+Fd*Cpd)
T(0) = 300

#Energy balance external counter-current flow
d(Tf)/d(V) = −(Ud*(T−Tf))/(mf*Cpf)
Tf(0) = 313.85

#Reaction rates
ra = −(((k1*k3*k5*Ca*Cb)/(k2*k4))−((k6*k8*k10*Cc*Cd)/(k7*k9)))*Cs
Cs = ((Cs0/(1+(k1*Ca/k2)+(k3*Cb/k4)+(k8*Cc/k7)+(k10*Cd/k9)))^2)
rb = ra
rc = −ra
rd = −ra

#Auxiliary equations
x = (Fa0−Fa)/Fa0
Ca = Fa/v
Cb = Fb/v
Cc = Fc/v
Cd = Fd/v
k1 = 95,000*EXP(−33,000/(8.3143*T))
k2 = 12,000*EXP(−29,000/(8.3143*T))
k3 = 11,000*EXP(−35,000/(8.3143*T))
k4 = 95,00*EXP(−31,000/(8.3143*T))
k5 = 95,000*EXP(−30,000/(8.3143*T))
k6 = 1,800*EXP(−32,000/(8.3143*T))
k7 = 10,000*EXP(−38,000/(8.3143*T))
k8 = 2,000*EXP(−35,000/(8.3143*T))
k9 = 12,000*EXP(−30,000/(8.3143*T))
k10 = 500*EXP(−18,000/(8.3143*T))

#Constants and initial values
Fa0 = 3
Fb0 = 5
Fc0 = 0.5
v = 1
Cs0 = 10
mf = 5

Cpf = 52
Ua = 50
Cpa = 120
Cpb = 140
Cpc = 150
Cpd = 110
DH = −7,200

#Independent variable
V(0) = 0
V(f) = 20

Running this script allow us to present the modification in the molar flow for all the components as well as the temperature profiles for the reaction temperature and for the external fluid. It is important to point out that the external fluid temperature flows in the opposite direction. The final conversion obtained with this approach is close to 0.93, which was the case for the adiabatic system. However, this method is more stable and easier to control due to the possibility to increase or reduce the flow from the cooling system.

The variations of the molar flow, conversion, and temperatures profiles, as a function of the volume, can be seen in Figures 4.27 to 4.29, respectively.

Due to the flow of the external cooling fluid going in the opposite direction, the temperature profile for the reaction itself is smoother and does not have such a straight up peak as before. It is important to notice that these plots are strongly

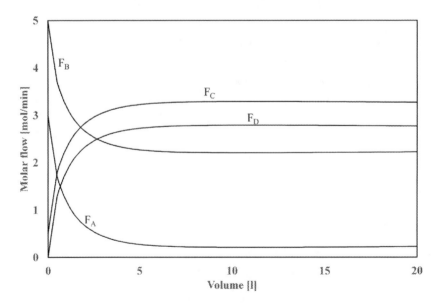

FIGURE 4.27 Variation of the molar flow as a function of the volume of the reactor.

Completely Solved Example

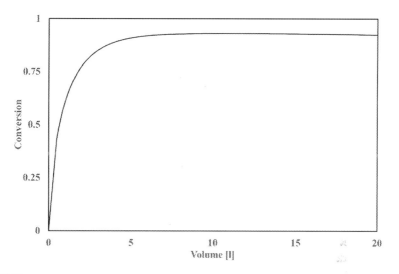

FIGURE 4.28 Variation of the conversion as a function of the volume of the reactor.

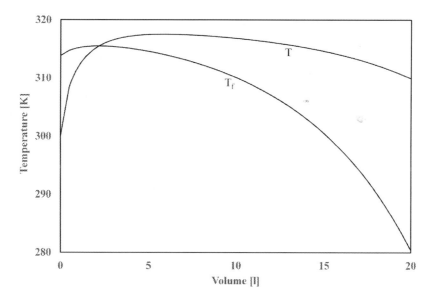

FIGURE 4.29 Temperature profiles as a function of the volume

affected by the flow of m_f. Therefore, we recommend that the student does more runs of the script with higher and lower fluid flows and discuss the effect.

Finally, as a matter of comparison, we want to present the variation of all conversion together and the same for all the temperature profiles. As shown in Figure 4.30, the conversion has some variations, especially from isothermal to all the rest, which is the largest gap. However, for all the other four cases, the conversion does not suffer much variation; so, it can be seen that all the lines are almost overlapping.

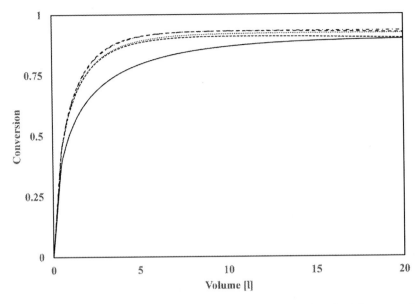

FIGURE 4.30 Variation of conversion. (———) Isothermal, (-•-) adiabatic, (--) constant fluid temperature, (•••) co-current flow, and (– –) counter-current flow.

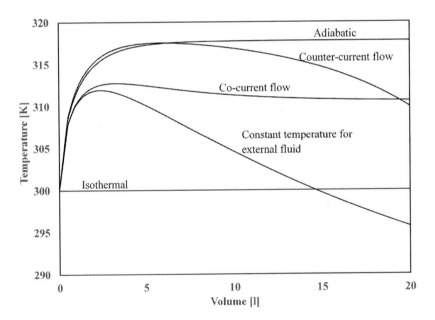

FIGURE 4.31 Temperature profiles.

Completely Solved Example

For the temperature profile, different profiles are generated based on the type of temperature control that we will have. The variations of temperature as a function of the volume can be seen in Figure 4.31. Here, only the reaction temperature is presented. It is important to point out that this profile of temperature allows the system to be controlled, whether easily or not. Having a cooling or warming fluid flowing outside the reactor is a very common technique to have a good temperature distribution inside the reactor. It can also help save energy since it can be used in different layouts, so that the energy is being produced in another part of the process. Additionally, it also provides better control with a valve that can be open or closed.

4.17 COMPARISON FOR A GAS PHASE SYSTEM WITH PRESSURE DROP

So far we have solved the problem for a liquid phase reaction. We will now solve a similar scenario, but for a gas phase reaction, with pressure drop within the solid catalysts. The kinetics obtained for all the previous models cannot be directly used since it was for a liquid reaction. However, we will use them, assuming that for this new reaction under consideration, they will be valid with some modifications to the value of the kinetics parameters.

Within this section, our major interest is not the kinetics; we want to show how the problem gets more and more complex under less and less ideal conditions, and in this case, by simply adding pressure, temperature, and change in the reaction volume.

We will solve this problem for an isothermal process, for an adiabatic scenario, for a constant outside temperature, for a co-current external flow, and for a counter-current external flow. We will present all the scripts accordingly and show the variation in the conversion of A, the pressure drops, and the reaction temperature profile, for all scenarios as well as the temperature profile for the outer fluids.

The problem to be solved is the same as the one previously solved where A and B produce C and D, and it follows the same mechanism with the same controlling step as before. The limiting step is the reaction step and therefore, we can use the kinetics expression developed in Equation 4.86.

Additional data is required for each case separately; we will provide a table in each case with the additional information required for each scenario.

4.17.1 Isothermal

First of all, we will solve the isothermal case which is the simplest case to solve compared with the next ones. To the typical PFR that we have solved before, we are adding the pressure drop to this case. Therefore, we have the following case to solve:

Mole balance:

$$\frac{dF_A}{dW} = r_A \qquad (4.125)$$

$$\frac{dF_B}{dW} = r_B \qquad (4.126)$$

$$\frac{dF_C}{dW} = r_C \tag{4.127}$$

$$\frac{dF_D}{dW} = r_D \tag{4.128}$$

Reaction rate:

$$r_A = \left(\frac{k_1 k_3 k_5 C_A C_B}{k_2 k_4} - \frac{k_6 k_8 k_{10} C_C C_D}{k_7 k_9}\right) \left(\frac{C_S^0}{\left(1 + \frac{k_1 C_A}{k_2} + \frac{k_3 C_B}{k_4} + \frac{k_8 C_C}{k_7} + \frac{k_{10} C_D}{k_9}\right)}\right)^2 \tag{4.129}$$

$$\text{With } \frac{r_A}{-1} = \frac{r_B}{-1} = \frac{r_C}{1} = \frac{r_D}{1}$$

Pressure drop:

$$\frac{dP}{dW} = -\frac{1}{A\rho_c(1-\theta)} \frac{U}{\rho_0 d_p} \left(\frac{1-\theta}{\theta^3}\right) \left(\frac{150(1-\theta)\mu}{d_p} + 1.75U\right) \frac{T}{T_0} \frac{P_0}{P} \frac{F_T}{F_{T0}} \tag{4.130}$$

Besides the differential equations presented above, some other auxiliary equations were also required to solve this problem; those are:

$$v = v^0 \frac{F_T}{F_T^0} \frac{P^0}{P} \frac{T}{T^0} \tag{4.131}$$

$$C_i = \frac{F_i}{v} \tag{4.132}$$

$$F_T = F_A + F_B + F_C + F_D \tag{4.133}$$

$$X_A = \frac{F_A^0 - F_A}{F_A^0} \tag{4.134}$$

$$F_{T0} = F_A^0 + F_B^0 + F_C^0 + F_D^0 \tag{4.135}$$

$$k_i = k_{0i} e^{\left(\frac{-E_{ai}}{R*T}\right)} \tag{4.136}$$

Energy balance
We are only missing the energy balance, for this case, it will be:

$$T = 300 [K] \tag{4.137}$$

Completely Solved Example

We can now solve the problem by solving Equations 4.125 to 4.137 simultaneously. To do that, we need some additional information for the constants involved as well as the initial and final values for the differential equations. Table 4.4 presents the values and units for those constants required for this problem.

TABLE 4.4
Values Used for Constants

Variable	Value	Units
k_{100}	95,000	l²/(mol*kg*min)
k_{200}	12,000	l/(kg*min)
k_{300}	11,000	l²/(mol*kg*min)
k_{400}	9,500	l/(kg*min)
k_{500}	95,000	l²/(mol*kg*min)
k_{600}	1,800	l²/(mol*kg*min)
k_{700}	10,000	l/(kg*min)
k_{800}	2,000	l²/(mol*kg*min)
k_{900}	12,000	l/(kg*min)
k_{1000}	500	l²/(mol*kg*min)
Ea_1	33,000	J/mol
Ea_2	29,000	J/mol
Ea_3	35,000	J/mol
Ea_4	31,000	J/mol
Ea_5	30,000	J/mol
Ea_6	32,000	J/mol
Ea_7	38,000	J/mol
Ea_8	35,000	J/mol
Ea_9	30,000	J/mol
Ea_{10}	18,000	J/mol
F_A^0	3	mol/min
F_B^0	5	mol/min
F_C^0	0.5	mol/min
F_D^0	0	mol/min
C_s^0	10	mol/l
$P0$	2,500	kPa
v_0	1	l/min
ρ_0	1.3	Kg/m³
ρ_c	2,800	Kg/m³
θ	0.6	-----
A	$\pi*D^2/4$	m²
D	0.2	m
μ	0.01	Kg/(m*s)
d_p	0.03	m
U	5.5	Kg/(m²*s)

Now we can put all this information into the software and solve the problem to obtain the variation of the molar flow, the change in pressure, temperature profile, and conversion of chemical A.

Script:

#Mole Balance

$d(Fa)/d(W) = ra$

$Fa(0) = 3$

$d(Fb)/d(W) = rb$

$Fb(0) = 5$

$d(Fc)/d(W) = rc$

$Fc(0) = 0.5$

$d(Fd)/d(W) = rd$

$Fd(0) = 0$

#Energy Balance

$T = 300$

#Pressure drop and its constants

$d(P)/d(W) = -1*(U/(dp*Rho0))*(((150*mu*(1-tita))/dp)+(1.75*U))*(1/(Ac*Rhoc*(1-tita)))*((1-tita)/((tita)^3))*((P0/P)*(FT/FT0)*(T/T0))$

$P(0) = 2,500$

#Reaction rates

$ra = -(((k1*k3*k5*Ca*Cb)/(k2*k4))-((k6*k8*k10*Cc*Cd)/(k7*k9)))*(((Cs0/(1+(k1*Ca/k2)+(k3*Cb/k4)+(k8*Cc/k7)+(k10*Cd/k9)))^2))$

$rb = ra$

$rc = -ra$

$rd = -ra$

#Auxiliary equations

$k1 = k001*EXP(-Ea1/(R*T))$

$k2 = k002*EXP(-Ea2/(R*T))$

$k3 = k003*EXP(-Ea3/(R*T))$

$k4 = k004*EXP(-Ea4/(R*T))$

$k5 = k005*EXP(-Ea5/(R*T))$

$k6 = k006*EXP(-Ea6/(R*T))$

$k7 = k007*EXP(-Ea7/(R*T))$

$k8 = k008*EXP(-Ea8/(R*T))$

Completely Solved Example

```
    k9 = k009*EXP(-Ea9/(R*T))
    k10 = k0010*EXP(-Ea10/(R*T))
    FT = Fa+Fb+Fc+Fd
    FT0 = Fa0+Fb0+Fc0+Fd0
    x = (Fa0-Fa)/Fa0
    Ca = Fa/v
    Cb = Fb/v
    Cc = Fc/v
    Cd = Fd/v
    v = v0*((P0/P)*(FT/FT0)*(T/T0))
Ac = 3.14159/4*D^2

#Initial values and constants

k001 = 95,000
    Ea1 = 33,000
    k002 = 12,000
    Ea2 = 29,000
    k003 = 11,000
    Ea3 = 35,000
    k004 = 9,500
    Ea4 = 31,000
    k005 = 95,000
    Ea5 = 30,000
    k006 = 1,800
    Ea6 = 32,000
    k007 = 10,000
    Ea7 = 38,000
    k008 = 2,000
    Ea8 = 35,000
    k009 = 12,000
    Ea9 = 30,000
    k0010 = 500
    Ea10 = 18,000
    R = 8.3143
    v0 = 1
    Fa0 = 3
    Fb0 = 5
    Fc0 = 0.5
```

Fd0 = 0
P0 = 2,500
Cs0 = 10
T0 = 300
tita = 0.6
D = 0.2
Rho0 = 1.3
Rhoc = 2,800
dp = 0.03
mu = 0.01
U = 5.5

#Independent variable
W(0) = 0
W(f) = 5

Solving the equations in Polymath allow us to then plot the variation of the different flows as well as the pressure and conversion profile. In this case, which is isothermal, the temperature profiles do not need to be plotted.

Figure 4.32 shows the variation of the different flows as a function of the amount of catalysts in the reactor.

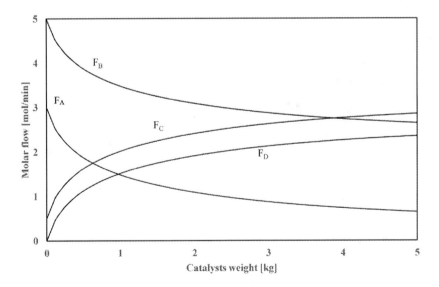

FIGURE 4.32 Flow profiles.

Completely Solved Example

As shown in Figure 4.32, the reactant flows are consumed while the products are generated. Chemical A is the limiting reactant and therefore, the conversion is based on this chemical. Figure 4.33 shows the drop in the pressure due to the presence of the catalysts and the variation on the conversion. The pressure drop is based on the amount of catalyst as well as in the porosity of it; how the porosity affects the pressure changes will be presented in Section 4.17.6.

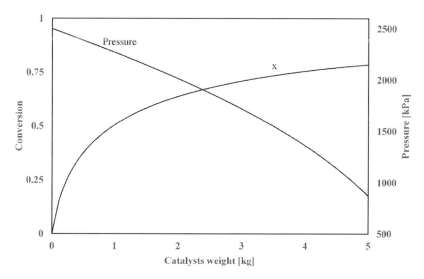

FIGURE 4.33 Pressure and conversion profiles.

4.17.2 Adiabatic

The second scenario to be solved is when the system is adiabatic. In this case, there is no exchange of heat with the outside of the reactor. Therefore, in this case, the only difference with the previous case will be the energy balance. However, we will include all the balances below, so that there is a clear overview of what is needed and how the script was developed.

Mole balance:

$$\frac{dF_A}{dW} = r_A \tag{4.138}$$

$$\frac{dF_B}{dW} = r_B \tag{4.139}$$

$$\frac{dF_C}{dW} = r_C \tag{4.140}$$

$$\frac{dF_D}{dW} = r_D \tag{4.141}$$

Reaction rate:

$$r_A = \left(\frac{k_1 k_3 k_5 C_A C_B}{k_2 k_4} - \frac{k_6 k_8 k_{10} C_C C_D}{k_7 k_9}\right) \left(\frac{C_S^0}{\left(1 + \frac{k_1 C_A}{k_2} + \frac{k_3 C_B}{k_4} + \frac{k_8 C_C}{k_7} + \frac{k_{10} C_D}{k_9}\right)}\right)^2 \quad (4.142)$$

$$\text{With } \frac{r_A}{-1} = \frac{r_B}{-1} = \frac{r_C}{1} = \frac{r_D}{1}$$

Pressure drop:

$$\frac{dP}{dW} = -\frac{1}{A\rho_c(1-\theta)}\frac{U}{\rho_0 d_p}\left(\frac{1-\theta}{\theta^3}\right)\left(\frac{150(1-\theta)\mu}{d_p} + 1.75U\right)\frac{T}{T_0}\frac{P_0}{P}\frac{F_T}{F_{T0}} \quad (4.143)$$

Besides the differential equations presented above, some other auxiliary equations were also required to solve this problem; those are:

$$v = v^0 \frac{F_T}{F_T^0}\frac{P^0}{P}\frac{T}{T^0} \quad (4.144)$$

$$C_i = \frac{F_i}{v} \quad (4.145)$$

$$F_T = F_A + F_B + F_C + F_D \quad (4.146)$$

$$x_A = \frac{F_A^0 - F_A}{F_A^0} \quad (4.147)$$

$$F_{T0} = F_A^0 + F_B^0 + F_C^0 + F_D^0 \quad (4.148)$$

$$k_i = k_{0i} e^{\left(\frac{-E_{ai}}{R*T}\right)} \quad (4.149)$$

In order to solve this problem, we need to have an energy balance to know the variation of the temperature as a function of the conversion.

Energy balance:
The complete energy balance can be seen in Equation 4.150, this expression gives, i.e., temperature as a function of conversion. This equation includes the terms needed when the heat of reaction is not considerably larger than the heat for heating and cooling the reactants and products.

Completely Solved Example

$$T = \frac{-xF_A^0 \Delta H_{RX}(T_R) + xF_A^0 \sum_{i=1}^{P} \frac{\upsilon_i}{|\upsilon_A|} C_{P_i} T_R + \sum_{i=1}^{P} F_i^0 C_{P_i} T_0}{\sum_{i=1}^{P} F_i^0 C_{P_i} + xF_A^0 \sum_{i=1}^{P} \frac{\upsilon_i}{|\upsilon_A|} C_{P_i}} \quad (4.150)$$

For this case, we will simplify the balance since in most cases, the heat of reactions is considerably larger than the term required to increase or decrease the reaction temperature to the reference temperature. Therefore, Equation 4.150 is reduced to:

$$T = \frac{-xF_A^0 \Delta H_{RX}(T_R) + \sum_{i=1}^{P} F_i^0 C_{P_i} T_0}{\sum_{i=1}^{P} F_i^0 C_{P_i}} \quad (4.151)$$

To solve the proposed problem we need to give some values to the constants involved; Table 4.5 will introduce all variables and their numerical value and units.

Now we can solve Equations 4.138 to 4.149 and 4.151 simultaneously in Polymath to obtain the different profiles for flows, pressure, conversion, and temperature. The script used is:

Script:

```
#Mole balance
    d(Fa)/d(W) = ra
    Fa(0) = 3
    d(Fb)/d(W) = rb
    Fb(0) = 5
    d(Fc)/d(W) = rc
    Fc(0) = 0.5
    d(Fd)/d(W) = rd
    Fd(0) = 0

#Energy balance
    T = (Fa0*Cpa*T0+Fb0*Cpb*T0+Fc0*Cpc*T0+Fd0*Cpd*T0-
    Fa0*x*DH)/(Fa0*Cpa+Fb0*Cpb+Fc0*Cpc+Fd0*Cpd)

#Pressure drop and its constants
```

$d(P)/d(W) = -1*(U/(dp*Rho0))*(((150*mu*(1-tita))/dp)+(1.75*U))*(1/(Ac*Rhoc*(1-tita)))*((1-tita)/((tita)^3))*((P0/P)*(FT/FT0)*(T/T0))$

$P(0) = 2,500$

#Reaction rates

$ra = -(((k1*k3*k5*Ca*Cb)/(k2*k4))-((k6*k8*k10*Cc*Cd)/(k7*k9)))*(((Cs0/(1+(k1*Ca/k2)+(k3*Cb/k4)+(k8*Cc/k7)+(k10*Cd/k9)))^2))$

$rb = ra$

$rc = -ra$

$rd = -ra$

#Auxiliary equations

$k1 = k001*EXP(-Ea1/(R*T))$

$k2 = k002*EXP(-Ea2/(R*T))$

$k3 = k003*EXP(-Ea3/(R*T))$

$k4 = k004*EXP(-Ea4/(R*T))$

$k5 = k005*EXP(-Ea5/(R*T))$

$k6 = k006*EXP(-Ea6/(R*T))$

$k7 = k007*EXP(-Ea7/(R*T))$

$k8 = k008*EXP(-Ea8/(R*T))$

$k9 = k009*EXP(-Ea9/(R*T))$

$k10 = k0010*EXP(-Ea10/(R*T))$

$FT = Fa+Fb+Fc+Fd$

$FT0 = Fa0+Fb0+Fc0+Fd0$

$x = (Fa0-Fa)/Fa0$

$Ca = Fa/v$

$Cb = Fb/v$

$Cc = Fc/v$

$Cd = Fd/v$

$v = v0*((P0/P)*(FT/FT0)*(T/T0))$

$Ac = 3.14159/4*D^2$

#Initial values and constants

$k001 = 95,000$

$Ea1 = 33,000$

$k002 = 12,000$

$Ea2 = 29,000$

$k003 = 11,000$

$Ea3 = 35,000$

Completely Solved Example 333

$\quad$ k004 = 9,500
$\quad$ Ea4 = 31,000
$\quad$ k005 = 95,000
$\quad$ Ea5 = 30,000
$\quad$ k006 = 1,800
$\quad$ Ea6 = 32,000
$\quad$ k007 = 10,000
$\quad$ Ea7 = 38,000
$\quad$ k008 = 2,000
$\quad$ Ea8 = 35,000
$\quad$ k009 = 12,000
$\quad$ Ea9 = 30,000
$\quad$ k0010 = 500
$\quad$ Ea10 = 18,000
$\quad$ R = 8.3143
$\quad$ v0 = 1
$\quad$ Fa0 = 3
$\quad$ Fb0 = 5
$\quad$ Fc0 = 0.5
$\quad$ Fd0 = 0
$\quad$ Cpa = 120
$\quad$ Cpb = 140
$\quad$ Cpc = 150
$\quad$ Cpd = 150
$\quad$ P0 = 2,500
$\quad$ Cs0 = 10
$\quad$ T0 = 300
$\quad$ DH = −7,200
$\quad$ tita = 0.6
$\quad$ D = 0.2
$\quad$ Rho0 = 1.3
$\quad$ Rhoc = 2,800
$\quad$ dp = 0.03
$\quad$ mu = 0.01
$\quad$ U = 5.5

#Independent variable
$\quad$ W(0) = 0
$\quad$ W(f) = 5

TABLE 4.5
Values Used for Constants

Variable	Value	Units
k_{100}	95,000	$l^2/(mol*kg*min)$
k_{200}	12,000	$l/(kg*min)$
k_{300}	11,000	$l^2/(mol*kg*min)$
k_{400}	9,500	$l/(kg*min)$
k_{500}	95,000	$l^2/(mol*kg*min)$
k_{600}	1,800	$l^2/(mol*kg*min)$
k_{700}	10,000	$l/(kg*min)$
k_{800}	2,000	$l^2/(mol*kg*min)$
k_{900}	12,000	$l/(kg*min)$
k_{1000}	500	$l^2/(mol*kg*min)$
Ea_1	33,000	J/mol
Ea_2	29,000	J/mol
Ea_3	35,000	J/mol
Ea_4	31,000	J/mol
Ea_5	30,000	J/mol
Ea_6	32,000	J/mol
Ea_7	38,000	J/mol
Ea_8	35,000	J/mol
Ea_9	30,000	J/mol
Ea_{10}	18,000	J/mol
F_A^0	3	mol/min
F_B^0	5	mol/min
F_C^0	0.5	mol/min
F_D^0	0	mol/min
Cp_A	120	$J/(mol*K)$
Cp_B	140	$J/(mol*K)$
Cp_C	150	$J/(mol*K)$
Cp_D	150	$J/(mol*K)$
ΔH	−7,200	J/mol
$T0$	300	K
R	8.3143	$J(mol*K)$
C_s^0	10	mol/l
$P0$	2,500	kPa
v_0	1	l/min
ρ_0	1.3	Kg/m^3
ρ_c	2,800	Kg/m^3
θ	0.6	-----
A	$\pi*D^2/4$	m^2
D	0.2	m
μ	0.01	$Kg/(m*s)$
d_p	0.03	m
U	5.5	$Kg/(m^2*s)$

Completely Solved Example 335

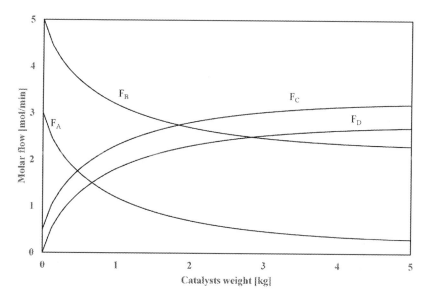

FIGURE 4.34 Molar flow profiles.

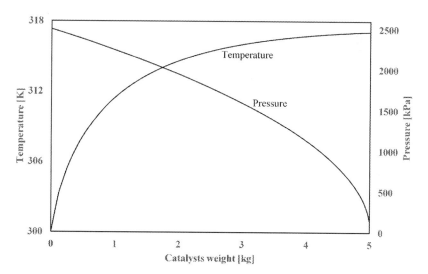

FIGURE 4.35 Pressure and temperature profiles.

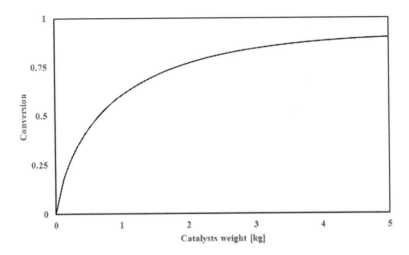

FIGURE 4.36 Conversion profile.

We can solve this problem in Polymath using the script above. The results shown in Figures 4.34 to 4.36 present the variations of the molar flow, pressure and temperature, and conversion, respectively. It can be seen that the pressure profile it has a higher slope closer to the end of the reactor; therefore, the pressure drops considerably more than in the previous case while the temperature increases around 18 K along the reactor. Conversion is improved 0.1% points in comparison with isothermal.

4.17.3 CONSTANT EXTERNAL TEMPERATURE

The third scenario we are going to solve is the energy balance considering a constant temperature in the outside flow. The balances that need to be solved are:

Mole balance:

$$\frac{dF_A}{dW} = r_A \qquad (4.152)$$

$$\frac{dF_B}{dW} = r_B \qquad (4.153)$$

$$\frac{dF_C}{dW} = r_C \qquad (4.154)$$

$$\frac{dF_D}{dW} = r_D \qquad (4.155)$$

Completely Solved Example

Reaction rate:

$$r_A = \left(\frac{k_1 k_3 k_5 C_A C_B}{k_2 k_4} - \frac{k_6 k_8 k_{10} C_C C_D}{k_7 k_9} \right) \left(\frac{C_S^0}{\left(1 + \frac{k_1 C_A}{k_2} + \frac{k_3 C_B}{k_4} + \frac{k_8 C_C}{k_7} + \frac{k_{10} C_D}{k_9}\right)} \right)^2 \quad (4.156)$$

$$\text{With } \frac{r_A}{-1} = \frac{r_B}{-1} = \frac{r_C}{1} = \frac{r_D}{1}$$

Pressure drop:

$$\frac{dP}{dW} = -\frac{1}{A\rho_c(1-\theta)} \frac{U}{\rho_0 d_p} \left(\frac{1-\theta}{\theta^3}\right)\left(\frac{150(1-\theta)\mu}{d_p} + 1.75U\right)\frac{T}{T_0}\frac{P_0}{P}\frac{F_T}{F_{T0}} \quad (4.157)$$

Besides the differential equations presented above, some other auxiliary equations are also required to solve this problem; those are:

$$v = v^0 \frac{F_T}{F_T^0} \frac{P^0}{P} \frac{T}{T^0} \quad (4.158)$$

$$C_i = \frac{F_i}{v} \quad (4.159)$$

$$F_T = F_A + F_B + F_C + F_D \quad (4.160)$$

$$x_A = \frac{F_A^0 - F_A}{F_A^0} \quad (4.161)$$

$$F_{T0} = F_A^0 + F_B^0 + F_C^0 + F_D^0 \quad (4.162)$$

$$k_i = k_{0i} e^{\left(\frac{-E_{ai}}{R*T}\right)} \quad (4.163)$$

In order to solve this problem, we need to have an energy balance to know the variation of the temperature as a function of the conversion.

Energy balance:
Since the external fluid temperature is constant, we then need to solve the energy balance following Equation 4.164:

$$\frac{dT}{dW} = \frac{(r_A \Delta H_{RX}) - (U\varpi(T - T_f))}{\sum F_i C p_i} \quad (4.164)$$

This expression, together with the previous one, will be solved using the data presented in Table 4.6.

We can now write Equations 4.152 to 4.164 into Polymath and solve them simultaneously. The script produce is as follows:

Script:

#Mole balance

d(Fa)/d(W) = ra
Fa(0) = 3
d(Fb)/d(W) = rb
Fb(0) = 5
d(Fc)/d(W) = rc
Fc(0) = 0.5
d(Fd)/d(W) = rd
Fd(0) = 0

#Energy balance

d(T)/d(W) = ((ra*DH)–(Ud*(T–Tf)))/((Fa*Cpa)+(Fb*Cpb)+(Fc*Cpc)+(Fd*Cpd))
T(0) = 300

#Pressure drop and its constants

d(P)/d(W) = –1*(U/(dp*Rho0))*(((150*mu*(1–tita))/dp)+(1.75*U))*(1/(Ac*Rhoc*(1–tita)))*((1–tita)/((tita)^3))*((P0/P)*(FT/FT0)*(T/T0))
P(0) = 2,500

#Reaction rates

ra = –(((k1*k3*k5*Ca*Cb)/(k2*k4))–((k6*k8*k10*Cc*Cd)/(k7*k9)))*(((Cs0/(1+(k1*Ca/k2)+(k3*Cb/k4)+(k8*Cc/k7)+(k10*Cd/k9)))^2))
rb = ra
rc = –ra
rd = –ra

#Auxiliary equations

k1 = k001*EXP(–Ea1/(R*T))
k2 = k002*EXP(–Ea2/(R*T))
k3 = k003*EXP(–Ea3/(R*T))
k4 = k004*EXP(–Ea4/(R*T))
k5 = k005*EXP(–Ea5/(R*T))
k6 = k006*EXP(–Ea6/(R*T))
k7 = k007*EXP(–Ea7/(R*T))

Completely Solved Example

k8 = k008*EXP(–Ea8/(R*T))
k9 = k009*EXP(–Ea9/(R*T))
k10 = k0010*EXP(–Ea10/(R*T))
FT = Fa+Fb+Fc+Fd
FT0 = Fa0+Fb0+Fc0+Fd0
x = (Fa0–Fa)/Fa0
Ca = Fa/v
Cb = Fb/v
Cc = Fc/v
Cd = Fd/v
v = v0*((P0/P)*(FT/FT0)*(T/T0))
Ac = 3.14159/4*D^2

#Initial values and constants
k001 = 95,000
Ea1 = 33,000
k002 = 12,000
Ea2 = 29,000
k003 = 11,000
Ea3 = 35,000
k004 = 9,500
Ea4 = 31,000
k005 = 95,000
Ea5 = 30,000
k006 = 1,800
Ea6 = 32,000
k007 = 10,000
Ea7 = 38,000
k008 = 2,000
Ea8 = 35,000
k009 = 12,000
Ea9 = 30,000
k0010 = 500
Ea10 = 18,000
R = 8.3143
v0 = 1
Cpa = 120
Cpb = 140

Cpc = 150
Cpd = 150
Fa0 = 3
Fb0 = 5
Fc0 = 0.5
Fd0 = 0
P0 = 2,500
Cs0 = 10
T0 = 300
DH = −7,200
tita = 0.6
D = 0.2
Rho0 = 1.3
Rhoc = 2,800
dp = 0.03
mu = 0.01
U = 5.5
Ud = 50
Tf = 260

#Independent variable
W(0) = 0
W(f) = 5

TABLE 4.6
Values Used for Constants

Variable	Value	Units
k_{100}	95,000	$l^2/(mol \cdot kg \cdot min)$
k_{200}	12,000	$l/(kg \cdot min)$
k_{300}	11,000	$l^2/(mol \cdot kg \cdot min)$
k_{400}	9,500	$l/(kg \cdot min)$
k_{500}	95,000	$l^2/(mol \cdot kg \cdot min)$
k_{600}	1,800	$l^2/(mol \cdot kg \cdot min)$
k_{700}	10,000	$l/(kg \cdot min)$
k_{800}	2,000	$l^2/(mol \cdot kg \cdot min)$
k_{900}	12,000	$l/(kg \cdot min)$
k_{1000}	500	$l^2/(mol \cdot kg \cdot min)$
Ea_1	33,000	J/mol
Ea_2	29,000	J/mol
Ea_3	35,000	J/mol
Ea_4	31,000	J/mol
Ea_5	30,000	J/mol

Completely Solved Example

Variable	Value	Units
Ea_6	32,000	J/mol
Ea_7	38,000	J/mol
Ea_8	35,000	J/mol
Ea_9	30,000	J/mol
Ea_{10}	18,000	J/mol
F_A^0	3	mol/min
F_B^0	5	mol/min
F_C^0	0.5	mol/min
F_D^0	0	mol/min
Cp_A	120	J/(mol*K)
Cp_B	140	J/(mol*K)
Cp_C	150	J/(mol*K)
Cp_D	150	J/(mol*K)
ΔH	−7,200	J/mol
$T0$	300	K
R	8.3143	J(mol*K)
C_s^0	10	mol/l
$P0$	2,500	kPa
v_0	1	l/min
ρ_0	1.3	Kg/m³
ρ_c	2,800	Kg/m³
θ	0.6	-----
A	$\pi*D^2/4$	m²
D	0.2	m
μ	0.01	Kg/(m*s)
d_p	0.03	m
U	5.5	Kg/(m²*s)
$U\varpi$	50	J/(min*dm³*K)
T_f	260	K

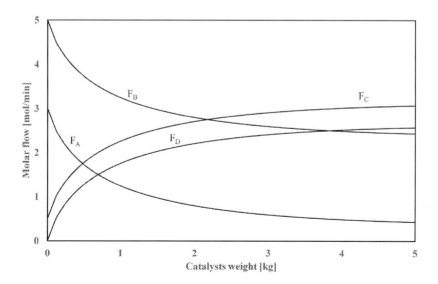

FIGURE 4.37 Molar flow profiles.

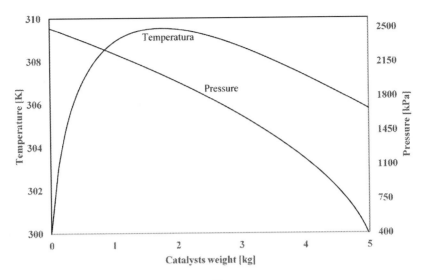

FIGURE 4.38 Pressure and temperature profiles.

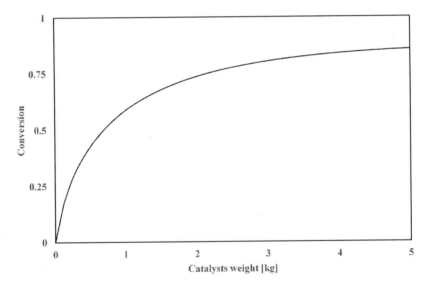

FIGURE 4.39 Conversion profile.

Figure 4.37 shows the variation of the molar flows of the chemicals involved in the reaction. These flows are similar to those presented for the adiabatic as well as the isothermal case. However, Figure 4.38 shows the variation of pressure and temperature. The profile of temperature shows a maximum value, this is due to the constant outside temperature that is constantly removing the heat being produced by the exothermic reaction. The pressure profile has a larger decay than the isothermal case but to a lesser extent than the adiabatic scenario. Figure 4.39 shows the variation of conversion that is slightly larger than the isothermal but lower than the adiabatic.

4.17.4 CO-CURRENT EXTERNAL FLOW

The fourth scenario that we are going to look into has an external fluid flowing co currently to the reactions flow. The balances that need to be solved are:

Mole balance:

$$\frac{dF_A}{dW} = r_A \tag{4.165}$$

$$\frac{dF_B}{dW} = r_B \tag{4.166}$$

$$\frac{dF_C}{dW} = r_C \tag{4.167}$$

$$\frac{dF_D}{dW} = r_D \tag{4.168}$$

Reaction rate:

$$r_A = \left(\frac{k_1 k_3 k_5 C_A C_B}{k_2 k_4} - \frac{k_6 k_8 k_{10} C_C C_D}{k_7 k_9}\right) \left(\frac{C_S^0}{\left(1 + \frac{k_1 C_A}{k_2} + \frac{k_3 C_B}{k_4} + \frac{k_8 C_C}{k_7} + \frac{k_{10} C_D}{k_9}\right)}\right)^2 \tag{4.169}$$

With $\dfrac{r_A}{-1} = \dfrac{r_B}{-1} = \dfrac{r_C}{1} = \dfrac{r_D}{1}$

Pressure drop:

$$\frac{dP}{dW} = -\frac{1}{A\rho_c(1-\theta)} \frac{U}{\rho_0 d_p} \left(\frac{1-\theta}{\theta^3}\right) \left(\frac{150(1-\theta)\mu}{d_p} + 1.75U\right) \frac{T}{T_0} \frac{P_0}{P} \frac{F_T}{F_{T0}} \tag{4.170}$$

Besides the differential equations presented above, some other auxiliary equations were also required to solve this problem:

$$v = v^0 \frac{F_T}{F_T^0} \frac{P^0}{P} \frac{T}{T^0} \tag{4.171}$$

$$C_i = \frac{F_i}{v} \tag{4.172}$$

$$F_T = F_A + F_B + F_C + F_D \tag{4.173}$$

$$x_A = \frac{F_A^0 - F_A}{F_A^0} \tag{4.174}$$

$$F_{T0} = F_A^0 + F_B^0 + F_C^0 + F_D^0 \tag{4.175}$$

$$k_i = k_{0i} e^{\left(\frac{-E_{ai}}{R*T}\right)} \tag{4.176}$$

In order to solve this problem, we need to have an energy balance to know the variation of the temperature as a function of the conversion.

Energy balance:
We then need to solve the energy balance that is as follows:

$$\frac{dT}{dW} = \frac{(r_A \Delta H_{RX}) - U\varpi(T - T_f)}{\Sigma F_i C p_i} \tag{4.177}$$

External energy balance:
In this case, we have a fluid that is on the outside of the reactor and it is either cooling or heating the reaction itself. Our reaction is exothermic; therefore, this is a cooling liquid. Since the external temperature changes constantly, we need to approach this with a differential equation, as presented in Equation 4.178, where the initial boundary condition is $T_f = 260$ [K].

$$\frac{dT_f}{dW} = \frac{U\varpi(T - T_f)}{\dot{m}_f C p_f} \tag{4.178}$$

In order to solve this problem, we will solve Equations 4.165 to 4.178 simultaneously. This is done in Polymath using the constants presented in Table 4.7.
We can now put them into Polymath to generate the desired script.

Script:
  #Mole balance
  d(Fa)/d(W) = ra
  Fa(0) = 3
  d(Fb)/d(W) = rb
  Fb(0) = 5
  d(Fc)/d(W) = rc
  Fc(0) = 0.5
  d(Fd)/d(W) = rd
  Fd(0) = 0

  #Energy balance
  d(T)/d(W) = ((ra*DH)−(Ud*(T−Tf)))/((Fa*Cpa)+(Fb*Cpb)+(Fc*Cpc)+(Fd*Cpd))
  T(0) = 300

Completely Solved Example   345

#External energy balance
d(Tf)/d(W) = (Ud*(T−Tf))/(mf*Cpf)
Tf(0) = 260

#Pressure drop and its constants
d(P)/d(W) = −1*(U/(dp*Rho0))*(((150*mu*(1−tita))/dp)+(1.75*U))* (1/(Ac*Rhoc*(1−tita)))*((1−tita)/((tita)^3))*((P0/P)*(FT/FT0)*(T/T0))
P(0) = 2,500

#Reaction rates
ra = −(((k1*k3*k5*Ca*Cb)/(k2*k4))−((k6*k8*k10*Cc*Cd)/(k7*k9)))*(((Cs0/(1+(k1*Ca/k2)+(k3*Cb/k4)+(k8*Cc/k7)+(k10*Cd/k9)))^2))
rb = ra
rc = −ra
rd = −ra

#Auxiliary equations
k1 = k001*EXP(−Ea1/(R*T))
k2 = k002*EXP(−Ea2/(R*T))
k3 = k003*EXP(−Ea3/(R*T))
k4 = k004*EXP(−Ea4/(R*T))
k5 = k005*EXP(−Ea5/(R*T))
k6 = k006*EXP(−Ea6/(R*T))
k7 = k007*EXP(−Ea7/(R*T))
k8 = k008*EXP(−Ea8/(R*T))
k9 = k009*EXP(−Ea9/(R*T))
k10 = k0010*EXP(−Ea10/(R*T))
FT = Fa+Fb+Fc+Fd
FT0 = Fa0+Fb0+Fc0+Fd0
x = (Fa0−Fa)/Fa0
Ca = Fa/v
Cb = Fb/v
Cc = Fc/v
Cd = Fd/v
v = v0*((P0/P)*(FT/FT0)*(T/T0))
Ac = 3.14159/4*D^2

#Initial values and constants

k001 = 95,000
Ea1 = 33,000
k002 = 12,000
Ea2 = 29,000
k003 = 11,000
Ea3 = 35,000
k004 = 9,500
Ea4 = 31,000
k005 = 95,000
Ea5 = 30,000
k006 = 1,800
Ea6 = 32,000
k007 = 10,000
Ea7 = 38,000
k008 = 2,000
Ea8 = 35,000
k009 = 12,000
Ea9 = 30,000
k0010 = 500
Ea10 = 18,000
R = 8.3143
v0 = 1
Cpa = 120
Cpb = 140
Cpc = 150
Cpd = 150
Fa0 = 3
Fb0 = 5
Fc0 = 0.5
Fd0 = 0
P0 = 2,500
Cs0 = 10
T0 = 300
DH = −7,200
tita = 0.6
D = 0.2
Rho0 = 1.3

Completely Solved Example

Rhoc = 2,800
dp = 0.03
mu = 0.01
U = 5.5
Ud = 50
mf = 1
Cpf = 150

#Independent variable
W(0) = 0
W(f) = 5

TABLE 4.7
Values Used for Constants

Variable	Value	Units
k_{100}	95.000	$l^2/(mol*kg*min)$
k_{200}	12,000	$l/(kg*min)$
k_{300}	11,000	$l^2/(mol*kg*min)$
k_{400}	9,500	$l/(kg*min)$
k_{500}	95,000	$l^2/(mol*kg*min)$
k_{600}	1,800	$l^2/(mol*kg*min)$
k_{700}	10,000	$l/(kg*min)$
k_{800}	2,000	$l^2/(mol*kg*min)$
k_{900}	12,000	$l/(kg*min)$
k_{1000}	500	$l^2/(mol*kg*min)$
Ea_1	33,000	J/mol
Ea_2	29,000	J/mol
Ea_3	35,000	J/mol
Ea_4	31,000	J/mol
Ea_5	30,000	J/mol
Ea_6	32,000	J/mol
Ea_7	38,000	J/mol
Ea_8	35,000	J/mol
Ea_9	30,000	J/mol
Ea_{10}	18,000	J/mol
F_A^0	3	mol/min
F_B^0	5	mol/min
F_C^0	0.5	mol/min
F_D^0	0	mol/min
Cp_A	120	$J/(mol*K)$
Cp_B	140	$J/(mol*K)$
Cp_C	150	$J/(mol*K)$
Cp_D	150	$J/(mol*K)$
ΔH	−7,200	J/mol
$T0$	300	K
R	8.3143	$J(mol*K)$

Variable	Value	Units
C_s^0	10	mol/l
$P0$	2,500	kPa
v_0	1	l/min
ρ_0	1.3	Kg/m³
ρ_c	2,800	Kg/m³
θ	0.6	-----
A	$\pi*D^2/4$	m²
D	0.2	m
μ	0.01	Kg/(m*s)
d_p	0.03	m
U	5.5	Kg/(m²*s)
$U\varpi$	50	J/(min*dm³*K)
T_f^0	260	K
m_f	1	mol/min
Cp_f	150	J/(mol*K)

The solution of this script in Polymath allows us to obtain the flow variations for each chemical as well as the change in pressure, reaction temperature, outside temperature, and conversion. Figure 4.40 shows the molar flow profiles for all chemicals; this is quite similar to the previous ones. Figure 4.41 presents the variation of pressure and conversion; here, the pressure drop is slightly higher than that for a constant ambient temperature, but not as much as for the adiabatic scenario. For the conversion, a similar tendency is shown where the final conversion is higher than then constant outside temperature scenario but lower than the adiabatic case. Figure 4.42 presents the temperature profile, which shows that the reaction temperature is higher than the previous scenario and its values does not decrease as much; this is due to the fact that the external fluid's temperature increases its value.

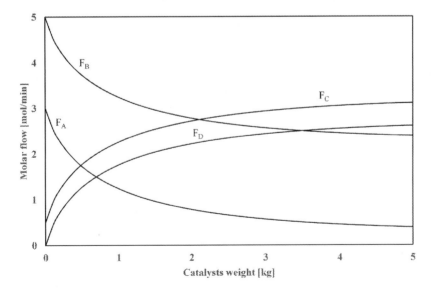

FIGURE 4.40 Molar flow profiles.

Completely Solved Example

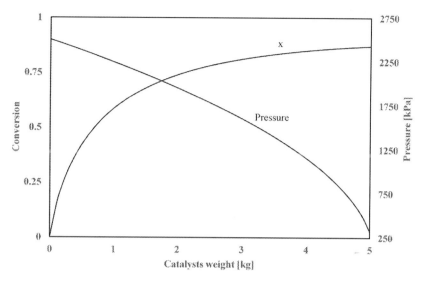

FIGURE 4.41 Pressure and conversion profiles.

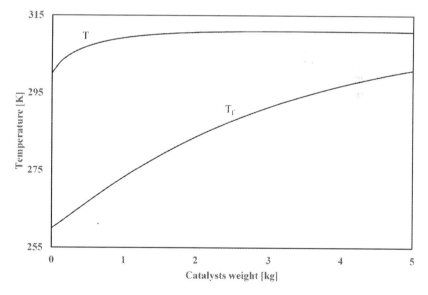

FIGURE 4.42 Temperature profiles.

4.17.5 Counter-Current External Flow

Finally, this problem will be solved with the external cooling/heating fluid going in counter-current flow. As mentioned before, this means that we have a final condition instead of initial boundary conditions and therefore, we need to apply the shooting methodology. The equations that we need to solve simultaneously are:

Mole balance:

$$\frac{dF_A}{dW} = r_A \tag{4.179}$$

$$\frac{dF_B}{dW} = r_B \tag{4.180}$$

$$\frac{dF_C}{dW} = r_C \tag{4.181}$$

$$\frac{dF_D}{dW} = r_D \tag{4.182}$$

Reaction rate:

$$r_A = \left(\frac{k_1 k_3 k_5 C_A C_B}{k_2 k_4} - \frac{k_6 k_8 k_{10} C_C C_D}{k_7 k_9}\right) \left(\frac{C_S^0}{\left(1 + \frac{k_1 C_A}{k_2} + \frac{k_3 C_B}{k_4} + \frac{k_8 C_C}{k_7} + \frac{k_{10} C_D}{k_9}\right)}\right)^2 \tag{4.183}$$

With $\dfrac{r_A}{-1} = \dfrac{r_B}{-1} = \dfrac{r_C}{1} = \dfrac{r_D}{1}$

Pressure drop:

$$\frac{dP}{dW} = -\frac{1}{A\rho_c(1-\theta)}\frac{U}{\rho_0 d_p}\left(\frac{1-\theta}{\theta^3}\right)\left(\frac{150(1-\theta)\mu}{d_p} + 1.75U\right)\frac{T}{T_0}\frac{P_0}{P}\frac{F_T}{F_{T0}} \tag{4.184}$$

Besides the differential equations presented above, some other auxiliary equations were also required to solve this problem; those are:

$$v = v^0 \frac{F_T}{F_T^0} \frac{P^0}{P} \frac{T}{T^0} \tag{4.185}$$

$$C_i = \frac{F_i}{v} \tag{4.186}$$

$$F_T = F_A + F_B + F_C + F_D \tag{4.187}$$

$$X_A = \frac{F_A^0 - F_A}{F_A^0} \tag{4.188}$$

Completely Solved Example

$$F_{T0} = F_A^0 + F_B^0 + F_C^0 + F_D^0 \tag{4.189}$$

$$k_i = k_{0i} e^{\left(\frac{-E_{ai}}{R*T}\right)} \tag{4.190}$$

In order to solve this problem, we need to have an energy balance to know the variation of the temperature as a function of the conversion.

Energy balance:
The energy balance, following Equation 4.164, is:

$$\frac{dT}{dW} = \frac{(r_A \Delta H_{RX}) - U\varpi(T - T_f)}{\Sigma F_i Cp_i} \tag{4.191}$$

External energy balance:
As in the previous case, we have a cooling fluid on the outside of the reactor which reduces the reaction temperature. In this case, the fluid has a counter-current flow with respect to the reaction mixture. Therefore, the only difference with Equation 4.178 is the presence of a minus sign. Furthermore, the value $T_f = 260$ [K] is still valid; however, for this reaction, this boundary condition is given when $W = 5$ [kg], *i.e.*, at the end of the process instead of at the beginning. Therefore, we will implement the shooting methodology to calculate the value of T_f when $W = 0$.

$$\frac{dT_f}{dW} = \frac{-U\varpi(T - T_f)}{\dot{m}_f Cp_f} \tag{4.192}$$

In order to solve this problem, we will solve Equations 4.179 to 4.192 simultaneously. This is done in Polymath using the constants presented in Table 4.8.

We can now put them into Polymath to generate the desired script.

Script:

#Mole balance
    d(Fa)/d(W) = ra
    Fa(0) = 3
    d(Fb)/d(W) = rb
    Fb(0) = 5
    d(Fc)/d(W) = rc
    Fc(0) = 0.5
    d(Fd)/d(W) = rd
    Fd(0) = 0

#Energy balance
$$d(T)/d(W) = ((ra*DH)-(Ud*(T-Tf)))/((Fa*Cpa)+(Fb*Cpb)+(Fc*Cpc)+(Fd*Cpd))$$
$$T(0) = 300$$

#External energy balance
$$d(Tf)/d(W) = -(Ud*(T-Tf))/(mf*Cpf)$$
$$Tf(0) = 301.015$$

#Pressure drop and its constants
$$d(P)/d(W) = -1*(U/(dp*Rho0))*(((150*mu*(1-tita))/dp)+(1.75*U))*(1/(Ac*Rhoc*(1-tita)))*((1-tita)/((tita)^3))*((P0/P)*(FT/FT0)*(T/T0))$$
$$P(0) = 2,500$$

#Reaction rates
$$ra = -(((k1*k3*k5*Ca*Cb)/(k2*k4))-((k6*k8*k10*Cc*Cd)/(k7*k9)))*((((Cs0/(1+(k1*Ca/k2)+(k3*Cb/k4)+(k8*Cc/k7)+(k10*Cd/k9)))^2))$$
$$rb = ra$$
$$rc = -ra$$
$$rd = -ra$$

#Auxiliary equations
$$k1 = k001*EXP(-Ea1/(R*T))$$
$$k2 = k002*EXP(-Ea2/(R*T))$$
$$k3 = k003*EXP(-Ea3/(R*T))$$
$$k4 = k004*EXP(-Ea4/(R*T))$$
$$k5 = k005*EXP(-Ea5/(R*T))$$
$$k6 = k006*EXP(-Ea6/(R*T))$$
$$k7 = k007*EXP(-Ea7/(R*T))$$
$$k8 = k008*EXP(-Ea8/(R*T))$$
$$k9 = k009*EXP(-Ea9/(R*T))$$
$$k10 = k0010*EXP(-Ea10/(R*T))$$
$$FT = Fa+Fb+Fc+Fd$$
$$FT0 = Fa0+Fb0+Fc0+Fd0$$
$$x = (Fa0-Fa)/Fa0$$
$$Ca = Fa/v$$
$$Cb = Fb/v$$
$$Cc = Fc/v$$

Completely Solved Example

$$Cd = Fd/v$$
$$v = v0*((P0/P)*(FT/FT0)*(T/T0))$$
$$Ac = 3.14159/4*D^2$$

#Initial values and constants
k001 = 95,000
Ea1 = 33,000
k002 = 12,000
Ea2 = 29,000
k003 = 11,000
Ea3 = 35,000
k004 = 9,500
Ea4 = 31,000
k005 = 95,000
Ea5 = 30,000
k006 = 1,800
Ea6 = 32,000
k007 = 10,000
Ea7 = 38,000
k008 = 2,000
Ea8 = 35,000
k009 = 12,000
Ea9 = 30,000
k0010 = 500
Ea10 = 18,000
R = 8.3143
v0 = 1
Cpa = 120
Cpb = 140
Cpc = 150
Cpd = 150
Fa0 = 3
Fb0 = 5
Fc0 = 0.5
Fd0 = 0
P0 = 2,500
Cs0 = 10
T0 = 300

DH = −7,200
tita = 0.6
D = 0.2
Rho0 = 1.3
Rhoc = 2,800
dp = 0.03
mu = 0.01
U = 5.5
Ud = 50
mf = 1
Cpf = 150

#Independent variable
W(0) = 0
W(f) = 5

TABLE 4.8
Values Used for Constants

Variable	Value	Units
k_{100}	95,000	$l^2/(mol*kg*min)$
k_{200}	12,000	$l/(kg*min)$
k_{300}	11,000	$l^2/(mol*kg*min)$
k_{400}	9,500	$l/(kg*min)$
k_{500}	95,000	$l^2/(mol*kg*min)$
k_{600}	1,800	$l^2/(mol*kg*min)$
k_{700}	10,000	$l/(kg*min)$
k_{800}	2,000	$l^2/(mol*kg*min)$
k_{900}	12,000	$l/(kg*min)$
k_{1000}	500	$l^2/(mol*kg*min)$
Ea_1	33,000	J/mol
Ea_2	29,000	J/mol
Ea_3	35,000	J/mol
Ea_4	31,000	J/mol
Ea_5	30,000	J/mol
Ea_6	32,000	J/mol
Ea_7	38,000	J/mol
Ea_8	35,000	J/mol
Ea_9	30,000	J/mol
Ea_{10}	18,000	J/mol
F_A^0	3	mol/min
F_B^0	5	mol/min
F_C^0	0.5	mol/min
F_D^0	0	mol/min
Cp_A	120	$J/(mol*K)$
Cp_B	140	$J/(mol*K)$
Cp_C	150	$J/(mol*K)$

Completely Solved Example

Variable	Value	Units
Cp_D	150	J/(mol*K)
ΔH	−7,200	J/mol
$T0$	300	K
R	8.3143	J(mol*K)
C_s^0	10	mol/l
$P0$	2,500	kPa
v_0	1	l/min
ρ_0	1.3	Kg/m³
ρ_c	2,800	Kg/m³
θ	0.6	-----
A	$\pi*D^2/4$	m²
D	0.2	m
μ	0.01	Kg/(m*s)
d_p	0.03	m
U	5.5	Kg/(m²*s)
$U\varpi$	50	J/(min*dm³*K)
T_f^0	301.015	K
m_f	1	mol/min
Cp_f	150	J/(mol*K)

By using Polymath, we can solve Equations 4.179 to 4.192 simultaneously; this will allow us to obtain the molar flow profiles as well as pressure, conversion, and temperature profiles. Figure 4.43 shows the molar flow profiles; these profiles show very similar tendencies as the previous ones. Figure 4.44 presents the variations for pressure as well as conversion. In comparison with the co-current flow, it can be seen that the pressure drop is higher for the counter-current flow.

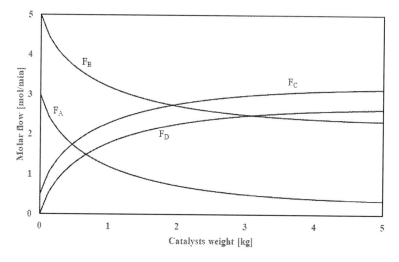

FIGURE 4.43 Molar flow profiles.

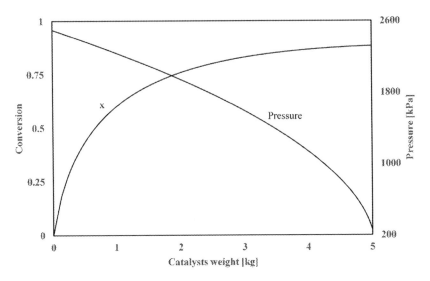

FIGURE 4.44 Pressure and conversion profiles.

Figure 4.45 presents the temperature profiles which shows that the reaction temperature reaches its highest value close to the highest for the co-current scenario. However, after this maximum, the reaction temperature drops sharply, unlike the co-current scenario.

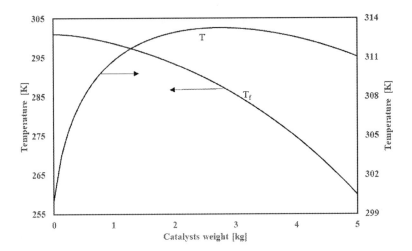

FIGURE 4.45 Temperature profiles.

4.17.6 Comparison of Previous Cases

In this section, we will try to compare the results from all the previous cases together to identify the differences, benefits, and disadvantages, of each scenario. We will

Completely Solved Example

focus on the conversion of chemical A, the pressure drop of the system, as well as the reaction temperature, since those are the main three key elements.

Figure 4.46 shows the variation of conversion as a function of the amount of catalysts present in the reactor.

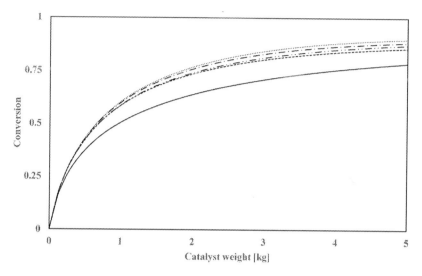

FIGURE 4.46 Conversion profiles. (———) isothermal, (···) adiabatic, (----), constant temperature on external fluid (-··-) co-current flow, and (-····-) counter-current flow.

It can be seen that the isothermal scenario gives the lower conversion, *i.e.*, almost 10% points lower than the adiabatic case. Among the possibilities, the adiabatic scenario is the best one in terms of conversion of chemical A, followed by the counter-current flow systems. Following the results in Figure 4.46, from best to worst, we have adiabatic, counter-current flow, co-current flow, constant fluid temperature, and isothermal.

However, we are only looking at the conversion profile, and in order to know what technology is most suitable for our case, we should look at more than just one variable. Therefore, we will also study the reaction temperature as shown in Figure 4.47, which presents the variation of the reaction temperature with the catalyst amount.

Figure 4.47 clearly shows that for the adiabatic case, the reaction temperature continuously increases, reaching the highest value at the end of the process and the highest of all study cases. It can be said that an increase of almost 18 [K] (5.6%) is not a large gap. However, the extent of temperature increase is not crucial but if that increase has any effect on the reaction, higher temperature could lead to deactivation of the catalyst, cracking of the reactants or products, or unwanted side reactions. Unwanted side reactions will still consume reactant A; therefore, the conversion plot might not be affected, but the yield of our process and eventually, the revenue of the technology might be severely affected by this. Beside the adiabatic scenario, counter-current flow and constant external temperature present a maximum value of temperature in their tendencies because the energy being removed is higher than the energy

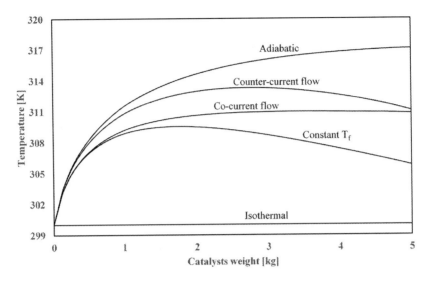

FIGURE 4.47 Temperature profiles.

produced after some time. This information can be useful if the reactor material or catalyst are highly sensitive to temperature, as the temperature could kill the catalyst, decrease the selectivity, or deactivate the catalysts by sintering.

Figure 4.48 shows the variation of pressure drop for all five cases, there is not much difference among all of them except the isothermal case, where the pressure drops slower and to a higher value. Since there is not much difference in these cases,

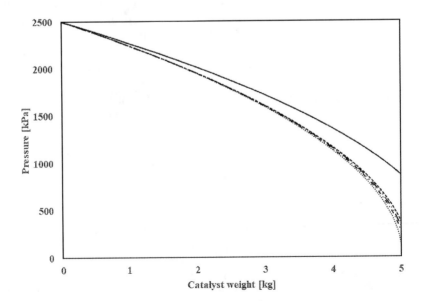

FIGURE 4.48 Pressure drop profiles.

Completely Solved Example

it is clear that the pressure has no effects on the conversion of the different reaction configurations. However, this problem was solved with $\theta = 0.6$. In order to see if the porosity of the catalyst has any effect on the pressure, temperature, and conversion profiles, we analyzed the scenario when the external fluids is at constant temperature ($T_f = 260$ [K]) for θ values of 0.6, 0.7, 0.8, 0.9, and 0.99. The final pressure, temperature, and conversion as a function of θ are presented in Table 4.9.

TABLE 4.9
Effect of θ on Conversion and Pressure

θ	Conversion	Pressure [kPa]	Temperature [K]
0.6	0.856	372.14	305.77
0.7	0.8622	1,746.55	305.88
0.8	0.8636	2,131.125	305.9
0.9	0.8642	2,314.17	305.9101
0.99	0.8644	2,405.18	305.9144

Evidently, the conversion and temperature do not show any mayor difference in their final value; however, pressure does. Figure 4.49 presents the variation of pressure along the reactor for the studied values of porosity. While the porosity increases toward 1, the system behaves more and more as a system with no particles in it, with an exit pressure closer and closer to the inlet pressure. When porosity is closer to 1, the behavior moves toward a plug flow with particles that have no mass, *i.e.*, so the tube is empty. This is an impossible physical scenario and therefore, cannot be simulated. Similarly, a porosity of zero corresponds to a completely plugged reactor and there is no flow in it.

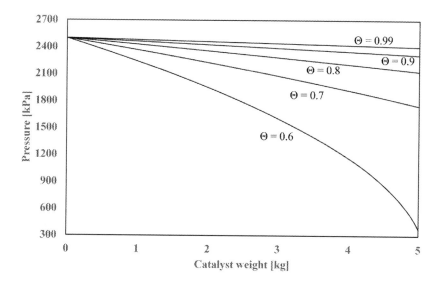

FIGURE 4.49 Pressure drop profiles.

REFERENCES

1. Sanchez, M., Marchetti, J.M., El Boulifi, N., Aracil, J., Martinez, M. "Kinetics of jojoba oil methanolysis using a waste from fish industry as catalyst". *Chemical Engineering Journal*. 262. (2015). 640–647.
2. Avhad, M.R., Sánchez, M., Peña, E., Bouaid, A., Martínez, M., Aracil, J., Marchetti, J.M. "Renewable production of value-added jojobyl alcohols and biodiesel using a naturally derived heterogeneous green catalyst". *Fuel*. 179. (2016). 332–338.
3. Sánchez, M., Avhad, M.R., Marchetti, J.M., Martínez, M., Aracil, J. "Enhancement of the jojobyl alcohols and biodiesel production using a renewable catalyst in a pressurized reactor". *Energy Conversion and Management*. 126. (2016). 1047–1053.
4. Avhad, M.R., Gangurde, L.S., Sánchez, M., Bouaid, A., Aracil, J., Martinez, M., Marchetti, J.M. "Enhancing biodiesel production using green glycerol enriched calcium oxide catalyst: an optimization study". *Catalysis Letters*. 148(4). (2018). 1169–1180.
5. Avhad, M.R., Sánchez, M., Bouaid, A., Martinez, M., Aracil, J., Marchetti, J.M. "Modeling chemical kinetics of avocado oil ethanolysis catalyzed by solid glycerol-enriched calcium oxide". *Energy Conversion and Management*. 126. (2016). 1168–1177.
6. Marchetti, J.M., Pedernera, M.N., Schbib, N.S. "Production of biodiesel from acid oil using sulphuric acid as catalyst: kinetics study". *International Journal of Low Carbon Technologies*. 6(1). (2011). 38–43.
7. Jasen, P., Marchetti, J.M. "Kinetic study of the esterification of free fatty acid and ethanol in the presence of triglycerides using solid resins as catalyst". *International Journal of Low Carbon Technologies*. 7. (2012). 325–330.

Index

A

activation energies, 1, 66
active phase, 5
    immobilization, 16
    impregnation, 12
adiabatic reactor, 122
adsorption, 18
ammonia production, 35, 86
Arrhenius equation, 88

B

backward reaction, reaction rate constant, 113
batch reactor (BR), 45, 57, 75, 153
    conversion profile, 75
    laboratory equipment, 260
    mole balance, 45
    scheme, 44, 46
biocatalysts, 6
biodiesel production, 37
BR, *see* batch reactor

C

catalysis, 1
catalyst characterization techniques, 19
catalyst deactivation, 225
    comparison, 228
catalyst immobilization, 16
catalysts, 1–2
    ammonia production, 35
    biodiesel production, 37
    density, 240
    Fischer-Tropsch process, 37
    homogeneous, 2, 16
        industry, 33
    site, 186
catalyst, surface area, 22
catalytic materials, 2, 8, 19
    Fischer-Tropsch, 36
chemical vapor deposition (CVD), 14
    scheme, 14
complex reactions, 49
    deactivation, 225
    generic method kinetics, 194
concentration profile
    adiabatic semi-batch, 158
    catalyst deactivation, 228
    comparison mass transfer limitations, 230
    internal mass transfer limitations, 231
    reaction in series, 102–103
    transesterification, 104
continuous reactor, *see* CSTR/PFR
continuous stirred tank reactor, 43, 57–58
    catalysts deactivation, 228
    energy balance, 147
    mole balance, 60
    scheme, 44, 58
        with heat exchanger, 147
        parallel reactors, 82
    volume calculation Levenspiel plot, 78
conversion, 73, 116
    adiabatic, 122
    concentration based, 73
    constant external fluids temperature, 137
    experimental data, 266
    external fluid, co-current flow, 142
conversion profile, 75
    comparison, 357
    first order reaction, 75
coprecipitation, 10, 12
covalent bonding, 17
    scheme, 17
CSTR, *see* Continuous stirred tank reactor

D

deactivation, 225
deactivation reaction rate constant, 226
desorption, 186, 203
    controlling, 206
differential method, 190

E

effectiveness factor, 232, 235, 238
    cylindrical pore, 234
    internal, 239
    spherical catalyst, 238
    Thiele modulus, 234
        first order reaction, 234
elementary reactions, 185–186
    activation energy scheme, 187
endothermic reaction, 134
energy, activation, 1
energy balance, 319
    adiabatic, 124, 131–132
    batch reactor, 153
    CSTR, 147
    multiple reactions
        batch reactor, 169
        constant external
            temperature, 162
        PFR, 159
    open system, 118

PFR
    constant external temperature, 137
    external fluid co-current, 141–142
    semi batch, 155
        adiabatic, 156
entrapment, 17
equilibrium, 114, 117, 165
equilibrium constant, 114
    concentration-based, 115–116
equilibrium conversion, 113
    temperature, 115, 131
equilibrium method, 196, 202
equilibrium reactions, 113
exothermic reaction, 115, 126
    equilibrium conversion, 127
experimental data, 188
    best fit, 297–298
    kinetics, 185
external energy balance
    co-current, 344
    counter-current, 351

F

FAEE, *see* fatty acid ethyl esters
fatty acid ethyl esters (FAEE), 103
FBR, *see* fluidized bed reactor
first-order elementary reaction, 76
first-order irreversible reaction, 47–48
    cylindrical porous, 236
    effectiveness factor, 234
    spherical porous, 236
fluidized bed reactor (FBR), 43, 57, 70
    scheme, 45

G

Gas chromatography (GC), 263
gas phase reaction, 82, 86, 89–91, 161, 178, 323
    non-elementary, 186
    pressure drop, 94
GC, *see* Gas chromatography

H

heat of reaction, 137
heterogeneous catalysis, 2–3
heterogeneous catalysts, 2
    industrial processes, 33
heterogeneous reactions, 186
    steps, 229
hydrogen, 34

I

impregnation, 12–13
    scheme, 12

infrared spectroscopy (IR), 25, 264
internal mass transfer limitations, 231
    Weisz-Prater criteria, 240
ion exchange, 15–16

K

kinetic expression, 185
kinetic modeling
    complex reaction, 209
    simple reaction, 194

L

Levenspiel plot, 77
    comparison CSTR vs. PFR, 78
limiting reactant, 67, 73

M

mass transfer coefficient, 239, 241
    overall, 245
mass transfer limitations, 228
    external, 238
membrane, 71
membrane reactor (MR), 43, 45, 57, 71
    scheme, 45, 71, 91
model selection criteria, 268
MR, *see* membrane reactor
MSC, *see* model selection criteria
multiple reactions, 100–101, 188
    semibatch, 166
Mussel shells, 26–27, 264–265

N

nanocatalysts, 2, 4
nanoscale catalytic materials, 4
non-elementary reactions, 186–187
    energy pathway representation, 187
    reversible, 202

O

overall effectiveness factor, 238

P

packed bed reactor (PBR), 43, 57, 70
    adiabatic, 134
    constant outside temperature, 134
    Ergun equation, 94
    mole balance, 71, 242
    pressure equation, 96
    scheme, 44, 70
parallel reactions, 104
PBR, *see* packed bed reactor

Index

PFR, *see* plug flow reactor
photocatalysis, 8–9
plug flow reactor (PFR), 43, 57, 62, 64–65, 137
   Levenspiel plot, 78
   mole balance, 71
   multiple reactions, 159
   scheme, 44, 62
porosity, 4, 95
porous materials, 4–5
precipitation, 10
pressure drop, 94
production of synthesis gas, 34
PSSH methodology, 199

R

Raman spectroscopy (RS), 19, 29
reaction kinetics, 1, 185, 188, 194
reaction order and reaction constant, interpolation, 193
reaction rate constant, 47, 188, 239, 241
reaction step, 110
   multiple reactions, 188
reactive system, scheme, 229
reactor volume, space time definition, 59
reproducibility, 262

S

selective oxidation, 36
selectivity, 101
semi-batch reactor (SMBR), 49, 154
   multiple reactions, 166
   output case, 57
series reactions, 101
space time, 59
steps, photocatalytic reaction, 9
support, catalysts, 6
surface area, 21–22, 239, 264
   catalyst deactivation, 225
sweep gas, 71

T

temperature profile
   comparison, 321–322
      gas phase, 358
   cooling/heating fluid flow effect, 318
   CSTR, 58
   energy balance, 137
   PBR, 70
   PFR, 62
temperature programed reduction (TPR), 26, 28
temperature programmed desorption (TPD), 26–27
Thiele modulus, cylindrical geometry, 231
TiO_2, 8
   scheme degradation, 9
TPD, *see* temperature programmed desorption
TPR, *see* temperature programed reduction
transesterification reaction
   profile, 104
   reaction steps, 104
transmission electron microscopy (TEM), 19, 24, 264
tubular reactor, *see* plug flow reactor

V

variation of conversion, 67
   gas phase reaction, 89
   mass of catalyst, 94
   temperature, 129

W

wet impregnation, *see* impregnation

X

XPS, *see* X-ray photoelectron spectroscopy
X-ray diffraction (XRD), 19
X-ray photoelectron spectroscopy (XPS), 30
   technique, 31
      scheme, 32
X-rays, 19
XRD, *see* X-ray diffraction